Remote Sensing Applications in Monitoring of Protected Areas

Remote Sensing Applications in Monitoring of Protected Areas

Special Issue Editors

Yeqiao Wang
Zhong Lu
Yongwei Sheng
Yuyu Zhou

MDPI • Basel • Beijing • Wuhan • Barcelona • Belgrade • Manchester • Tokyo • Cluj • Tianjin

Special Issue Editors

Yeqiao Wang
University of Rhode Island
USA

Zhong Lu
Southern Methodist University
USA

Yongwei Sheng
University of California
USA

Yuyu Zhou
Iowa State University
USA

Editorial Office
MDPI
St. Alban-Anlage 66
4052 Basel, Switzerland

This is a reprint of articles from the Special Issue published online in the open access journal *Remote Sensing* (ISSN 2072-4292) (available at: https://www.mdpi.com/journal/remotesensing/special_issues/Protected_Areas).

For citation purposes, cite each article independently as indicated on the article page online and as indicated below:

LastName, A.A.; LastName, B.B.; LastName, C.C. Article Title. *Journal Name* **Year**, *Article Number*, Page Range.

ISBN 978-3-03936-368-1 (Hbk)
ISBN 978-3-03936-369-8 (PDF)

Cover image courtesy of Yeqiao Wang.

© 2020 by the authors. Articles in this book are Open Access and distributed under the Creative Commons Attribution (CC BY) license, which allows users to download, copy and build upon published articles, as long as the author and publisher are properly credited, which ensures maximum dissemination and a wider impact of our publications.

The book as a whole is distributed by MDPI under the terms and conditions of the Creative Commons license CC BY-NC-ND.

Contents

About the Special Issue Editors vii

Preface to "Remote Sensing Applications in Monitoring of Protected Areas" ix

Yeqiao Wang, Zhong Lu, Yongwei Sheng and Yuyu Zhou
Remote Sensing Applications in Monitoring of Protected Areas
Reprinted from: *Remote Sens.* **2020**, *12*, 1370, doi:10.3390/rs12091370 1

Liangxian Fan, Jianjun Zhao, Yeqiao Wang, Zhoupeng Ren, Hongyan Zhang and Xiaoyi Guo
Assessment of Night-Time Lighting for Global Terrestrial Protected and Wilderness Areas
Reprinted from: *Remote Sens.* **2019**, *11*, 2699, doi:10.3390/rs11222699 17

Xiangyi Bei, Yunjun Yao, Lilin Zhang, Tongren Xu, Kun Jia, Xiaotong Zhang, Ke Shang, Jia Xu and Xiaowei Chen
Long-Term Spatiotemporal Dynamics of Terrestrial Biophysical Variables in the Three-River Headwaters Region of China from Satellite and Meteorological Datasets
Reprinted from: *Remote Sens.* **2019**, *11*, 1633, doi:10.3390/rs11141633 37

Chunyan Lu, Chunying Ren, Zongming Wang, Bai Zhang, Weidong Man, Hao Yu, Yibin Gao and Mingyue Liu
Monitoring and Assessment of Wetland Loss and Fragmentation in the Cross-Boundary Protected Area: A Case Study of Wusuli River Basin
Reprinted from: *Remote Sens.* **2019**, *11*, 2581, doi:10.3390/rs11212581 63

Zeinab Shirvani
A Holistic Analysis for Landslide Susceptibility Mapping Applying Geographic Object-Based Random Forest: A Comparison between Protected and Non-Protected Forests
Reprinted from: *Remote Sens.* **2020**, *12*, 434, doi:10.3390/rs12030434 85

Lei Fang, Ellen V. Crocker, Jian Yang, Yan Yan, Yuanzheng Yang and Zhihua Liu
Competition and Burn Severity Determine Post-Fire Sapling Recovery in a Nationally Protected Boreal Forest of China: An Analysis from Very High-Resolution Satellite Imagery
Reprinted from: *Remote Sens.* **2019**, *11*, 603, doi:10.3390/rs11060603 107

Bo Li, Fang Huang, Lijie Qin, Hang Qi and Ning Sun
Spatio-Temporal Variations of Carbon Use Efficiency in Natural Terrestrial Ecosystems and the Relationship with Climatic Factors in the Songnen Plain, China
Reprinted from: *Remote Sens.* **2019**, *11*, 2513, doi:10.3390/rs11212513 129

Lin Chen, Chunying Ren, Bai Zhang, Zongming Wang and Yeqiao Wang
Mapping Spatial Variations of Structure and Function Parameters for Forest Condition Assessment of the Changbai Mountain National Nature Reserve
Reprinted from: *Remote Sens.* **2019**, *11*, 3004, doi:10.3390/rs11243004 149

Yang Han, Ziying Li, Chang Huang, Yuyu Zhou, Shengwei Zong, Tianyi Hao, Haofang Niu and Haiyan Yao
Monitoring Droughts in the Greater Changbai Mountains Using Multiple Remote Sensing-Based Drought Indices
Reprinted from: *Remote Sens.* **2020**, *12*, 530, doi:10.3390/rs12030530 175

Jian Xu, Chen Gao and Yeqiao Wang
Extraction of Spatial and Temporal Patterns of Concentrations of Chlorophyll-a and Total Suspended Matter in Poyang Lake Using GF-1 Satellite Data
Reprinted from: *Remote Sens.* **2020**, *12*, 622, doi:10.3390/rs12040622 . . . . **193**

**Emilio Guirado, Javier Blanco-Sacristán, Juan Pedro Rigol-Sánchez,
Domingo Alcaraz-Segura and Javier Cabello**
A Multi-Temporal Object-Based Image Analysis to Detect Long-Lived Shrub Cover Changes in Drylands
Reprinted from: *Remote Sens.* **2019**, *11*, 2649, doi:10.3390/rs11222649 . . . . **215**

Anthony Campbell and Yeqiao Wang
High Spatial Resolution Remote Sensing for Salt Marsh Mapping and Change Analysis at Fire Island National Seashore
Reprinted from: *Remote Sens.* **2019**, *11*, 1107, doi:10.3390/rs11091107 . . . . **233**

Peili Duan, Yeqiao Wang and Peng Yin
Remote Sensing Applications in Monitoring of Protected Areas: A Bibliometric Analysis
Reprinted from: *Remote Sens.* **2019**, *12*, 772, doi:10.3390/rs12050772 . . . . **251**

About the Special Issue Editors

Yeqiao Wang is a professor at the Department of Natural Resources Science, College of the Environment and Life Sciences, University of Rhode Island. He earned his B.Sc. from the Northeast Normal University in 1982 and his M.Sc. degree from the Chinese Academy of Sciences in 1987. He earned the M.Sc. and Ph.D. degrees in natural resources management and engineering from the University of Connecticut in 1992 and 1995, respectively. From 1995 to 1999, he held the position of assistant professor in Department of Geography and Department of Anthropology, University of Illinois at Chicago. He has been on the faculty of the University of Rhode Island since 1999. Among his awards and recognitions, Dr. Wang was a recipient of the prestigious Presidential Early Career Award for Scientists and Engineers (PECASE) in 2000 by former U.S. President William J. Clinton. Dr. Wang's specialties and research interests are in terrestrial remote sensing and its applications in natural resources analysis and mapping. Dr. Wang has published over 170 refereed articles. He was the editor of *Remote Sensing of Coastal Environments and Remote Sensing of Protected Lands*, published by CRC Press in 2009 and 2011, respectively. He served as the Editor-in-Chief for the *Encyclopedia of Natural Resources*, a three-volume set of *Land, Air,* and *Water*, published by CRC Press in 2014. He was the editor of *The Handbook of Natural Resources*, Second Edition, a six-volume set, including *Terrestrial Ecosystems and Biodiversity* (Vol. 1), *Landscape and Land Capacity* (Vol. 2), *Wetlands and Habitats* (Vol. 3), *Fresh Water and Watersheds* (Vol. 4), *Coastal and Marine Environments* (Vol. 5), and *Atmosphere and Climate* (Vol. 6), published by CRC Press in 2020.

Zhong Lu received the B.Sc. and M.Sc. degrees from Peking University, Beijing, China, in 1989 and 1992, respectively, and a Ph.D. from the University of Alaska Fairbanks, Fairbanks, Alaska in 1996. He was a physical scientist with United States Geological Survey during 1997–2013, and he is now a professor and endowed Shuler-Foscue chair at the Roy M. Huffington Department of Earth Sciences, Southern Methodist University, Dallas, Texas, USA (www.smu.edu/dedman/lu). His research interests include technique developments of interferometric synthetic aperture radar (InSAR) processing and their applications to the study of volcano, landslide, and human-induced geohazards. He has produced more than 200 peer-reviewed journal articles and book chapters focused on InSAR techniques and applications, along with a book titled *InSAR Imaging of Aleutian Volcanoes: Monitoring a Volcanic Arc from Space* (Springer, 2014). He is a member of NASA-ISRO SAR (NISAR) Science Team (2012 to present); Senior Associate Editor of the journal *Remote Sensing* and the journal *Frontier in Earth Sciences*; and a member of editorial boards of *International Journal of Image and Data Fusion* and *Geomatics, Natural Hazards and Risk*.

Yongwei Sheng is a scientist in the field of Geospatial Information Systems and Technologies and their applications in large-area environmental monitoring and assessment, with over 100 journal publications. He graduated from Earth Science Department, Zhejiang University with B.Sc. and M.E. in 1988 and 1991, respectively, and obtained his Ph.D. in Environmental Science, Policy, and Management in 2000 from University of California, Berkeley. He is currently a Professor at Department of Geography, University of California, Los Angeles (UCLA), with primary research interests in lake dynamics at regional and global scales. He has been a member of NASA SWOT Mission Science Team and USGS/NASA Landsat Science Team.

Yuyu Zhou is an associate professor at the Department of Geological and Atmospheric Sciences, Iowa State University. He received his B.Sc. degree in geography and M.Sc. degree in remote sensing from Beijing Normal University and his Ph.D. degree in environmental sciences from University of Rhode Island. Before moving to Iowa State University, He worked as a geography scientist in the Joint Global Change Research Institute, Pacific Northwest National Laboratory. His research interests lie in the applications of geospatial technologies, including remote sensing, GIS, geovisualization, spatial analytic tools, and integrated assessment modeling, to understanding the problems of regional and global environmental change (e.g., urbanization, urban heat islands, ecosystem phenology, energy supply and demand, and greenhouse gas emissions) and their potential solutions. His research focus has always been in quantifying spatiotemporal patterns of environmental change and developing modeling mechanisms to bridge the driving forces (both natural and socioeconomic factors) and consequences of environmental change so that the impacts of human activities on environment can be effectively measured, modeled, and evaluated.

Preface to "Remote Sensing Applications in Monitoring of Protected Areas"

Protected areas have been established worldwide for achieving long-term goals in conservation of nature and the associated ecosystem services and cultural values. Globally, 15% of the world's terrestrial lands and inland waters, excluding Antarctica, are designated as protected areas. About 4.12% of the global ocean and 10.2% of coastal and marine areas under national jurisdiction are set as marine protected areas. Protected lands and waters serve as the fundamental building blocks of virtually all national and international conservation strategies, supported by governments and international institutions. Some of the protected areas are the only places that contain undisturbed landscape, seascape, and ecosystems on the planet Earth. With intensified impacts from climate and environmental change, protected areas have become more important to serve as indicators of ecosystem status and functions. Earth's remaining wilderness areas are becoming increasingly important buffers against changing environmental and ecological conditions.

The development of remote sensing platforms and sensors and the improvement in science and technology provide crucial support for monitoring of protected areas across the world. Remote sensing monitoring can provide essential information for the efficient, transparent, repeatable, and defensible decision-making in management and governance of protected areas. Time-series remote sensing data have allowed for reconstruction of histories of disturbances induced by anthropogenic and natural impacts. Remote sensing has unique advantages in monitoring frontier lands, which are always in remote and difficult-to-reach locations. Remote sensing technologies have profoundly changed the practice of monitoring and understanding the dynamics of protected areas and the surrounding environments. Integration of ground-based observations and remote sensing data has been practiced in monitoring the change of protected areas.

Remote sensing is among the most fascinating frontiers of science and technology that are constantly improving. Protected areas are by no means uniform entities. Protected areas have a wide range of management aims and are governed by stakeholders at different administrative levels and across spatial scales. Advances in remote sensing have helped gather and share information about protected areas at unprecedented rates. There are many new and exciting remote sensing applications that contribute to better informing management of protected areas. The achievements, challenges, lessons learned, and recommendations for remote sensing of protected areas deserve further attention.

The articles collected in this book reflect the subjects and contents of a published Special Issue of the open access journal *Remote Sensing* under the theme of remote sensing applications in monitoring of protected areas. We hope that this book can provide a snapshot of examples about remote sensing applications to address issues in inventory, monitoring, and management of protected areas. We also hope that this book can inspire a broader scope of interests in scientific research and management of protected lands and waters as the natural treasures on the planet Earth.

As the Guest Editors of the Special Issue, we appreciate the professionalism and support from all contributors, editors, and reviewers for their dedication and efforts toward the advancement of remote sensing applications in monitoring of protected areas. In particular, we express our gratitude to Ms. Sharon Fan, Section Managing Editor of the MDPI AG *Remote Sensing* Editorial Office, for her visionary suggestion in the preparation of the Special Issue and this book.

Yeqiao Wang, Zhong Lu, Yongwei Sheng, Yuyu Zhou
Special Issue Editors

Editorial

Remote Sensing Applications in Monitoring of Protected Areas

Yeqiao Wang [1,*], Zhong Lu [2], Yongwei Sheng [3] and Yuyu Zhou [4]

1. Department of Natural Resources Science, University of Rhode Island, Kingston, RI 02881, USA
2. Huffington Department of Earth Sciences, Southern Methodist University, Dallas, TX 75275, USA; zhonglu@smu.edu
3. Department of Geography, University of California, Los Angeles (UCLA), Los Angeles, CA 90095, USA; ysheng@geog.ucla.edu
4. Department of Geological & Atmospheric Sciences, Iowa State University, Ames, IA 50011, USA; yuyuzhou@iastate.edu
* Correspondence: yqwang@uri.edu

Received: 21 April 2020; Accepted: 24 April 2020; Published: 26 April 2020

Abstract: Protected areas (PAs) have been established worldwide for achieving long-term goals in the conservation of nature with the associated ecosystem services and cultural values. Globally, 15% of the world's terrestrial lands and inland waters, excluding Antarctica, are designated as PAs. About 4.12% of the global ocean and 10.2% of coastal and marine areas under national jurisdiction are set as marine protected areas (MPAs). Protected lands and waters serve as the fundamental building blocks of virtually all national and international conservation strategies, supported by governments and international institutions. Some of the PAs are the only places that contain undisturbed landscape, seascape and ecosystems on the planet Earth. With intensified impacts from climate and environmental change, PAs have become more important to serve as indicators of ecosystem status and functions. Earth's remaining wilderness areas are becoming increasingly important buffers against changing conditions. The development of remote sensing platforms and sensors and the improvement in science and technology provide crucial support for the monitoring and management of PAs across the world. In this editorial paper, we reviewed research developments using state-of-the-art remote sensing technologies, discussed the challenges of remote sensing applications in the inventory, monitoring, management and governance of PAs and summarized the highlights of the articles published in this Special Issue.

Keywords: protected areas (PAs); biodiversity conservation; spatiotemporal dynamics; climate change; human disturbances; management and governance

1. Introduction

The World Commission on Protected Areas adopted a definition that describes a protected area (PA) as a clearly defined geographical space, recognized, dedicated and managed, through legal or other effective means, to achieve the long-term conservation of nature with the associated ecosystem services and cultural values [1]. In general, protected areas (PAs) include national parks (NPs), national forests, national seashores, all levels of natural reserves, wildlife refuges and sanctuaries, designated areas for the conservation of native biological diversity and natural and cultural heritage and significance. PAs also include some of the last frontiers that have unique landscape characteristics and ecosystem functions in wilderness conditions [2]. Along shorelines and over ocean and sea, the International Union for the Conservation of Nature (IUCN) has defined marine protected areas (MPAs) as any area of intertidal or subtidal terrain, together with its overlying water and associated flora, fauna, historical and cultural features, which has been reserved by law or other effective means to protect part or all of the enclosed environment [3]. As reported by the World Database on Protected Areas (WDPA, https://www.protectedplanet.net/), 15% of the world's terrestrial lands and inland waters, excluding Antarctica, is under protection. About 4.12% of the global ocean and 10.2% of the coastal and marine areas under national jurisdiction are set as MPAs. About 19.2% of key biodiversity areas are completely covered as PAs [4]. Protected lands and waters serve as the fundamental building blocks of virtually all national and international conservation strategies, supported by governments and international institutions. These policies and their implementations provide the protection of threatened species around the world. The IUCN has categorized PAs into seven types, namely the strict nature reserve (Ia), wilderness area (Ib), national park (II), natural monument or feature (III), habitat/species management area (IV), protected landscape/seascape (V) and the protected area with a sustainable use of natural resources (VI) [1]. PAs are increasingly recognized as essential providers of ecosystem services and biological resources, key components in climate change mitigation strategies, as well as vehicles for protecting threatened human communities or sites of great cultural and spiritual value.

PAs have been created over past millennia for a multitude of reasons [5]. The establishment of the Yellowstone National Park in 1872 by the United States (U.S.) Congress ushered in the modern era of the governmental protection of natural areas, which catalyzed a global movement [6,7]. The 1916 National Park Service Organic Act of the United States established the purpose of national parks, including to conserve the scenery and the natural and historic objects and the wild life therein, and to provide for the enjoyment of the same in such a manner and by such means that will leave them unimpaired for the enjoyment of future generations [8]. As the National Parks Omnibus Management Act of 1998, the agency undertook a program of inventory and monitoring of National Park System resources to establish the baseline information and to provide information on the long-term trends in the condition of National Park System resources [9]. Remote sensing applications have contributed greatly in such inventory and monitoring efforts [10–12].

Even with the implementation of a tremendous variety of monitoring programs and conservation efforts with achievements, wild species' population decline, biodiversity loss, extinction, system degradation, pathogen spread and state change events are occurring at unprecedented rates [13,14]. The effects are augmented by continued changes in land use, invasive spread, alongside the direct, indirect and interactive effects of climate change and disruption. PAs become more important in serving as indicators of ecosystem conditions and functions either by their status and/or by contrasting to their adjacent unprotected areas. PAs are highly prized by the society with diversified representative characteristics. Earth's remaining wilderness areas are becoming increasingly important buffers against the changing environmental conditions. However, they are not an explicit target in many international policy frameworks [15]. The most recent United Nation's report concluded that up to one million animal and plant species were facing extinction, for which humans were to blame [16]. The most impacting drivers on global biodiversity scenarios toward the year 2100 include human-induced changes in land use, climate, nitrogen deposition, biotic exchange and atmospheric CO_2 [17].

The WDPA data showed that the Latin American and the Caribbean regions have 4.85 million km^2 (24%) of protected land. Brazil has half (2.47 million km^2) of the entire region protected, making it the largest national terrestrial PA network in the world [WDPA, https://protectedplanet.net/]. Worldwide, 77% of land, excluding Antarctica, and 87% of the ocean has been modified by the direct effects of human activities [18]. PAs in China, for example, have typically incorporated core and buffer zones with human habitation. A study mapped and analyzed the human footprint index at 1-km scale for 1834 terrestrial nature reserves of mainland China and concluded that the reserves designated at higher levels of governance were more pristine than those at lower levels. This was significant as China started to consider the reclassification of some reserves as NPs [19]. Another nationwide assessment quantified the provision of threatened species habitats and key regulating services in natural reserves in China. The study illuminated a strategy for strengthening PAs through creating the first comprehensive national park system of China [20]. As a strategic movement, in June 2019, the Chinese government announced a guideline for the establishment of a new NP-centered system for the protection of natural areas with the implementation plan in 2020. The crown jewels on the list of the first 10 designated NPs included the Three-River-Source NP, Giant Panda NP, Northeastern China Tiger and Leopard NP, among others. The Three-River-Source NP covers an area of about 363,000 km^2 and encompasses the headwaters of three major rivers, i.e., the Yellow, Yangtze and Lancang rivers, in the eastern Tibetan Plateau. The comprehensive system of NPs aims to protect the lands and waters with key natural resources and biodiversity.

Remote sensing provides a comprehensive geospatial capacity to map and study PAs in different spatial details and contexts, e.g., pixel size, area coverage, immediate adjacent areas of PAs and the broader background of the land and water that support the PAs; temporal frequencies, e.g., daily, weekly and monthly observations; and spectral properties. Remote sensing observations, in combination with field-based measurements, create new and exciting opportunities to meet the needs of monitoring PAs [21].

2. Remote Sensing Applications in Monitoring of Protected Areas

It has long been recognized that the on-the-ground monitoring of PA ecosystems is expensive, primarily due to the size and logistical constraints of national parks, designated wilderness, wildlife refuges and other large PAs. Remote sensing monitoring can provide essential information for the efficient, transparent, repeatable and defensible decision making in ecological systems [22]. The integration of ground-based data (e.g., focal species populations) and remote sensing has been practiced in monitoring and modeling environmental change in many PAs [5,23–25].

Remote sensing has unique advantages in monitoring the landscape dynamics of PAs around the world. The temporal depth of remote sensing can be used to provide monitoring with the continuity of deployments of new satellites and sensor systems and image acquisition capability. Multispectral optical sensors, e.g., the Landsat Thematic Mapper (TM), Enhanced Thematic Mapper Plus (ETM+) and Operational Land Imager (OLI), Advanced Spaceborne Thermal Emission and Reflection Radiometer (ASTER), SPOT High Resolution Visible (HRV) and High Resolution Visible and Infrared (HRVIR), Moderate Resolution Imaging Spectroradiometer (MODIS), Visible Infrared Imaging Radiometer Suite (VIIRS), Advanced Very-High-Resolution Radiometer (AVHRR) and Sentinel-2 MultiSpectral Instrument (MSI), and the derivatives of data products, have been routinely applied in PA inventory and monitoring research and applications. The approaches translated an ecologically based view of change into the spectral domain when archives of multispectral images were considered. Spectral indices have been used as the proxy for ecological attributes and have been tracked as time-series trajectories. The developed algorithms use statistical fitting rules to identify periods of consistent progression in the spectral trajectory (segments) and the turning points (vertices) that separate these periods. The change detection methods capture a wide range of processes affecting vegetation, such as the decline and mortality, growth and recovery and the combination of other driving factors [18,26,27]. Active sensors, such as the synthetic aperture radar (SAR), and satellites including the European Remote Sensing (ERS-1/-2) and Envisat, the Japanese Earth Resources Satellite 1 (JERS-1), the Phased Array type L-band Synthetic Aperture Radar (PALSAR-1/-2), RADARSAT-1/-2, the Constellation of Small Satellites for Mediterranean Basin Observation (COSMO-SkyMed), TerraSAR-X and TanDEM-X, and Sentinel-1A/B, have been proven effective in monitoring the changing environments at the local, regional and global scales [28–33]. The interferometric synthetic aperture radar (InSAR) has been used to construct a global digital elevation model (DEM), to map characteristics of the Earth's surface and measure land surface deformation at an unprecedented precision and spatial resolution under all-weather conditions [34].

Time-series remote sensing data have allowed for the reconstruction of the histories of disturbances induced by anthropogenic and natural impacts. Typical examples have included: the inventory and monitoring studies in NPs and PAs in a landscape context, such as in the Acadia NP and other northeastern U.S. NPs [12,35], the Yellowstone NP [36] and the Olympic NP of the Pacific Northwest of the U.S. [37]; for monitoring the interannual variability in snowpack and lake ice in southwest

Alaska [38]; in the assessment of national forests of eastern U.S. [39]; in monitoring the land cover change and ecological integrity of Canada's national parks [40], such as the wildlife habitat changes in Kejimkujik NP and the national historic site in southern Nova Scotia of the Canadian Atlantic Coastal Uplands Natural Region [41]; in operational active fire mapping and burnt area identification to Mexican nature PAs [42]; in Tibetan Plateau [43] and Changbai Mountain National Nature Reserve [44,45] of China.

Remote sensing has unique advantages in monitoring frontier lands, which are always in remote and difficult-to-reach locations. Examples have included: satellite-observed dynamics of lake-rich regions across the Tibetan Plateau and the Arctic; forest disturbance and dynamics in Siberia; the assessment of the complex Amur tiger and Far Eastern leopard habitats in the Russian Far East; in the landscape and ecosystem characterizations in China and Southeast Asia; in conservation efforts of tree kangaroos in Papua New Guinea; and in PAs in the Albertine Rift of Africa [46–62]. Remote sensing has advantages in monitoring vast habitats both inside and surrounding the PAs. This is particularly true when ecological functioning and habitats within NPs and PAs are influenced by natural resources outside of their borders [63–65]. Remote sensing applications have been among the critical approaches in the assessment of landscape contexts and the conversion risks of PAs surrounded by accelerated human population growth [66–70].

MPAs are among the critical components of protected waters. Important factors that affect the way plants and animals respond to MPAs include the distribution of habitat types, the level of connectivity to nearby fish habitats, wave exposure, depth distribution, prior level of resource extraction and regulations. Conservation benefits are evident through increased habitat heterogeneity at the seascape level, the increased abundance of threatened species and habitats and the maintenance of a full range of genotypes [71]. Remote sensing data that quantify spatial patterns in habitat type, oceanographic conditions, and benthic complexity can be integrated with in situ ecological data for the design, evaluation and the monitoring of MPA networks to design, assess and monitor MPAs [72,73]. Combining remote sensing products with in situ ecological and physical data can support the development of a statistically robust monitoring program of the living marine resources within and adjacent to marine protected areas [74]. Individual MPAs need to be networked in order to provide large-scale ecosystem benefits and to have the greatest chance of protecting all species, life stages and ecological linkages if they encompass representative portions of all ecologically relevant habitat types in a replicated manner. High-resolution remote sensing data are capable of mapping the physical and biological features of a benthic habitat, such as the monitoring of the coral reef in the Hawaii Archipelago and near-shore PAs in California and New England [75].

Coastal habitats, such as sand dunes, barrier islands, tidal wetlands, marshes, mangrove forests and submerged aquatic vegetation provide foods, shelters and breeding grounds for terrestrial and marine species. Coastal habitats also provide irreplaceable services such as filtering pollutants and retaining nutrients, maintaining water quality, protecting shorelines and absorbing flood waters. As coastal habitats are facing intensified natural and anthropogenic disturbances through direct impacts such as hurricanes, tsunamis, harmful algae blooms and cumulative and secondary impacts such as climate change, sea level rise, oil spill and urban development, the inventory and monitoring of coastal environments has become one of the most challenging tasks of the society in resource management and humanity administration. Remote sensing technologies with space-borne and airborne sensor systems in data acquisition and observation have profoundly changed the practice of monitoring and understanding the dynamics of coastal environments. Remote sensing applications have greatly enhanced the monitoring capacity of coastal PAs and practical implementations across spatial scales [76,77]. Very high resolution (VHR) imageries from airborne and satellite sensors, unmanned aerial vehicles (UAV), light detection and ranging (LiDAR), hyperspectral sensors, ground-based sensor networks and wireless geospatial service web systems have been increasingly applied with local focused interests on coastal PAs [78–86].

The improved capacity of data science and infrastructure, e.g., cloud computing, Google Earth Engine (GEE) and big Earth data approaches, facilitates data sharing and the integration and modeling processes [87–89]. For example, the capacity and service from GEE open opportunities for explorations that benefit from decades of data acquisition from remote sensing [90–96].

3. Challenges of Remote Sensing Monitoring of Protected Areas

The impacts of climate and human-induced environmental changes will continue to disrupt ecosystem functions and services, as well as the habitats and biodiversity. The future projections indicate a potentially catastrophic loss of global biodiversity [97–102]. Earth's remaining wilderness areas are becoming increasingly important buffers against changing conditions. Protected lands and waters are becoming more important, serving as indicators of ecosystem status and functions and as the barometer for guiding the national and international strategies in collaborated mitigation and conservation efforts.

PAs are functional from many forms of direct human intervention. The landscapes and seascapes of PAs are dynamic rather than static. Vegetation is changing continuously in response to both endogenous and exogenous pressures. PAs and their networks provide critical habitats for biodiversity conservation and yet their performances are challenged under the changing climate and shifting resource patterns [103]. Monitoring the dynamics of PAs requires tools that capture a wide range of processes over large areas. The evaluation of management effectiveness is a vital component of responsive, pro-active PA management [104,105]. Ecosystem indicators, whether process-based (e.g., productivity), pattern-based (e.g., land-use activities), or component-based (specie populations), vary in space and time. A major limiting factor in comprehensive ecological models is the lack of explanatory geospatial data. The issues conspire against the ready, standardized integration of remote sensing into ecological research for the management and governance of PAs.

Remote sensing is a universal tool for scientists and land managers. New developments of remote sensing platforms, sensors and improvements in science and technology provide crucial support for monitoring PAs across the world. Remote sensing data products, coupled with user-friendly data exploration, analyses, and accessible modeling tools, allow scientists and practitioners to gain a better understanding of how environmental changes affect specie populations, ecosystem functions and the services that sustain them. The lessons learned and the recommendations put forward for the remote sensing of PAs include: the allocation of sufficient time to develop a genuine science–management partnership; the communication of results in a management-relevant context; the confirmation or embellishment of existing frameworks and processes; plans for persistence and change; and to build on existing, widely used data analysis tools and software frameworks [10,21].

Field survey and in situ observations are essential to identify protected habitats through remote sensing. Almost every remote sensing exercise requires a field survey to define the habitats, to calibrate remote sensing imagery and to evaluate the accuracy out of remote sensing outputs [106]. With precise and accurate positioning and field survey becoming a routine operation, challenges remain for the incorporation of data from ground-based sensor networks and wireless geospatial service web systems with remote sensing observations for the comprehensive analyses and assessment of PAs.

The monitoring of landscape dynamics of PAs is among the primary advantage of remote sensing. The link between the pattern and the process, however, has been identified as a seminal challenge in landscape ecology. Disturbance is an important process that creates and responds to a pattern. The integration of remote sensing-based and in situ monitoring, including the consideration of scaling site observations, to the ecosystem level and the explicit link through ecosystem-based modeling to management options and recommendations, present the practical challenges and opportunities in the variety of PAs [23,26,39].

Remote sensing science is effective for managers and researchers across many domains. The lack of standardized protocols, workflow architecture, guidelines, training and software tools has led into a complexity. When evaluating the trends in resource and ecological conditions, the resource managers of PAs pursue analyses that use all the available information. Thus, they seek remote sensing change detection analyses that may include historical aerial photography, combined with more recent satellite images acquired in different spectral bands at various spatial and temporal resolutions. In addition, many resource problems must be evaluated at multiple spatial scales [12,69]. These practical issues result in unusually complex requirements and procedures that can be worked out only through the sustained collaboration between remote sensing scientists and PA managers. A key lesson is about the importance, difficulty and time-consumption of the mutual learning process [11]. In the management perspectives, there is a considerable potential to expand the operational use of remote sensing to monitor PAs among routine implementations. The uses of such information in operational monitoring present difficulties in designing and implementing a program that provides useful information at management levels and at an affordable cost [18,107]. The integration of remote sensing data into a framework for the data assimilation, processing, modeling and reporting is becoming essential [108–110].

It is worthy to point out that one of the most important limitations to the use of remote sensing data for the monitoring of PAs is the variant mapping accuracy and the cost of acquiring ground-based data for verification and validation. This is a common challenge of obtaining and integrating traditional in situ measurements and approaches with remote sensing mapping and modeling. It also shows that remote sensing cannot always meet the entire information collection needs. Whereas remote sensing-based techniques address spatial and temporal domains inaccessible to traditional approaches, remote sensing cannot match the accuracy, precision and the thematic richness of in situ measurements and monitoring at the plot scale. Therefore, the design of remote sensing-based monitoring methods needs to be carefully integrated with a very efficient protocol for the inclusion of field observations and survey data [10,111].

As the amenity values of PAs attract the rapid developments and impacts of human-induced land use change, remote sensing has to meet an increasingly essential requirement to address a range of monitoring across spatial scales and from terrestrial to coastal and open waters [112,113]. Challenges and uncertainties remain for the data continuity and systematic technology improvements toward consistent long-term monitoring applications in the future [114].

4. Highlights of the Special Issue Articles

With the rapid development of remote sensing science and technologies, this Special Issue aims to publish original manuscripts of the latest innovative research and advancement in the remote sensing of PAs. The articles in this Special Issue include applications of using data from multiple sensor systems in the monitoring of PAs from global to local interests.

The Defense Meteorological Satellite Program/Operational Linescan System (DMSP/OLS) night-time stable light (NTL) has been proven to be an effective indicator of the intensity and change of human-induced urban development over a long time span and at a larger spatial scale [115]. The study by Fan et al. [116] used the NTL data from 1992 to 2013 to characterize the human-induced urban development and studied the spatial and temporal variation of the NTL of global terrestrial PAs. The study selected seven types of PAs defined by the IUCN, including the strict nature reserve (Ia), the wilderness area (Ib), the national park (II), the natural monument or feature (III), the habitat/species management area (IV), the protected landscape/seascape (V), and the protected area with a sustainable use of natural resources (VI). The study evaluated the NTL magnitudes in PAs and their surrounding buffer zones. The results revealed the level, growth rate, trend and distribution pattern of the NTL in global PAs.

Terrestrial biophysical variables play an essential role in quantifying the amount of energy budget, water cycle and carbon sink over the Three-River Headwaters Region of China (TRHR). Bei et al. [117] evaluated the spatiotemporal dynamics of the biophysical variables including meteorological variables, vegetation and evapotranspiration (ET) over the TRHR and analyzed the response of the vegetation and the ET to climate change in the period from 1982 to 2015 using the China Meteorological Forcing Dataset (CMFD) and the Global Inventory Modeling and Mapping Studies (GIMMS) NDVI3g product, among others. The main input gridded datasets included meteorological reanalysis data, a satellite-based vegetation index dataset and the ET product developed by a process-based Priestley–Taylor algorithm.

The study suggested a 'dryer warming' and a 'wetter warming' tendency in different areas of the TRHR. The study revealed that more than 56.8% of the areas in the TRHR presented a significant increment in vegetation. The analysis noted that the ET was governed by the terrestrial water supply in the arid region of the western TRHR.

Salt marshes are changing due to natural and anthropogenic stressors such as sea level rise, nutrient enrichment, herbivory, storm surge and coastal development. A study by Campbell and Wang [105] analyzed the salt marsh change at the Fire Island National Seashore, a nationally protected area in New York, using the object-based image analysis (OBIA) to classify a combination of data from Worldview-2 and Worldview-3 satellites, Topobathymetric LiDAR, and National Agricultural Imagery Program (NAIP) aerial imageries. The salt marsh classification was trained and tested with the vegetation plot data. In October 2012, Hurricane Sandy caused extensive overwash and breached a section of the island. This study quantified the continuing effects of the breach on the surrounding salt marsh. The tidal inundation at the time of image acquisition was analyzed using the LiDAR-derived DEM to create a bathtub model at the target tidal stage. The study revealed the geospatial distribution and rates of change within the salt marsh interior and the salt marsh edge. The Worldview imagery was able to classify the salt marsh environments accurately at an overall accuracy of 92.75%. The study suggested that the NAIP data were adequate for determining the rates of salt marsh change with a high accuracy. The cost and revisit time of the NAIP imagery created an ideal open data source for high spatial resolution monitoring and the change analysis of salt marsh environments.

Anticipating how boreal forest landscapes will change in response to fire regimes requires disentangling the effects of various spatial controls on the recovery process of tree saplings. The spatially explicit monitoring of post-fire vegetation recovery through moderate resolution Landsat imagery is a popular technique but is filled with ambiguous information due to mixed pixel effects. On the other hand, very-high resolution satellite imagery accurately measures the crown size of tree saplings but has gained little attention. Its utility for estimating leaf area index (LAI) and tree sapling abundance (TSA) in post-fire landscapes remains untested. A study by Fang et al. [118] compared the explanatory power of the Landsat imagery with 0.5-m WorldView-2 VHR imagery for the LAI and TSA based on the field-sampling data and subsequently mapped the distribution of the LAI and TSA based on the most predictive relationships. The results showed that the pixel percentage of the canopy trees (PPCT) derived from VHR imagery outperformed all the Landsat-derived spectral indices for explaining the variance of the LAI and TSA. The analyses concluded that mitigating wildfire severity and size may increase forest resilience to wildfire damage. Given the easily damaged seed banks and relatively short seed dispersal distance of coniferous trees, reasonable human help for the natural recovery of coniferous forests was necessary for severe burns with a large patch size, particularly in certain areas. The research showed that WorldView-2 VHR imagery better resolved the key characteristics of forest landscapes, providing a valuable tool to land managers and researchers alike.

Climate change and human activities alter the spatial distribution and structure of vegetation, especially in drylands. In this context, the object-based image analysis (OBIA) has been used to monitor changes in vegetation, but only a few studies have related them to anthropogenic pressure. Guirado et al. [119] assessed changes in the cover, number and shape of *Ziziphus lotus* shrub individuals in a coastal groundwater-dependent ecosystem in Spain over a period of 60 years and related them to human activities in the area. In particular, the study evaluated how sand mining, groundwater extraction and the protection of the area affected the shrubs. To do this, the study developed an object-based methodology to create accurate maps of the vegetation patches and compared the cover changes in the individuals. The changes in shrub size and shape were related to soil loss, seawater intrusion and the legal protection of the area measured by the average minimum distance and average random distance analysis. It was found that both the sand mining and seawater intrusion had a negative effect on individuals; on the contrary, the protection of the area had a positive effect on the size of the individuals' coverage. The findings supported the use of the OBIA for monitoring scattered vegetation patches in drylands, key to any monitoring program aimed at vegetation preservation.

Forest condition is the baseline information for ecological evaluation and management. A study by Chen et al. [120] mapped the structure and function parameters for forest condition assessment in the Changbai Mountain National Nature Reserve (CMNNR). Various mapping algorithms, including statistical regression, random forests, and random forest kriging were employed with predictors from Advanced Land Observing Satellite (ALOS)-2, Sentinel-1, Sentinel-2 satellite sensors, digital surface model of ALOS and 1803 field-sampled forest plots. The combined predicted parameters and weights from the principal component analysis as well as the forest conditions were assessed. The models explained the spatial dynamics and characteristics of forest parameters based on the independent validation. The mean assessment score suggested that forest conditions in the CMNNR were mainly the result of spatial variations of function parameters such as stand volume and soil fertility. This study provided a methodology on forest condition assessment at regional scales, as well as the up-to-date information for the forest ecosystem in the CMNNR.

Han et al. [121] reported on the monitoring of droughts in the Greater Changbai Mountains (GCM) region by six drought indices, i.e., the precipitation condition index (PCI), temperature condition index (TCI), vegetation condition index (VCI), vegetation health index (VHI), scaled drought condition index (SDCI) and the temperature–vegetation dryness index (TVDI), between 2001 and 2018. This study provided a reference for the selection of drought indices for monitoring droughts to gain a better understanding of the ecosystem conditions and the environment.

The Songnen Plain (SNP) is an important grain production base and a designated red-line protection in China. The understanding of carbon use efficiency (CUE) of natural ecosystems in protected farmland areas is vital to predicting the impacts of natural and anthropogenic disturbances on carbon budgets and evaluating ecosystem functions. An article by Li et al. [122] studied variations in the ecosystem CUE in the SNP using MODIS data products and the Carnegie–Ames–Stanford approach (CASA) model. The relationships revealed between the CUE and the phenological and climate factors helped explain the CUE of the natural ecosystems in the protected farmland areas and improved the understanding about the dynamics of ecosystem carbon allocation in temperate semi-humid to semi-arid transitional regions under climate and phenological fluctuations.

The comparative evaluation of cross-boundary wetland PAs is essential to underpin knowledge-based bilateral conservation policies and funding decisions by governments and managers. The article by Lu et al. [123] reported on a study of monitoring wetland change in the Wusuli River Basin in the crossboundary zone of China and Russia from 1990 to 2015 using Landsat images. The spatial-temporal distribution of wetlands was identified using a rule-based object-oriented classification method. The wetland dynamics were determined by combining the annual land change area (ALCA), the annual land change rate (ALCR), the landscape metrics and the spatial analysis. The study revealed the changes of the natural wetlands in the Wusuli River Basin and the patterns of change. The study provided critical information for the conservation and management of ecological conditions in cross-boundary wetlands.

Despite recent progress in landslide susceptibility mapping, a holistic method is still needed to integrate and customize influential factors with a focus on forest regions. A study by Shirvani [124] tested the performance of geographic object-based random forest modeling in the susceptibility of protected and non-protected forests to landslides in northeast Iran using Landsat 8 multispectral images and DEM data. The study derived features of conditioning factors. The study confirmed that some anthropogenic activities such as forest fragmentation and logging significantly intensified the susceptibility of the non-protected forests to landslides.

As the largest freshwater lake in China, Poyang Lake provides tremendous services and functions to its surrounding ecosystem, such as water conservation and the sustaining of biodiversity, and has significant impacts on the security and sustainability of the regional ecology. The lake and associated wetlands are among the protected aquatic ecosystems with global significance. The Poyang Lake region has recently experienced increased urbanization and anthropogenic disturbances, which has greatly impacted the lake environment. The concentrations of chlorophyll-a (Chl-a) and total suspended matter

(TSM) are important indicators for assessing the water quality of lakes. The study by Xu et al. [125] used data from the Gaofen-1 (GF-1) satellite, in situ measurements of the reflectance of the lake water and the analysis of the Chl-a and TSM concentrations of the lake water samples to investigate the spatial and temporal variation and distribution patterns. The study analyzed the measured reflectance spectra and conducted a correlation analysis to identify the spectral bands that were sensitive to the concentration of Chl-a and TSM, respectively. The modeling results revealed the spatial and temporal variations of the water quality in Poyang Lake and demonstrated the capacities of the GF-1 satellite data in the monitoring of lake water quality.

The article by Duan et al. [126] presented an analysis of research publications, from a bibliometric perspective, on the remote sensing of PAs. The analysis focused on the time period from 1991 to 2018. The study extracted 4546 academic publications from the Web of Science database. Using VOSviewer software, the study evaluated the co-authorships among countries and institutions, as well as the co-occurrences of the keywords. The results indicated an increasing trend of annual publications in the remote sensing of PAs. This analysis revealed the major topical subjects, leading countries and most influential institutions around the world that have conducted relevant research in scientific publications. The study also revealed the journals that published the most articles in the subject of interests and the collaborative patterns related to the remote sensing of PAs. The analysis provided insights for understanding the intellectual structure of the field and identifying the future research directions.

5. Conclusion Remarks

Remote sensing is among the most fascinating frontiers of science and technology that are constantly improving our understanding of PAs. PAs are by no means uniform entities and have a wide range of management aims and are governed by many stakeholders. Advances in remote sensing have helped gather and share information about PAs at unprecedented rates and scales. There are many new and exciting applications for remotely sensed data that contribute to the better informing management of PAs. The achievements through the applications of science and technologies, the challenges, the lessons learned and the recommendations for the remote sensing of PAs deserve further attention [127].

The subjects and contents of the articles collected in this Special Issue reflect the state-of-the-art of remote sensing technologies for: capturing the dynamics of ecosystem variations; the evaluations of available sensors, data and the new development of integrated approaches; the methods for processing advanced remote sensing and time series data; and the integration of multisource and open source data. These studies contributed in the monitoring of PAs from the perspectives of in situ measurements, habitat assessments, socio-economic development, policy and management factors, and in inventory and practical implementations. The applications of monitoring from biospheric, atmospheric, hydrospheric and societal dimensions reflect the advantages of remote sensing in habitat mapping and biodiversity conservation, in the detection of effects from natural and anthropogenic disturbances, as well as in revealing uncertainties for the assessment of the resilience and sustainability of PAs and the mitigation approaches under the changing environments.

Funding: This research received no external funding.

Acknowledgments: As the Guest Editors, we appreciate the professionalism and support from all contributors, editors, and reviewers for their efforts and dedication toward the advancement of remote sensing in monitoring of protected areas with the publication of this Special Issue.

Conflicts of Interest: The authors declare no conflict of interest.

References

1. Dudley, N. *Guidelines for Applying Protected Area Management Categories*; IUCN: Gland, Switzerland, 2008.
2. Wang, Y. *Remote Sensing of Protected Lands*; CRC Press: Boca Raton, FL, USA, 2011; ISBN 978-1-4398-4187-7.
3. Kelleher, G. *Guidelines for Marine Protected Areas*; IUCN: Gland, Switzerland, 1999.
4. UNEP-WCMC and IUCN. *Protected Planet Report 2016*; UNEP-WCMC: Cambridge, UK; IUCN: Gland, Switzerland, 2016.
5. Crabtree, R.; Sheldon, J. Monitoring and modeling environmental change in protected areas: Integration of focal species populations and remote sensing. In *Remote Sensing of Protected Lands*; CRC Press: Boca Raton, FL, USA, 2011; pp. 495–524.
6. IUCN. Shaping a sustainable future. In *The IUCN Programme 2009–2012*; IUCN: Gland, Switzerland, 2008.
7. Heinen, J.; Hite, K. Protected natural areas. In *Encyclopedia of Earth*; Cutler, J., Cleveland, C.J., Eds.; Environmental Information Coalition, National Council for Science and the Environment: Washington, DC, USA, 2007.
8. National Park Service. Organic Act of 1916. Available online: https://www.nps.gov/grba/learn/management/organic-act-of-1916.htm (accessed on 20 April 2020).
9. National Park Service. National Parks Omnibus Management Act of 1998. Available online: https://www.nps.gov/gis/data_standards/omnibus_management_act.html (accessed on 20 April 2020).
10. Gross, J.E.; Goetz, S.J.; Cihlar, J. Application of remote sensing to parks and protected area monitoring: Introduction to the special issue. *Remote Sens. Environ.* **2009**, *113*, 1343–1345. [CrossRef]
11. Kennedy, R.E.; Townsend, P.A.; Gross, J.E.; Cohen, W.B.; Bolstad, P.; Wang, Y.Q.; Adams, P. Remote sensing change detection and natural resource monitoring for managing natural landscapes. *Remote Sens. Environ.* **2009**, *113*, 1382–1396. [CrossRef]
12. Wang, Y.; Mitchell, B.R.; Nugranad-Marzilli, J.; Bonynge, G.; Zhou, Y.; Shriver, G.W. Remote sensing of land-cover change and landscape context of the national parks: A case study of the Northeast Temperate Network. *Remote Sens. Environ.* **2009**, *113*, 1453–1461. [CrossRef]
13. Hoffmann, M.; Hilton-Taylor, C.; Angulo, A.; Böhm, M.; Brooks, T.M.; Butchart, S.H.; Carpenter, K.E.; Chanson, J.; Collen, B.; Cox, N.A.; et al. The impact of conservation on the status of the World's vertebrates. *Science* **2010**, *330*, 1503–1509. [CrossRef]
14. Pereira, H.M.; Leadley, P.W.; Proença, V.; Alkemade, R.; Scharlemann, J.P.; Fernandez-Manjarrés, J.F.; Araújo, M.B.; Balvanera, P.; Biggs, R.; Cheung, W.W.; et al. Scenarios for global biodiversity in the 21st century. *Science* **2010**, *330*, 1496–1501. [CrossRef]
15. Watson, J.E.M.; Allan, J.R. Protected the Last of the Wild. *Nature* **2018**, *563*, 27. [CrossRef] [PubMed]
16. IPBES. *Intergovernmental Science-Policy Platform on Biodiversity and Ecosystem Services (IPBES) Media Release, Nature's Dangerous Decline 'Unprecedented' Species Extinction Rates 'Accelerating'*; IPBES: Bonn, Germany, 2019.
17. Sala, O.E.F.S.; Chapin, J.J., III; Armesto, E.; Berlow, J.; Bloomfield, R.; Dirzo, E.; Huber-Sanwald, L.F.; Huenneke, R.B.; Jackson, A.; Kinzig, R.; et al. Global Biodiversity Scenarios for the Year 2100. *Science* **2010**, *287*, 1770–1774. [CrossRef] [PubMed]
18. Kennedy, R.E.; Yang, Z.; Cohen, W.B. Detecting trends in forest disturbance and recovery using yearly Landsat time series: 1. LandTrendr—Temporal segmentation algorithms. *Remote Sens. Environ.* **2010**, *114*, 2897–2910. [CrossRef]
19. Buckley, R.; Zhou, R.; Zhong, L. How Pristine Are China's Parks? *Front. Ecol. Evol.* **2016**, *4*, 136. [CrossRef]
20. Xu, W.; Xiao, Y.; Zhang, J.; Yang, W.; Zhang, L.; Hull, W.; Wang, Z.; Zheng, H.; Liu, J.; Polasky, S.; et al. Strengthening protected areas for biodiversity and ecosystem services in China. *Proc. Natl. Acad. Sci.* **2017**, *114*, 1601–1606. [CrossRef]
21. Fancy, S.G.; Gross, J.E.; Carter, S.L. Monitoring the condition of natural resources in U.S. National Parks. *Environ. Monit. Assess.* **2009**, *151*, 161–174. [CrossRef]
22. Gross, J.E.; Nemani, R.R.; Turner, W.; Melton, F. Remote sensing for the national parks. *Park Sci.* **2006**, *24*, 30–36.
23. Nagler, P.L.; Glenn, E.P.; Hinojosa-Huerta, O. Synthesis of ground and remote sensing data for monitoring ecosystem functions in the Colorado River Delta, Mexico. *Remote Sens. Environ.* **2009**, *113*, 1473–1485. [CrossRef]

24. Clark, J.; Wang, Y.; August, P. Assessing current and projected suitable habitats for tree-of-heaven along the Appalachian Trail. *Philos. Trans. R. Soc. B* **2012**, *369*, 20130192. [CrossRef] [PubMed]
25. Fu, B.; Li, Y.; Wang, Y.; Campbell, A.; Zhang, B.; Yin, S.; Zhu, H.; Xing, Z.; Jin, X. Evaluation of riparian condition of Songhua River by integration of remote sensing and field measurements. *Sci. Rep.* **2017**, *7*, 2565. [CrossRef] [PubMed]
26. Dennison, P.E.; Nagler, P.L.; Hultine, K.R.; Glenn, E.P.; Ehleringer, J.R. Remote monitoring of tamarisk defoliation and evapotranspiration following saltcedar leaf beetle attack. *Remote Sens. Environ.* **2009**, *113*, 1462–1472. [CrossRef]
27. Wang, Y.; Moskovits, D.K. Tracking Fragmentation of Natural Communities and Changes in Land Cover: Applications of Landsat Data for Conservation in an Urban Landscape (Chicago Wilderness). *Conserv. Biol.* **2001**, *15*, 835–843. [CrossRef]
28. Sippel, S.; Hamilton, S.; Melack, J. Inundation area and morphometry of lakes on the Amazon river floodplain, Brazil. *Arch. Fur Hydrobiol. Stuttg.* **1992**, *123*, 385–400.
29. Birkett, C. Synergistic remote sensing of Lake Chad: Variability of basin inundation. *Remote Sens. Environ.* **2000**, *72*, 218–236. [CrossRef]
30. Zhang, J.; Xu, K.; Yang, Y.; Qi, L.; Hayashi, S.; Watanabe, M. Measuring water storage fluctuations in lake Dongting, China, by topex/poseidon satellite altimetry. *Environ. Monit. Assess.* **2006**, *115*, 23–37. [CrossRef]
31. Schlaffer, S.; Matgen, P.; Hollaus, M.; Wagner, W. Flood detection from multi-temporal SAR data using harmonic analysis and change detection. *Int. J. Appl. Earth Obs. Geoinf.* **2015**, *38*, 15–24. [CrossRef]
32. Smith, L.C.; Sheng, Y.; MacDonald, G.M. A first pan-arctic assessment of the influence of glaciation, permafrost, topography and peatlands on northern hemisphere lake distribution. *Permafr. Periglac. Process.* **2007**, *18*, 201–208. [CrossRef]
33. Matta, E.; Giardino, C.; Boggero, A.; Bresciani, M. Use of satellite and in situ reflectance data for lake water color characterization in the Everest Himalayan region. *Mt. Res. Dev.* **2017**, *37*, 16–23. [CrossRef]
34. Lu, Z.; Zhang, L. Frontiers of Radar Remote Sensing. *Photogram. Eng. Remote Sens.* **2014**, *80*, 5–13.
35. Goetz, S.J.; Jantz, P.; Jantz, C.A. Connectivity of core habitat in the northeastern United States: Parks and protected areas in a landscape context. *Remote Sens. Environ.* **2009**, *113*, 1421–1429. [CrossRef]
36. Crabtree, R.L.; Potter, C.S.; Mullen, R.S.; Sheldon, J.W.; Huang, S.; Harmsen, J.A. A modeling and spatiotemporal analysis framework for monitoring environmentalchange using NPP as an ecosystem indicator. *Remote Sens. Environ.* **2009**, *113*, 1486–1496. [CrossRef]
37. Huang, C.; Schleerweis, K.; Thomas, N.; Goward, S.N. Forest Dynamics within and around Olympic National Park Assessed Using Time Series Landsat Observations. In *Remote Sensing of Protected Lands*; CRC Press: Boca Raton, FL, USA, 2011; pp. 75–94.
38. Reed, B.; Budde, M.; Spencer, P.; Miller, A.E. Integration of MODIS-derived metrics to assess interannual variability in snowpack, lake ice, and NDVI in southwest Alaska. *Remote Sens. Environ.* **2009**, *113*, 1443–1452. [CrossRef]
39. Huang, C.; Goward, S.N.; Schleeweis, K.; Thomas, N.; Masek, J.G.; Zhu, Z. Dynamics of national forests assessed using the Landsat record: Case studies in eastern United States. *Remote Sens. Environ.* **2009**, *113*, 1430–1442. [CrossRef]
40. Fraser, R.; Olthof, I.; Pouliot, D. Monitoring land cover change and ecological integrity in Canada's national parks. *Remote Sens. Environ.* **2009**, *113*, 1397–1409. [CrossRef]
41. Zorn, P.; Ure, D.; Sharma, R.; O'Grady, S. Using earth observation to monitor species-specific habitat change in the Greater Kejimkujik National Park Region of Canada. In *Remote Sensing of Protected Lands*; CRC Press: Boca Raton, FL, USA, 2011; pp. 95–110.
42. Ressl, R.; Lopez, G.; Cruz, I.; Ressl, S.; Schmidt, M.I.; Jiménez, R. Operational active fire mapping and burnt area identification applicable to Mexican nature protection areas using MODIS-DB data. *Remote Sens. Environ.* **2009**, *113*, 1113–1126. [CrossRef]
43. Gillespie, T.W.; Madson, A.; Cusack, C.F.; Xue, Y. Changes in NDVI and human population in protected areas on the Tibetan Plateau. *Arct. Antarct. Alp. Res.* **2019**, *51*, 428–439. [CrossRef]
44. Guo, X.; Zhang, H.; Wang, Y.; Clark, J. Mapping and assessing typhoon-induced forest disturbance in Changbai Mountain National Nature Reserve using time series Landsat imagery. *J. Mt. Sci.* **2015**, *12*, 404–416. [CrossRef]

45. Chi, H.; Sun, G.; Huang, J.; Li, R.; Ren, X.; Ni, W.; Fu, A. Estimation of Forest Aboveground Biomass in Changbai Mountain Region Using ICESat/GLAS and Landsat/TM Data. *Remote Sens.* **2017**, *9*, 707. [CrossRef]
46. Sheng, Y.; Alsdorf, D. Automated ortho-rectification of Amazon basin-wide SAR mosaics using SRTM DEM data. *IEEE Trans. Geosci. Remote Sens.* **2005**, *43*, 1929–1940. [CrossRef]
47. Arima, E.Y.; Walker, R.T.; Sales, M.; Souza, C., Jr.; Perz, S.G. The fragmentation of space in the Amazon basin: Emergent road networks. *Photogram. Eng. Remote Sens.* **2008**, *74*, 699–709. [CrossRef]
48. Walsh, S.J.; Messina, J.P.; Brown, D.G. Mapping & modeling land use/land cover dynamics in frontier settings. *Photogram. Eng. Remote Sens.* **2008**, *74*, 677–679.
49. Mena, C.F. Trajectories of land-use and land-cover in the northern Ecuadorian Amazon: Temporal composition, spatial configuration, and probability of change. *Photogram. Eng. Remote Sens.* **2008**, *74*, 737–751. [CrossRef]
50. Wang, C.; Qi, J.; Cochrane, M. Assessment of tropical forest degradation with canopy fractional cover from Landsat ETM+ and IKONOS imagery. *Earth Interact.* **2005**, *9*, 1–18. [CrossRef]
51. Wang, C.; Qi, J. Biophysical estimation in tropical forests using JERS-1 SAR and VNIR Imagery: II-aboveground woody biomass. *Int. J. Remote Sens.* **2008**, *29*, 6827–6849. [CrossRef]
52. Sun, G.; Ranson, K.J.; Kharuk, V.I. Radiometric slope correction for forest biomass estimation from SAR data in Western Sayani mountains, Siberia. *Remote Sens. Environ.* **2002**, *79*, 279–287. [CrossRef]
53. Bergen, K.M.; Zhao, T.; Kharuk, V.; Blam, Y.; Brown, D.G.; Peterson, L.K.; Miller, N. Changing regimes: Forested land cover dynamics in central Siberia 1974–2001. *Photogram. Eng. Remote Sens.* **2008**, *74*, 787–798. [CrossRef]
54. Kharuk, V.I.; Ranson, K.J.; Im, S.T. Siberian silkmoth outbreak pattern analysis based on SPOT VEGETATION data. *Int. J. Remote Sens.* **2009**, *30*, 2377–2388. [CrossRef]
55. Stow, D.A.; Hope, A.; McGuire, D.; Verbyla, D.; Gamon, J.; Huemmrich, F.; Houston, S.; Racine, C.; Sturm, M.; Tape, K.; et al. Remote sensing of vegetation and land-cover change in Arctic tundra ecosystems. *Remote Sens. Environ.* **2004**, *89*, 281–308. [CrossRef]
56. Sheng, Y.; Shah, C.A.; Smith, L.C. Automated image registration for hydrologic change detection in the lake-rich arctic. *IEEE Geosci. Remote Sens. Lett.* **2008**, *5*, 414–418. [CrossRef]
57. Sheng, Y.; Li, J. Satellite-observed endorheic lake dynamics across the Tibetan plateau between circa 1976 and 2000. In *Remote Sensing of Protected Lands*; CRC Press: Boca Raton, FL, USA, 2011; pp. 305–319.
58. Ranson, K.J.; Sun, G.; Kharuk, V.I.; Howl, J. Multisensor Remote Sensing of Forest Dynamics in Central Siberia. In *Remote Sensing of Protected Lands*; CRC Press: Boca Raton, FL, USA, 2011; pp. 321–377.
59. Sherman, N.J.; Loboda, T.V.; Sun, G.; Shugart, H.H. Remote sensing and modeling for assessment of complex Amur (Siberian) Tiger and Amur (Far Eastern) Leopard Habitats in the Russian Far East. In *Remote Sensing of Protected Lands*; CRC Press: Boca Raton, FL, USA, 2011; pp. 379–407.
60. Chen, L.; Ren, C.; Li, L.; Wang, Y.; Zhang, B.; Wang, Z.; Li, L. A Comparative Assessment of Geostatistical, Machine Learning, and Hybrid Approaches for Mapping Topsoil Organic Carbon Content. *ISPRS Int. J. Geo-Inf.* **2019**, *8*, 174. [CrossRef]
61. Stabach, J.A.; Dabek, L.; Jensen, R.; Wang, Y.Q. Discrimination of dominant forest types for Matschie's tree kangaroo conservation in Papua New Guinea using high-resolution remote sensing data. *Int. J. Remote Sens.* **2009**, *30*, 405–422. [CrossRef]
62. Ayebare, S.; Moyer, D.; Plumptre, A.J.; Wang, Y. Remote sensing for biodiversity conservation of the Albertine Rift in Eastern Africa. In *Remote Sensing of Protected Lands*; CRC Press: Boca Raton, FL, USA, 2011; pp. 183–201.
63. GAO (U.S. General Accounting Office). *Activities Outside Park Borders Have Caused Damage to Resources and Will Likely Cause More*; U.S. Government Printing Office GAO/RCED-94-59; GAO: Washington, DC, USA, 1994.
64. Hansen, A.J.; DeFries, R. Ecological mechanisms linking protected areas to surrounding lands. *Ecol. Appl.* **2007**, *17*, 974–988. [CrossRef]
65. Svancara, L.; Scott, J.M.; Loveland, T.R.; Pidgorna, A.B. Assessing the landscape context and conversion risk of protected areas using remote-sensing derived data. *Remote Sens. Environ.* **2009**, *113*, 1357–1369. [CrossRef]
66. Wittemyer, G.; Elsen, P.; Bean, W.T.; Burton, A.C.O.; Brashares, J.S. Accelerated human population growth at protected area edges. *Science* **2008**, *321*, 123–126. [CrossRef]
67. Townsend, P.A.; Lookingbill, T.R.; Kingdon, C.C. Spatial pattern analysis for monitoring protected areas. *Remote Sens. Environ.* **2009**, *113*, 1410–1420. [CrossRef]

68. Wiens, J.A.; Sutter, R.D.; Anderson, M.; Blanchard, J.; Barnett, A.; Aguilar-Amuchastegui, N. Selecting and conserving lands for biodiversity: The role of remote sensing. *Remote Sens. Environ.* **2009**, *113*, 1370–1381. [CrossRef]
69. Jones, D.A.; Hansen, A.J.; Bly, K.; Doherty, K.; Verschuyl, J.P.; Paugh, J.I.; Carle, R.; Story, S.J. Monitoring land use and cover around parks: A conceptual approach. *Remote Sens. Environ.* **2009**, *113*, 1346–1356. [CrossRef]
70. Lu, X.; Zhou, Y.; Liu, Y.; Yannick, L.P. The role of protected areas in land use/land cover change and the carbon cycle in the conterminous United States. *Glob. Chang. Biol.* **2017**. [CrossRef] [PubMed]
71. Edgar, G.J.; Russ, G.R.; Babcock, R.C. Marine protected areas. In *Marine Ecology*; Oxford University Press: Oxford, UK, 2007; pp. 533–555.
72. Wedding, L.; Friedlander, A.M. Determining the influence of seascape structure on coral reef fishes in Hawaii using a geospatial approach. *Mar. Geod.* **2008**, *31*, 246–266. [CrossRef]
73. Wedding, L.; Friedlander, A.; McGranaghan, M.; Yost, R.; Monaco, M.E. Using bathymetric Lidar to define nearshore benthic habitat complexity: Implications for management of reef fish assemblages in Hawaii. *Remote Sens. Environ.* **2008**, *112*, 4159–4165. [CrossRef]
74. Friedlander, A.M.; Brown, E.K.; Monaco, M.E. Coupling ecology and GIS to evaluate efficacy of màine protected areas in Hawaii. *Ecol. Appl.* **2007**, *17*, 715–730. [CrossRef]
75. Friedlander, A.M.; Wedding, L.M.; Caselle, J.E.; Costa, B.M. Integration of remote sensing and in situ ecology for the design and evaluation of marine protected areas: Examples from tropical and temperate ecosystems. In *Remote Sensing of Protected Lands*; CRC Press: Boca Raton, FL, USA, 2011; pp. 245–280.
76. Wang, Y.; Tobey, J.; Bonynge, G.; Nugrand, J.; Makota, V.; Ngusaru, N.; Traber, M. Involving Geospatial Information in the Analysis of Land Cover Change along Tanzania Coast. *Coast. Manag.* **2005**, *33*, 89–101. [CrossRef]
77. Wang, Y.; Traber, M.; Milestead, B.; Stevens, S. Terrestrial and submerged aquatic vegetation mapping in Fire Island National Seashore using high spatial resolution remote sensing data. *Mar. Geod.* **2007**, *30*, 77–95. [CrossRef]
78. Li, X.; Cheng, X.; Gong, P.; Yan, K. Design and Implementation of a Wireless Sensor Network-Based Remote Water-Level Monitoring System. *Sensors* **2011**, *11*, 1706–1720. [CrossRef] [PubMed]
79. Laliberte, A.S.; Goforth, M.A.; Steele, C.M.; Rango, A. Multispectral Remote Sensing from Unmanned Aircraft: Image Processing Workflows and Applications for Rangeland Environments. *Remote Sens.* **2011**, *3*, 2529–2551. [CrossRef]
80. D' Oleire-Oltmanns, S.; Marzolff, I.; Peter, K.D.; Ries, J.B. Unmanned Aerial Vehicle (UAV) for Monitoring Soil Erosion in Morocco. *Remote Sens.* **2012**, *4*, 3390–3416. [CrossRef]
81. Hruska, R.; Mitchell, J.; Anderson, M.; Glenn, N.F. Radiometric and Geometric Analysis of Hyperspectral Imagery Acquired from an Unmanned Aerial Vehicle. *Remote Sens.* **2012**, *4*, 2736–2752. [CrossRef]
82. Harwin, S.; Lucieer, A. Assessing the Accuracy of Georeferenced Point Clouds Produced via Multi-View Stereopsis from Unmanned Aerial Vehicle (UAV) Imagery. *Remote Sens.* **2012**, *4*, 1573–1599. [CrossRef]
83. Wallace, L.; Lucieer, A.; Watson, C.; Turner, D. Development of a UAV-LiDAR System with Application to Forest Inventory. *Remote Sens.* **2012**, *4*, 1519–1543. [CrossRef]
84. Campbell, A.; Wang, Y. Examining the Influence of Tidal Stage on Salt Marsh Mapping using High Spatial Resolution Satellite Remote Sensing and Topobathymetric LiDAR. *IEEE Trans. Geosci. Remote Sens.* **2018**, *56*, 5169–5176. [CrossRef]
85. Campbell, A.; Wang, Y. Assessment of salt marsh change on Assateague Island National Seashore between 1962 and 2016. *Photogram. Eng. Remote Sens.* **2020**, *86*, 187–194. [CrossRef]
86. Campbell, A.; Wang, Y.; Christiano, M.; Stevens, S. Salt Marsh Monitoring in Jamaica Bay, New York from 2003 to 2013: A Decade of Change from Restoration to Hurricane Sandy. *Remote Sens.* **2017**, *9*, 131. [CrossRef]
87. Hansen, M.C.; Potapov, P.V.; Moore, R.; Hancher, M.; Turubanova, S.A.; Tyukavina, A.; Thau, D.; Stehman, S.V.; Goetz, S.J.; Loveland, T.R.; et al. High-Resolution Global Maps of 21st-Century Forest Cover Change. *Science* **2013**, *342*, 850–853. [CrossRef]
88. Gorelick, N.; Hancher, M.; Dixon, M.; Ilyushchenko, S.; Thau, D.; Moore, R. Google Earth Engine: Planetary-scale geospatial analysis for everyone. *Remote Sens. Environ.* **2016**, *202*, 18–27. [CrossRef]
89. Pekel, J.-F.; Cottam, A.; Gorelick, N.; Belward, A.S. High-resolution mapping of global surface water and its long-term changes. *Nature* **2016**, *540*, 418–422. [CrossRef]

90. Sengupta, D.; Chen, R.; Meadows, M.E.; Choi, Y.R.; Banerjee, A.; Zilong, X. Mapping Trajectories of Coastal Land Reclamation in Nine Deltaic Megacities using Google Earth Engine. *Remote Sens.* **2019**, *11*, 2621. [CrossRef]
91. Li, Q.; Qiu, C.; Ma, L.; Schmitt, M.; Zhu, X.X. Mapping the Land Cover of Africa at 10 m Resolution from Multi-Source Remote Sensing Data with Google Earth Engine. *Remote Sens.* **2020**, *12*, 602. [CrossRef]
92. Banerjee, A.; Chen, R.E.; Meadows, M.; Singh, R.; Mal, S.; Sengupta, D. An Analysis of Long-Term Rainfall Trends and Variability in the Uttarakhand Himalaya Using Google Earth Engine. *Remote Sens.* **2020**, *12*, 709. [CrossRef]
93. Campbell, A.; Wang, Y. Salt marsh monitoring along the Mid-Atlantic coast by Google Earth Engine enabled time series. *PLoS ONE* **2020**, *15*, e0229605. [CrossRef] [PubMed]
94. Richards, D.R.; Belcher, R.N. Global Changes in Urban Vegetation Cover. *Remote Sens.* **2020**, *12*, 23. [CrossRef]
95. Stromann, O.; Nascetti, A.; Yousif, O.; Ban, Y. Dimensionality Reduction and Feature Selection for Object-Based Land Cover Classification based on Sentinel-1 and Sentinel-2 Time Series Using Google Earth Engine. *Remote Sens.* **2020**, *12*, 76. [CrossRef]
96. Zhang, K.; Dong, X.; Liu, Z.; Gao, W.; Hu, Z.; Wu, G. Mapping Tidal Flats with Landsat 8 Images and Google Earth Engine: A Case Study of the China 's Eastern Coastal Zone circa 2015. *Remote Sens.* **2019**, *11*, 924. [CrossRef]
97. Barnosky, A.D.; Hadly, E.A.; Bascompte, J.; Berlow, E.L.; Brown, J.H.; Fortelius, M.; Getz, W.M.; Harte, J.; Hastings, A.; Marquet, P.A.; et al. Approaching a state shift in Earth's biosphere. *Nature* **2012**, *486*, 52–58. [CrossRef]
98. Urban, M.C. Accelerating extinction risk from climate change. *Science* **2015**, *348*, 571–573. [CrossRef]
99. Wernberg, T.; Bennett, S.; Babcock, R.C.; de Bettignies, T.; Cure, K.; Depczynski, M.; Dufois, F.; Fromont, J.; Fulton, C.J.; Hovey, R.K.; et al. Climate-driven regime shift of a temperate marine ecosystem. *Science* **2016**, *353*, 169–172. [CrossRef]
100. Newbold, T. Future effects of climate and land-use change on terrestrial vertebrate community diversity under different scenarios. *Proc. R. Soc. B* **2018**, *285*, 20180792. [CrossRef] [PubMed]
101. Warren, R.; Price, J.; Graham, E.; Forstenhaeusler, N.; VanDerWal, J. The projected effect on insects, vertebrates, and plants of limiting global warming to 1.5 °C rather than 2 °C. *Science* **2018**, *360*, 791–795. [CrossRef] [PubMed]
102. Trisos, C.H.; Merow, C.; Pigot, A.L. The projected timing of abrupt ecological disruption from climate change. *Nature* **2020**. [CrossRef] [PubMed]
103. Thomas, C.; Gillingham, P.K. The performance of protected areas for biodiversity under climate change. *Biol. J. Linn. Soc.* **2015**, *115*, 718–730. [CrossRef]
104. Hockings, M.; Stolton, S.; Leverington, F.; Dudley, N.; Courrau, J. *Evaluating Effectiveness: A Framework for Assessing Management Effectiveness of Protected Areas*, 2nd ed.; IUCN: Gland, Switzerland; Cambridge, UK, 2006; p. 105.
105. Campbell, A.; Wang, Y. High Spatial Resolution Remote Sensing for Salt Marsh Mapping and Change Analysis at Fire Island National Seashore. *Remote Sens.* **2019**, *11*, 1107. [CrossRef]
106. Wang, Y.; Ngusaru, A.; Tobey, J.; Makota, V.; Bonynge, G.; Nugranad, J.; Traber, M.; Hale, L.; Bowen, R. Remote Sensing of Mangrove Change Along the Tanzania Coast. *Mar. Geod.* **2003**, *26*, 35–48. [CrossRef]
107. Gross, J.E.; Hansen, A.J.; Goetz, S.J.; Theobald, D.M.; Melton, F.M.; Piekielek, N.B.; Nemani, R.R. Remote Sensing for Inventory and Monitoring of U.S. National Parks. In *Remote Sensing of Protected Lands*; CRC Press: Boca Raton, FL, USA, 2011; pp. 29–56.
108. Nemani, R.R.; Hashimoto, H.; Votava, P.; Melton, F.; White, M.; Wang, W.; Michaelis, A.; Mutch, L.; Milesi, C.; Hiatt, S.; et al. Monitoring and forecasting ecosystem dynamics using the using the Terrestrial Observation and Prediction System (TOPS). *Remote Sens. Environ.* **2009**, *113*, 1497–1509. [CrossRef]
109. Zhao, J.; Wang, Y.; Hashimoto, H.; Melton, F.S.; Hiatt, S.H.; Zhang, H.; Nemani, R.R. The variation of land surface phenology from 1982 to 2006 along the Appalachian Trail. *IEEE Trans. Geosci. Remote Sens.* **2012**, *51*, 2087–2095. [CrossRef]
110. Meng, L.; Zhou, Y.; Lia, X.; Asrar, G.R.; Mao, J.; Alan, D.; Wanamaker, A.D., Jr.; Wang, Y. Divergent responses of spring phenology to daytime and nighttime warming. *Agric. For. Meteorol.* **2020**, *281*, 107832. [CrossRef]

111. Mao, D.; Wang, Z.; Dua, B.; Li, L.; Tian, Y.; Jia, M.; Zeng, Y.; Song, K.; Jiang, M.; Wang, Y. National wetland mapping in China: A new product resulting from object based and hierarchical classification of Landsat 8 OLI images. *ISPRS J. Photogramm. Remote Sens.* **2020**, *164*, 11–25. [CrossRef]
112. Wang, Y. Coastal Environments: Remote Sensing. In *Coastal and Marine Environments, the Handbook of Natural Resources*, 2nd ed.; CRC Press: Boca Raton, FL, USA, 2020; pp. 267–276.
113. Wang, Y.; Yésou, H. Remote Sensing of Floodpath Lakes and Wetlands: A Challenging Frontier in the Monitoring of Changing Environments. *Remote Sens.* **2018**, *10*, 1955. [CrossRef]
114. Popkin, G. U.S. government considers charging for popular earth-observing data. *Nature* **2018**, *556*, 417–418. [CrossRef]
115. Zhou, Y.; Smith, S.J.; Zhao, K.; Imhoff, M.; Thomson, A.; Bond-Lamberty, B.; Asrar, G.R.; Zhang, X.; He, C.; Elvidge, C.D. A global map of urban extent from nightlights. *Environ. Res. Lett.* **2015**, *10*, 054011. [CrossRef]
116. Fan, L.; Zhao, J.; Wang, Y.; Ren, Z.; Zhang, H.; Guo, X. Assessment of Night-Time Lighting for Global Terrestrial Protected and Wilderness Areas. *Remote Sens.* **2019**, *11*, 2699. [CrossRef]
117. Bei, X.; Yao, Y.; Zhang, L.; Xu, T.; Jia, K.; Zhang, X.; Shang, K.; Xu, J.; Chen, X. Long-Term Spatiotemporal Dynamics of Terrestrial Biophysical Variables in the Three-River Headwaters Region of China from Satellite and Meteorological Datasets. *Remote Sens.* **2019**, *11*, 1633. [CrossRef]
118. Fang, L.; Crocker, E.V.; Yang, J.; Yan, Y.; Yang, Y.; Liu, Z. Competition and Burn Severity Determine Post-Fire Sapling Recovery in a Nationally Protected Boreal Forest of China: An Analysis from Very High-Resolution Satellite Imagery. *Remote Sens.* **2019**, *11*, 603. [CrossRef]
119. Guirado, E.; Blanco-Sacristán, J.; Rigol-Sánchez, J.P.; Alcaraz-Segura, D.; Cabello, J. A Multi-Temporal Object-Based Image Analysis to Detect Long-Lived Shrub Cover Changes in Drylands. *Remote Sens.* **2019**, *11*, 2649. [CrossRef]
120. Chen, L.; Ren, C.; Zhang, B.; Wang, Z.; Wang, Y. Mapping Spatial Variations of Structure and Function Parameters for Forest Condition Assessment of the Changbai Mountain National Nature Reserve. *Remote Sens.* **2019**, *11*, 3004. [CrossRef]
121. Han, Y.; Li, Z.; Huang, C.; Zhou, Y.; Zong, S.; Hao, T.; Niu, H.; Yao, H. Monitoring Droughts in the Greater Changbai Mountains Using Multiple Remote Sensing-Based Drought Indices. *Remote Sens.* **2020**, *12*, 530. [CrossRef]
122. Li, B.; Huang, F.; Qin, L.; Qi, H.; Sun, N. Spatio-Temporal Variations of Carbon Use Efficiency in Natural Terrestrial Ecosystems and the Relationship with Climatic Factors in the Songnen Plain, China. *Remote Sens.* **2019**, *11*, 2513. [CrossRef]
123. Lu, C.; Ren, C.; Wang, Z.; Zhang, B.; Man, W.; Yu, H.; Gao, Y.; Liu, M. Monitoring and Assessment of Wetland Loss and Fragmentation in the Cross-Boundary Protected Area: A Case Study of Wusuli River Basin. *Remote Sens.* **2019**, *11*, 2581. [CrossRef]
124. Shirvani, Z. A Holistic Analysis for Landslide Susceptibility Mapping Applying Geographic Object-Based Random Forest: A Comparison between Protected and Non-Protected Forests. *Remote Sens.* **2020**, *12*, 434. [CrossRef]
125. Xu, J.; Gao, C.; Wang, Y. Extraction of Spatial and Temporal Patterns of Concentrations of Chlorophyll-a and Total Suspended Matter in Poyang Lake Using GF-1 Satellite Data. *Remote Sens.* **2020**, *12*, 622. [CrossRef]
126. Duan, P.; Wang, Y.; Yin, P. Remote Sensing Applications in Monitoring of Protected Areas: A Bibliometric Analysis. *Remote Sens.* **2020**, *12*, 772. [CrossRef]
127. Wang, Y. Protected Areas: Remote Sensing. In *Landscape and Land Capacity, the Handbook of Natural Resources*, 2nd ed.; CRC Press: Boca Raton, FL, USA, 2020; pp. 75–84.

© 2020 by the authors. Licensee MDPI, Basel, Switzerland. This article is an open access article distributed under the terms and conditions of the Creative Commons Attribution (CC BY) license (http://creativecommons.org/licenses/by/4.0/).

Article

Assessment of Night-Time Lighting for Global Terrestrial Protected and Wilderness Areas

Liangxian Fan [1,2], Jianjun Zhao [1,2,*], Yeqiao Wang [3,*], Zhoupeng Ren [4], Hongyan Zhang [1,2] and Xiaoyi Guo [1,2]

1. Key Laboratory of Geographical Processes and Ecological Security in Changbai Mountains, Ministry of Education, School of Geographical Sciences, Northeast Normal University, Changchun 130024, China; fanlx202@nenu.edu.cn (L.F.); Zhy@nenu.edu.cn (H.Z.); Guoxy914@nenu.edu.cn (X.G.)
2. Urban Remote Sensing Application Innovation Center, School of Geographical Sciences, Northeast Normal University, Changchun 130024, China
3. Department of Natural Resources Science, University of Rhode Island, Kingston, RI 02881, USA
4. State Key Laboratory of Resources and Environmental Information System (LREIS), Institute of Geographic Sciences and Natural Resources Research, Chinese Academy of Sciences, Beijing 100101, China; renzp@lreis.ac.cn
* Correspondence: zhaojj662@nenu.edu.cn (J.Z.); yqwang@uri.edu (Y.W.)

Received: 1 October 2019; Accepted: 15 November 2019; Published: 18 November 2019

Abstract: Protected areas (PAs) play an important role in biodiversity conservation and ecosystem integrity. However, human development has threatened and affected the function and effectiveness of PAs. The Defense Meteorological Satellite Program/Operational Linescan System (DMSP/OLS) night-time stable light (NTL) data have proven to be an effective indicator of the intensity and change of human-induced urban development over a long time span and at a larger spatial scale. We used the NTL data from 1992 to 2013 to characterize the human-induced urban development and studied the spatial and temporal variation of the NTL of global terrestrial PAs. We selected seven types of PAs defined by the International Union for Conversation of Nature (IUCN), including strict nature reserve (Ia), wilderness area (Ib), national park (II), natural monument or feature (III), habitat/species management area (IV), protected landscape/seascape (V), and protected area with sustainable use of natural resources (VI). We evaluated the NTL digital number (DN) in PAs and their surrounding buffer zones, i.e., 0–1 km, 1–5 km, 5–10 km, 10–25 km, 25–50 km, and 50–100 km. The results revealed the level, growth rate, trend, and distribution pattern of NTL in PAs. Within PAs, areas of types V and Ib had the highest and lowest NTL levels, respectively. In the surrounding 1–100 km buffer zones, type V PAs also had the highest NTL level, but type VI PAs had the lowest NTL level. The NTL level in the areas surrounding PAs was higher than that within PAs. Types Ia and III PAs showed the highest and lowest NTL growth rate from 1992 to 2013, respectively, both inside and outside of PAs. The NTL distributions surrounding the Ib and VI PAs were different from other types. The areas close to Ib and VI boundaries, i.e., in the 0–25 km buffer zones, showed lower NTL levels, for which the highest NTL level was observed within the 25–100 km buffer zone. However, other types of PAs showed the opposite NTL patterns. The NTL level was lower in the distant buffer zones, and the lowest night light was within the 1–25 km buffer zones. Globally, 6.9% of PAs are being affected by NTL. Conditions of wilderness areas, e.g., high latitude regions, Tibetan Plateau, Amazon, and Caribbean, are the least affected by NTL. The PAs in Europe, Asia, and North America are more affected by NTL than South America, Africa, and Oceania.

Keywords: Protected areas; Night-time light; Global lightscape assessment; Human disturbance

1. Introduction

A protected area (PA) is defined as a geographical space that is recognized, dedicated, and managed, through legal or other effective means, to achieve the long-term conservation of nature with associated ecosystem services and cultural values in mind [1]. In general, protected areas include national parks, national forests, national seashores, all levels of natural reserves, wildlife refuges and sanctuaries, and designated areas for the conservation of native biological diversity and natural and cultural heritage and significance. Protected areas also include some of the last frontiers that have unique landscape characteristics and ecosystem functions in wilderness conditions [2].

PAs reflect the efforts to protect the world's threatened species and their habitats. PAs are increasingly recognized as essential providers of ecosystem services and biological resources, key components in climate change mitigation strategies, and in some cases, are vehicles for protecting threatened human communities or sites of great cultural and spiritual value.

Humans have created protected areas over the past millennia for a multitude of reasons [3,4]. The establishment of Yellowstone National Park in 1872 by the US Congress ushered in the modern era of governmental protection of natural areas, which catalyzed a global movement [5,6]. However, even with the implementation of a tremendous variety of monitoring programs, as well as conservation planning efforts and achievements, species' population decline, biodiversity loss, extinction, system degradation, pathogen spread, and state change events are occurring at unprecedented rates [7,8]. The effects are augmented by continued changes in land use, urbanization, and invasive spread alongside the direct, indirect, and interactive effects of climate change and disruption [3,4]. Protected areas become more important in serving as indicators of ecosystem conditions and functions, either by their status and/or by comparison with unprotected adjacent areas. Protected areas are highly prized by society with diversified representative characteristics. Earth's remaining wilderness areas are becoming increasingly important buffers against the changing environment. However, they are not yet an explicit target in international policy frameworks [9].

Over the past four decades, more PAs have been established around the world. PAs play a vital role in conserving biodiversity; specifically, PAs provide a paradise for endangered and species from declining habitats and a sanctuary for over-harvested and poached species [10–12]. PAs have a positive impact on the local environment, such as maintaining water resources, regulating climate, and preventing forest damage [13], among other benefits [14,15].

On the other hand, ecosystems of PAs have been disturbed by human development, in particular, urbanization, such that biodiversity and habitats in those PAs have been reduced [16–21]. The effectiveness of PAs has been affected by the land use of the surrounding areas. Nightlight may impose problems to PAs [22,23]. The threat from human development to PAs is a concern. Currently, assessment studies on the impact of human development on PAs are mainly based on data, such as human population [14] and housing [24] at regional or local scales. However, accurate, reliable, and comprehensive population or housing statistics are often not available in global scale monitoring.

The Defense Meteorological Satellite Program/Operational Linescan System (DMSP/OLS) night-time stable light (NTL) dataset is a timely and effective data source for monitoring human development and mapping the dynamics of land cover on a regional or global scale [25–28]. The NTL data have a wide time span and are suitable for dynamic analysis on long-term sequences. The NTL can be measured and used as a proxy of human development. NTL imposes impacts on several taxa in terrestrial and aquatic ecosystems, including mammals, birds, reptiles, amphibians, fish, invertebrates, and plants. It is necessary to study the lighting conditions of reserves [29–31]. DMSP/OLS NTL is an excellent data source for analyzing the impact of human development on various ecosystems [32], vegetation [33], and biodiversity [34] over long-term sequences and over large geographical areas and in remote locations [35].

DMSP/OLS NTL data have been used as an effective indicator to evaluate the conservation of protected areas [36–38]. Reported studies have investigated the difference in NTL between the interior and the surrounding areas and have revealed that the NTL level of the PAs is lower than that of the surrounding areas [29]. Studies suggested that the NTL level of the boundary of PAs is particularly high [39]. In addition, some studies have combined NTL with other data to conduct research, for example, Geldmann et al. used the NTL and population data to construct a temporal human pressure index (THPI) on the time series, which quantifies the human pressure on the protected area. However, the spatial resolution of the study was 10 km^2, and only attempted comparisons between 1995 and 2010 [40].

In this study, we used NTL data as an indicator of the intensity of human development to portray the characteristics of NTL in various types of PAs defined by the International Union for Conversation of Nature (IUCN). The objectives of this study were to: (1) reveal human development surrounding different types of terrestrial PAs, growth rate, and trend characteristics measured using the NTL level; and (2) make an assessment about the NTL effects on global terrestrial PAs.

2. Materials and Methods

2.1. Protected Areas and Data

We used the 2019 World Database on Protected Areas (WDPA, https://protectedplanet.net/), as the most updated data source for obtaining the coverage of global protected areas. We selected seven types of terrestrial PAs defined by the IUCN, including Ia, Ib, and II–VI, as defined in Table 1. In order to explore the spatial distribution and change of NTL within PAs and their vicinity, we examined buffer zones with distances of 0–1 km, 1–5 km, 5–10 km, 10–25 km, 25–50 km, and 50–100 km from PA boundaries. This range of buffer distances was chosen to capture different types of NTL effects. Human activities within PA boundaries and within a 1 km distance exert a direct and significant influence on protected areas, e.g., habitat loss, noise, and light pollution [41]. At further distances, artificial surfaces contribute to the landscape disturbances and effects, such as the isolation of protected areas, disruption of connectivity, and introduction and spread of invasive species. Even as far as 50 km from the protected area, human development can impose effects on PAs. PAs may be impacted even if the source of lighting lies kilometers away owing to skyglow [24].

Table 1. Seven types of terrestrial protected areas defined by IUCN.

IUCN PAs Type	IUCN PAs Code	Number of PAs	Area (km^2)	Description
Strict Nature Reserve	Ia	11,921	3,874,328	PAs that are strictly set aside to protect biodiversity and also possibly geological/geomorphological features, where human visitation, use, and impacts are strictly controlled and limited to ensure the protection of the conservation values.
Wilderness Area	Ib	3114	1,152,403	PAs that are usually large unmodified or slightly modified areas, retaining their natural character and influence, without permanent or significant human habitation.
National Park	II	5523	6,178,074	Large natural or near natural areas set aside to protect large-scale ecological processes, along with the complement of species and ecosystems characteristic of the area, which also provide a foundation for environmentally and culturally compatible spiritual, scientific, educational, recreational, and visitor opportunities.
Natural Monument or Feature	III	14,317	434,460	PAs set aside to protect specific natural monuments, which can be a landform; sea mount; submarine cavern; geological feature, such as a cave; or even a living feature, such as an ancient grove. They are generally quite small protected areas and often have a high visitor value.

Table 1. Cont.

IUCN PAs Type	IUCN PAs Code	Number of PAs	Area (km^2)	Description
Habitat/Species Management Area	IV	56,836	7,250,215	PAs aiming to protect particular species or habitats and management reflects this priority. Many category IV PAs will need regular, active interventions to address the requirements of particular species or to maintain habitats, but this is not a requirement of the category.
Protected Landscape/Seascape	V	44,915	4,351,409	PAs where the interaction of people and nature over time have produced the area of distinct character with significant ecological, biological, cultural, and scenic value.
Protected Area with Sustainable use of Natural Resources	VI	7177	12,757,697	PAs that conserve ecosystems and habitats, together with associated cultural values and traditional natural resource management systems. They are generally large, with most of the area in a natural condition, where a proportion is under sustainable natural resource management and where low-level, non-industrial use of natural resources compatible with nature conservation is seen as one of the main aims of the area.

2.2. DMSP/OLS NTL Data

The Operational Linescan System (OLS) is one of the sensors that is carried by the Defense Meteorological Satellite Program (DMSP). The DMSP/OLS NTL data came from the National Geophysical Data Center (NGDC) under the National Oceanic and Atmospheric Administration (NOAA), which eliminates accidental noise effects, such as clouds and flaring, with a 0.008333 degrees spatial resolution. The number of each pixel is not the actual light level on the ground, but rather the relative brightness level recorded as a digital number (DN) from 0 to 63. Currently, NTL data from 1992 to 2013 are available through online access. DMSP/OLS is different from Landsat, Satellite Pour l'Observation de la Terre (SPOT), and the Advanced Very High Resolution Radiometer (AVHRR) sensors that use the ground object to monitor the reflected radiation characteristics of sunlight. The DMSP/OLS sensor can work at night and capture the low intensity of urban lights and even small-scale residential areas and traffic lights, providing powerful data support for monitoring human development research on a large scale [42].

2.3. Data Processing

For origin data, discrepancies appeared between the DN values and the number of lit pixels from different satellites in the same year, and abnormal fluctuations appeared in the DN values for different years derived from the same satellite. It was necessary to calibrate the original NTL between different years and satellites. In this study, we assumed that all regions of the world that developed positive reflection in NTL would keep the DN value and the number of lit pixels either consequently increased or remained the same. The NTL were corrected using four main steps as described below.

2.3.1. Inter-Annual Correction

As the economy and population continue to develop over time, the assumption was that all PA regions would be positively developing and increasing in terms of the amount of NTL. Therefore, the DN values of each pixel in the time series would either increase or remain unchanged. With this assumption, the light pixels detected in the early image would remain in the latter image, and the DN value should not be larger than the DN value detected in their subsequent images. If the pixel that detected the light in the earlier image disappeared in the later year, the pixel with light would be considered as an unstable light pixel, and its DN value would be replaced with a value of 0. Second, the DN value of each stable light pixel was corrected to ensure that the DN value in the early image was smaller than the DN value of the corresponding position pixel in the later image. Due to the replacement of satellites between 1992 and 2013, the light sensitivity of each sensor was different, leading to the difference in the number of lit pixels and the DN values of the corresponding position pixels from different sensors. Thus, each satellite was interannually corrected using the following step [43]:

$$DN_{(n,i)} = \begin{cases} 0, & DN_{(n+1,i)} = 0 \\ DN_{(n-1,i)}, & DN_{(n+1,i)} > 0 \text{ nd } DN_{(n-1,i)} > D\, DN_{(n,i)} \\ DN_{(n,i)}, & \text{otherwise} \end{cases} \quad (n = 1992, 1993, \ldots, 2013), \quad (1)$$

where $DN_{(n-1,i)}$, $DN_{(n,i)}$, and $DN_{(n+1,i)}$ are the DN values of the ith lit pixel of the NTL data in the ($n-1$)th, the nth, and ($n+1$)th years, respectively.

2.3.2. Inter-Satellite Correction

The NTL from 1992 to 2013 were derived from multiple sensors. The data were not continuous and the data between different sensors could not be directly compared. It was necessary to correct the sensors to improve the continuity and comparability of the NTL. Although the image of satellite F18 did not intersect with those of the other satellites, there was an intersection between F10 and F12, F12 and F14, F14 and F5, and F15 and F16. First, based on F10, using the data of the intersection year, 100,000 pixels were randomly selected on each continent to establish a minimum linear-square regression model between two satellites; thus, the data of F12 could be corrected, which could then be used to correct the data of F14. The same process was applied to data from other satellite pairs.

The regression formulas are as follows:
Correct F12 based on F10:

$$F12 = 0.870 \times F10 - 0.078 \left(R^2 = 0.925\right) \quad (2)$$

Correct F14 based on F12:

$$F14 = 0.927 \times F12 - 1.709 \left(R^2 = 0.965\right) \quad (3)$$

Correct F15 based on F14:

$$F15 = 0.961 \times F14 + 1.408 \left(R^2 = 0.951\right) \quad (4)$$

Correct F16 based on F15:

$$F16 = 1.062 \times F15 + 1.672 \left(R^2 = 0.960\right) \quad (5)$$

2.3.3. Intra-Annual Correction

The intra-annual composition correction used images from two satellites in the same year to remove any intra-annually unstable lit pixels. There were two images from two different satellites in 1994 and 1997–2008. We used the average DN values of the two NTL images to calculate the annual NTL data for these years. First, we examined all the lit pixels to determine whether the pixel was an unstable light pixel during the year. If only one satellite was detected, the light pixel was defined as an unstable light pixel during the year. Second, in intra-annual composites, the DN values of intra-annually unstable lit pixels were replaced with values of 0, and the DN value of each intra-annually stable lit pixel was replaced by the average DN value of two NTL images from the same year. This produced one intra-annual composite for each year [43].

$$DN_{(n,i)} = \begin{cases} 0, & DN^a_{(n,i)} = 0 | DN^b_{(n,i)} = 0 \\ DN^a_{(n,i)} + DN^b_{(n,i)}, & \text{otherwise} \end{cases} \quad (n = 1992, 1993, \ldots, 2013), \qquad (6)$$

where $DN^a_{(n,i)}$ and $DN^b_{(n,i)}$ are the DN values of the ith lit pixel from two NTL data in the nth year, and $DN_{(n,i)}$ is the DN value of the ith lit pixel of the intra-annual composite in the nth year.

2.3.4. Correction for Different Sensors

We performed another first-step correction on this series of data after the first three steps, which resulted in a dataset that grew positively and with continuity.

2.3.5. Evaluation of the Calibrated NTL Time Series

A number of studies have shown that the DMSP/OLS NTL data are correlated with economic activities [44–48] and have used gross domestic product (GDP) and electricity consumption (EC) to evaluate the performance of intercalibration techniques [28,49–53]. We applied this assessment method in this study to evaluate the calibrated NTL time series. We obtained country-level GDP data (1992–2013) from the World Bank (http://data.worldbank.org/) and EC records derived from International Energy Statistics (http://www.eia.gov/beta/international/data/browser/#). We compared the raw NTL and the calibrated NTL with GDP and EC data (Figure 1) to reduce the inconsistency of the data and enhanced correlation with GDP and EC. We calculated the Pearson correlation between the raw NTL with GDP and EC, and between the calibrated NTL and the two ancillary datasets (Figure 2). The calibrated method improved the correlations between the NTL time series and GDP and EC.

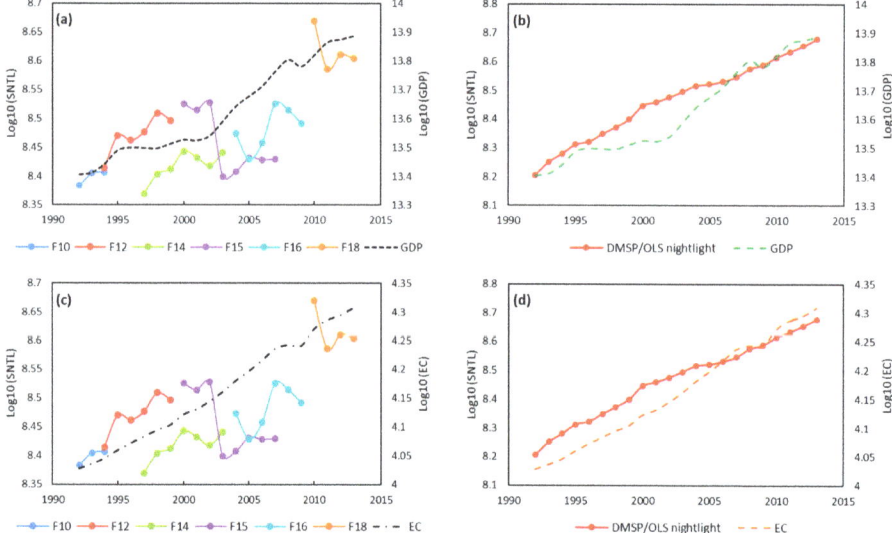

Figure 1. Gross domestic product (GDP; unit: trillions of dollars) and global the sum of NTL (SNTL) (1992–2013) from (**a**) the raw NTL and (**b**) the calibrated NTL. Electricity consumption (EC; unit: billion kWh) and global SNTL (1992–2013) from (**c**) the raw NTL and (**d**) the calibrated NTL. DMSP/OLS—Defense Meteorological Satellite Program/ Operational Linescan System.

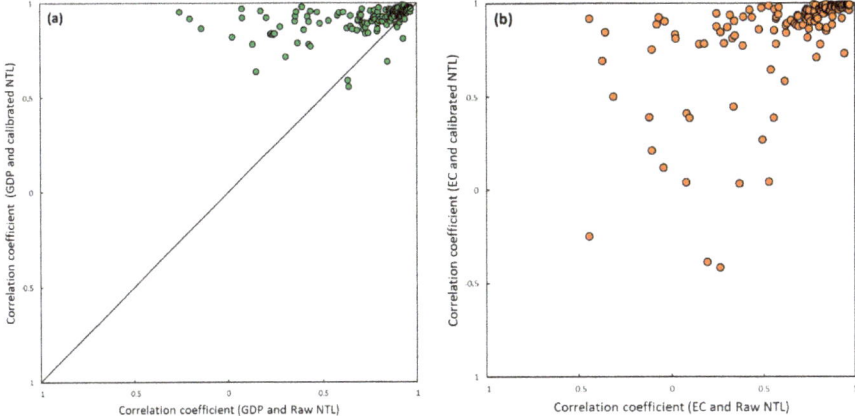

Figure 2. The correlation of GDP (**a**) and EC (**b**) with the raw NTL and the calibrated NTL. The black line is the 1:1 line.

2.4. Calculation of the Trend Using Sen's Slope

The Sen's slope method can be used to avoid the loss of time series data and the influence of data distribution on the analysis results to eliminate the interference of outliers in a time series [54]. Sen's slope has been used in the analysis of long-term sequence data sets to detect the magnitude of the change trend. We used Sen's slope to calculate the trend of the NTL from 1992 to 2013. The calculation formula is:

$$Q = \text{median} \frac{X_j - X_i}{j - i} \qquad 0 < i < j \leq n \qquad (7)$$

where Q is the Sen's slope; and Xj and Xi are the average DN value corresponding to j and i year, respectively. If the length of the time series is n, the number of Qi is $N = n(n-1)/2$; and the Q is determined by N.

$$Q = \begin{cases} Q_{[(N-1)/2]} & N \text{ is odd} \\ \left(Q_{(N/2)} + Q_{[(N+2)/2]}\right)/2 & N \text{ is even} \end{cases} \qquad (8)$$

where $Q > 0$, $Q < 0$ and $Q = 0$ indicate that there is a rising trend, a downward trend, and no obvious trend, respectively.

3. Results

3.1. Spatial Distribution and Trend of NTL for Global PAs

We superimposed the global PAs and the global NTL, except for Antarctica (Figure 3). The result revealed that the DN values for most of the PAs were lower than that of the surrounding areas. Further assessment ranked the average NTL DN values of all PAs on each continent as: Europe > Asia > North America > South America > Africa > Oceania. The DN values in Europe was much higher than those in the other continents (Figure 4). The ranking of the average change trend on all continents was consistent with the average DN value, i.e., the highest was in Europe and the lowest was in Oceania (Figure 5).

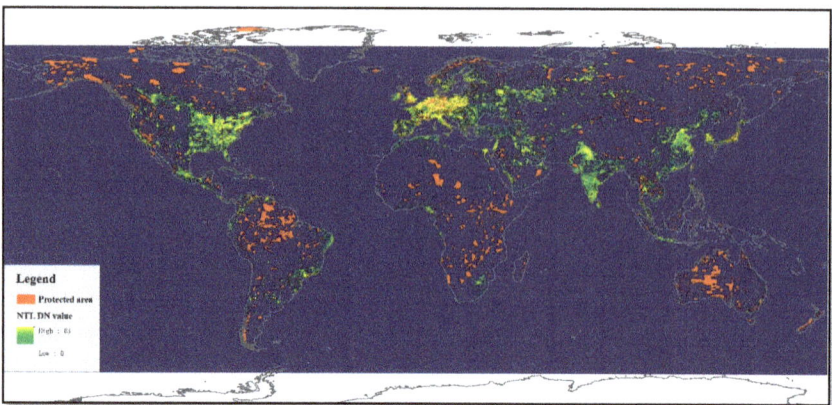

Figure 3. Spatial distribution of NTL in association with global terrestrial protected areas (PAs). The orange-colored areas represent the PAs; the dark blue-colored area (including oceans) were in darkness; the yellow-colored areas had a higher NTL digital number (DN) value than the green-colored areas; the white-colored area had no data.

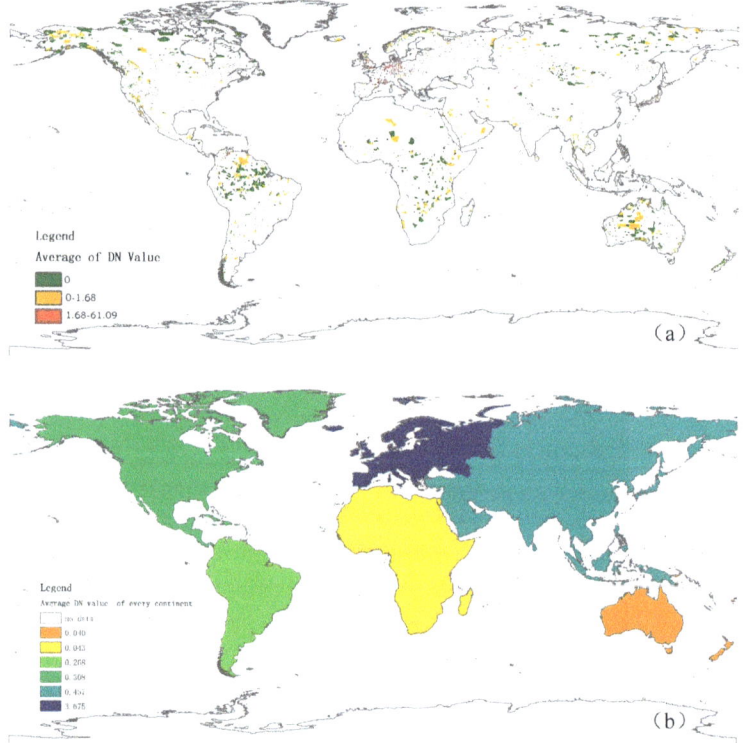

Figure 4. (**a**) Average NTL DN value for every PA and (**b**) average NTL DN value for every continent. For (**a**), we used the natural breaks (Jens) method to divide the PAs into three levels according to the average DN value. The green-colored PAs were those with the average value of 0, the yellow-colored PAs had a lower than average DN value, and the red-colored PAs had a higher than average DN value. For (**b**), the ranking of the average NTL DN value of PAs in each continent gave the order as Europe > Asia > North America > South America > Africa > Oceania.

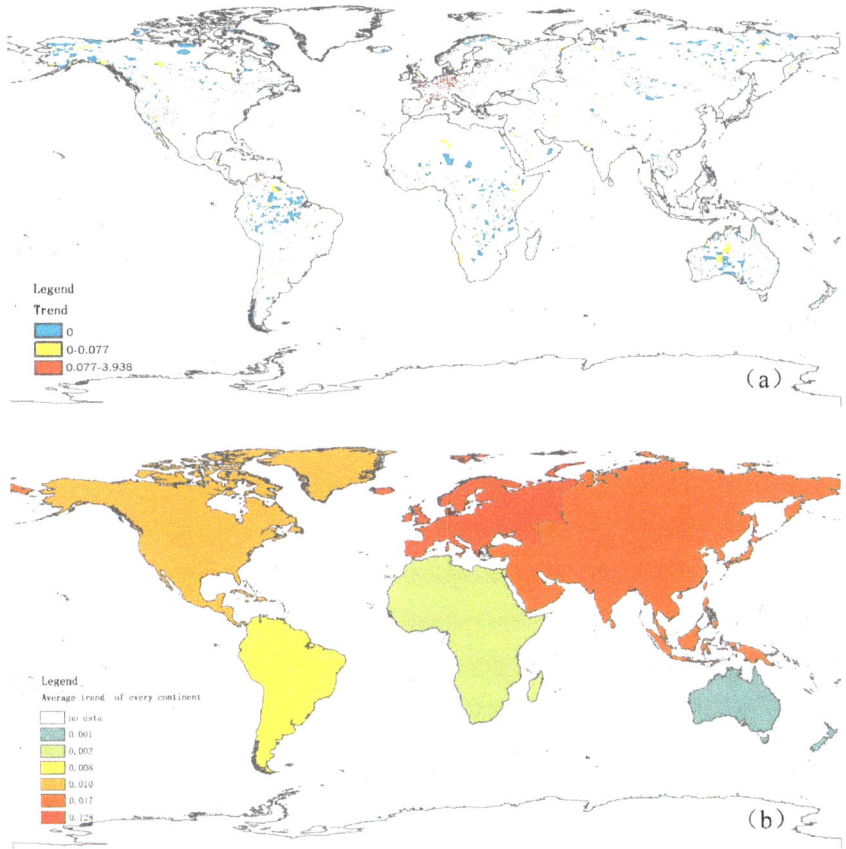

Figure 5. (**a**) Average trend of NTL of the global terrestrial PAs and (**b**) average trend of every continent. For (**a**), we used the natural breaks (Jens) method to divide the PAs into three levels according to the average trend of NTL. The blue-colored PAs had an average trend of 0, the yellow-colored PAs had a lower than average trend, and the red-colored PAs had a higher than average trend. For (**b**), the ranking of the average NTL trend of PAs in each continent gave the order as Europe > Asia > North America > South America > Africa > Oceania.

3.2. Average NTL Level in Different PAs and Buffers

Figure 6 illustrates that NTL DN values in types III and V PAs were significantly higher than that of the other five types of PAs in the interior and buffer zones. With the buffer radius increased, the average NTL DN values of all PAs increased first and then decreased (Figure 7). The highest DN values of different types of PAs appeared in the range of 1–10 km.

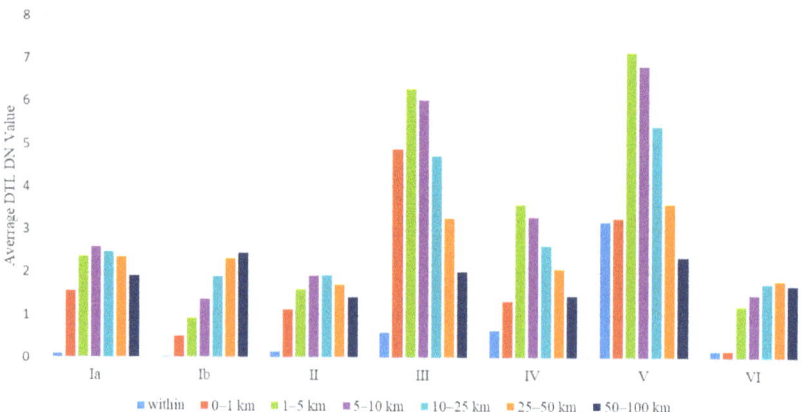

Figure 6. NTL level on different buffers for each type of PA. Columns with different colors represent the interior of PAs and different buffer zones.

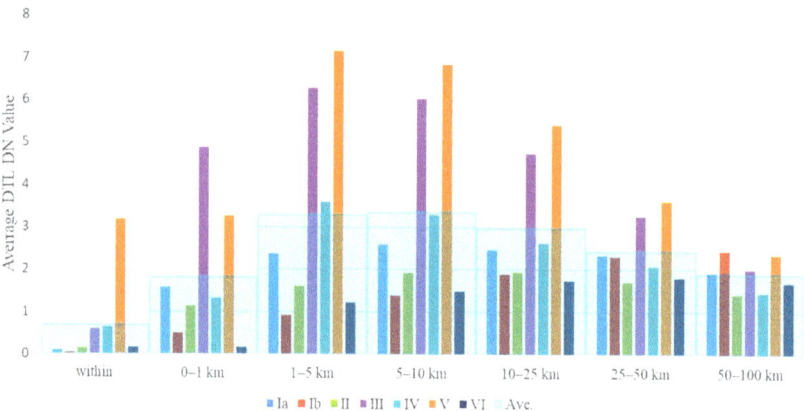

Figure 7. NTL level for each type of PA in buffer zones. Columns with different colors represent different types of PAs. The light blue shadowed areas represents the mean DN values of PAs and buffers.

From the time series, the NTL level within and outside PAs increased without exception, but the NTL distributions among seven types of PAs had different characteristics (Figure 8).

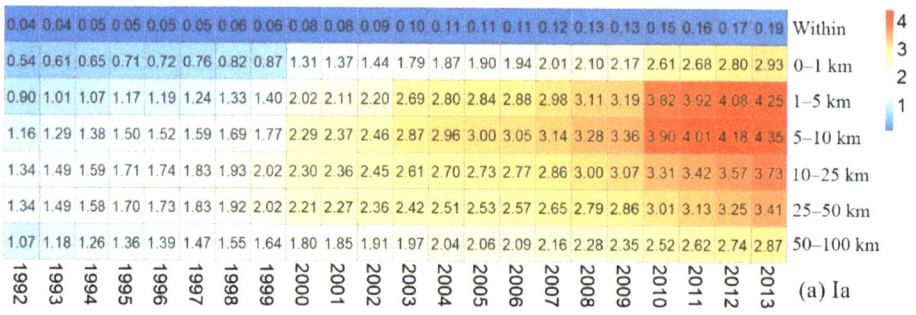

Figure 8. *Cont.*

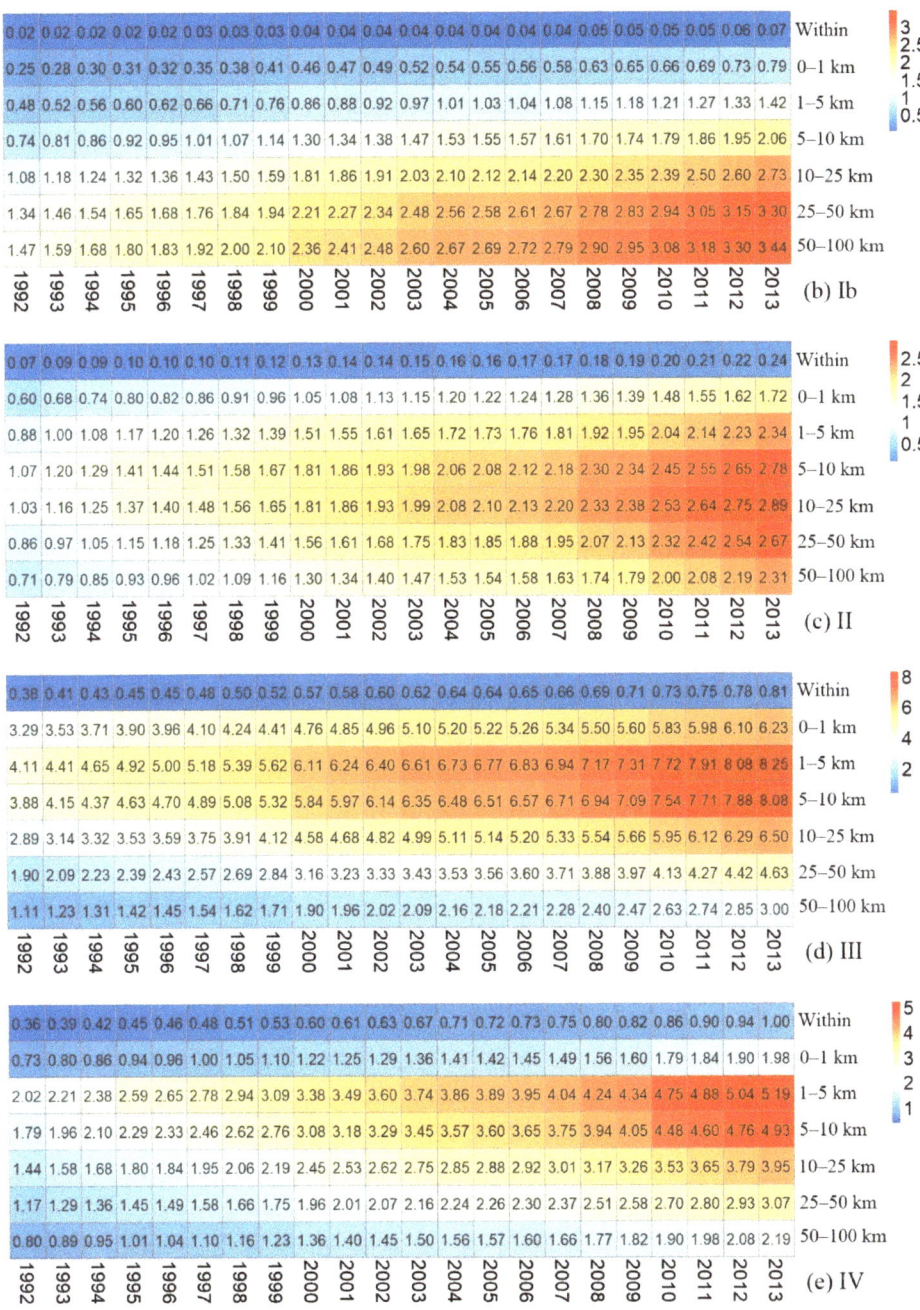

Figure 8. Cont.

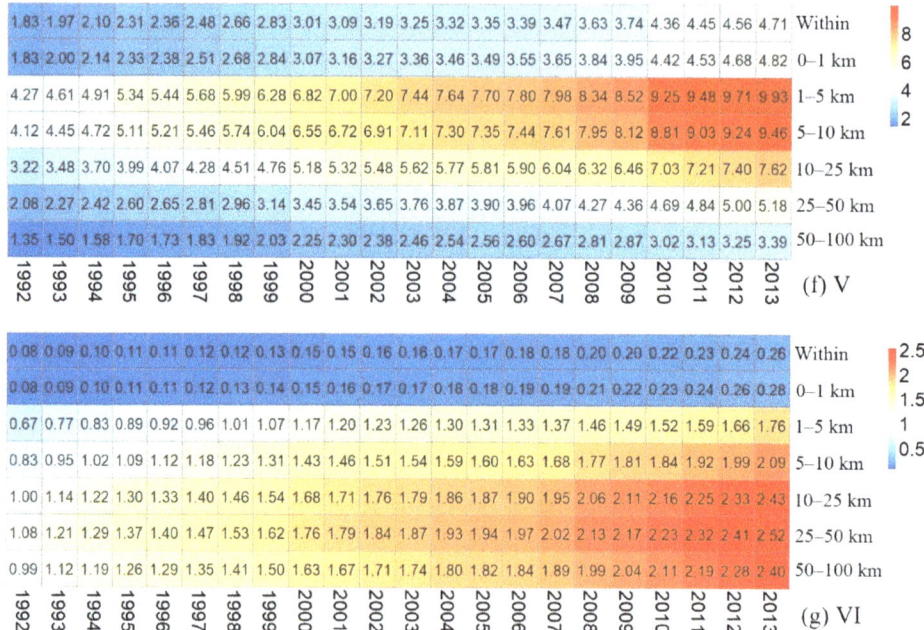

Figure 8. Changes within PAs and in different buffer zones in the NTL level of different types of PAs in the time series. The numbers in the grid represent the average NTL DN values. The color of the grids is from blue to red, and the corresponding values are from small to large. (**a**) strict nature reserve (Ia), (**b**) wilderness area (Ib), (**c**) national park (II), (**d**) natural monument or feature (III), (**e**) habitat/species management area (IV), (**f**) protected landscape/seascape (V), and (**g**) protected area with sustainable use of natural resources (VI).

Type Ia PAs represent strict nature reserves. As shown in Figure 8a, the average NTL DN value in buffer zones of type Ia PAs increased in the 0–10 km buffer zone and then decreased with over the 10–100 km buffer zone, reaching the maximum DN value in the 5–10 km buffer zone. The average DN value in the 0–100 km buffer zone of type Ia was much higher than the average DN value within the boundaries of the PAs.

Type Ib PAs represent wilderness area with large unmodified or slightly modified areas (Figure 8b). The average DN value within boundaries of type Ib was lower than that of type Ia PAs with the lowest NTL value among all types of PAs. The average DN value of the type Ib buffer increased from the 0–1 km to the 50–100 km buffer zones.

Type II PAs represent national parks (Figure 8c). The change trend of the DN value of the type II PAs buffer was similar to that of Ia (Figure 8a), suggesting a trend of increasing in the close buffer zones (0–10 km) and decreasing in distant buffer zones (10–100 km), with the maximum DN value in the 5–10 km buffer areas. The fluctuation of the NTL DN value of type II was the lowest among all types of PAs.

Type III PAs represent natural monuments or features (Figure 8d). The difference between DN values within PAs and 0–100 km buffer zones were the highest among all types of PAs. The NTL outside of type III PAs increased in the 1–5 km buffer zone and then decreased, indicating more human development in this range.

Type IV PAs represent habitat/species management areas (Figure 8e). The nearest (0–1 km) buffer zone had the lowest DN value. The area with the highest NTL level was concentrated in the 1–10 km buffer zone.

Type V PAs represent protected landscapes/seascapes (Figure 8f). The DN value of type V PAs was much higher than that of the other types of PAs (Figure 9). The average DN value and the range of the buffer zones was also the highest among all PA types (Figure 9). The brighter areas around the PAs were concentrated within 1–25 km, and the 0–1 km and 25–100 km areas were darker.

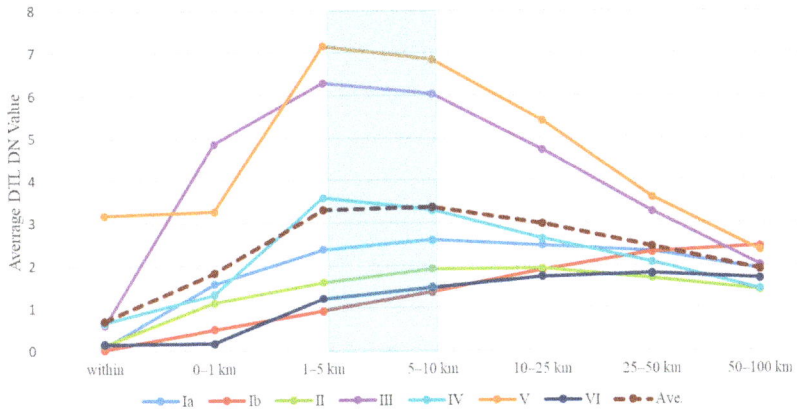

Figure 9. The NTL level of different types of PAs in their interior and surrounding buffer zones. The blue shadowed area represents the fact that the 1–10 km buffer zone had the highest NTL level.

The average DN values in type VI PAs buffers were the lowest among all PAs (Figure 8g). The DN values of the 0–1 km buffer zone were the lowest among all types of PAs. As distance from the boundaries of the PAs increased, the DN value in the buffer zones increased, and reached a maximum in the 25–50 km buffer zone.

The results showed that the NTL levels within PAs were lower than that of surrounding areas, indicating that human development was limited and controlled by the boundary of the PAs. The NTL level within and outside the PAs increased from 1992 to 2013 (Figure 8). The areas with the highest NTL level in the buffer zone constantly approached the boundary of the type Ia PAs (Figure 8a).

In general, the ranking of the NTL level among the seven types of PAs was V > IV > III > VI > II > Ia > Ib (Table 2). The ranking of NTL level in the 0–100 km buffer zones outside the boundaries of types of PAs was V > III > IV > Ia > II > Ib > VI (Figure 7). The ranking of the NTL level of buffer zones and the interior was 5–10 km > 1–5 km > 10–25 km > 25–50 km > 50–100 km > 0–1 km. The NTL level within the PAs was significantly lower than that observed outside the PAs (Table 2). For most types of PAs, e.g., Ia, II, III, IV, and V, the brightest areas around the PAs were concentrated in the 1–25 km buffer zone. However, for types Ib and VI, the brightest areas around PAs were concentrated at further distances (Figure 9).

Table 2. NTL level of interior and buffer zones for the seven types of PAs.

PA Type	Interior	0–1 km	1–5 km	5–10 km	10–25 km	25–50 km	50–100 km
Ia	0.098	1.572	2.372	2.598	2.478	2.344	1.916↑
Ib	0.039	0.496	0.920	1.380	1.897	2.317	2.452↑
II	0.147	1.129	1.602	1.920	1.932	1.702	1.427
III	0.594	4.868↑	6.289↑	6.038	4.734↑	3.272↑	2.013↑
IV	0.652	1.317	3.593↑	3.301	2.631	2.078	1.456
V	3.184↑	3.270↑	7.151↑	6.840↑	5.418↑	3.610↑	2.357↑
VI	0.160	0.169	1.217	1.483	1.739	1.811	1.692
Ave.	0.696	1.832	3.307	3.366	2.975	2.448	1.902

3.3. NTL Growth Rate in Different PAs and Buffers

The 1992–2013 NTL growth rates of every type of PAs and their buffer zones were doubled, except within the 0–1 km for type III PAs (Figure 10), indicating enhanced human development within and around PAs. The NTL growth rate for type Ia PAs was the highest both within and the surrounding outside among all types. Especially, NTL growth rates within the PAs and in the 0–10 km buffer zones were much higher than that of other types. The DN value of the interior Ia increased from 0.04 in 1992 to 0.19 in 2013, representing an increase of 378%. The 0–1 km buffer zone increased by 441% from 0.54 in 1992 to 2.93 in 2013. The growth rates of the 1–5 km and 5–10 km buffer zones were 372% and 274%, respectively. The NTL growth rate from 1992 to 2013 for type Ib PAs decreased from 217% for the internal area to 135% for the furthest buffer zone. The growth rate in the 0–1 km buffer zone was 215%, which was much greater than that of other buffer zones. NTL growth rates for type Ib PAs were much lower than that of Ia. The NTL growth rate for type II PAs ranged from 160% in the 5–10 km buffer zone to 226% in the outermost 50–100 km buffer zone. The growth rate within the type II PAs was 222%, ranking second. For type III PAs, the NTL growth rate was 114% within the protected boundary area. The average growth rate of the buffer zones was the lowest among all types of PAs. The growth rate increased with distance, from 90% in the 0–1 km buffer zone to 170% in the 50–100 km buffer zone. The NTL growth rate within type IV PAs (178%) was slightly higher than that of its surrounding areas (157–175%). The difference in growth rates between buffer zones was small, with a minimum of 157% in the 1–5 km buffer zone and a maximum of 175% in the 5–25 km buffer zone. The NTL growth rate within the type V PAs was 157%. The growth rate of the 0–1 km buffer zone was 164%, which was the highest among all type V PA buffers. The growth rate of the remaining buffer zones decreased with distance from the boundary, i.e., from 133% in the 1–5 km buffer zone to 151% in the 50–100 km buffer zone. From 1992 to 2013, the NTL growth rate of type VI PAs was 245%. The buffers near the boundary had a higher growth rate than that of the buffers farther away. The highest growth rate (263%) occurred in the 0–1 km buffer zone, and the minimum growth rate was 142% in the 50–100 km buffer zone.

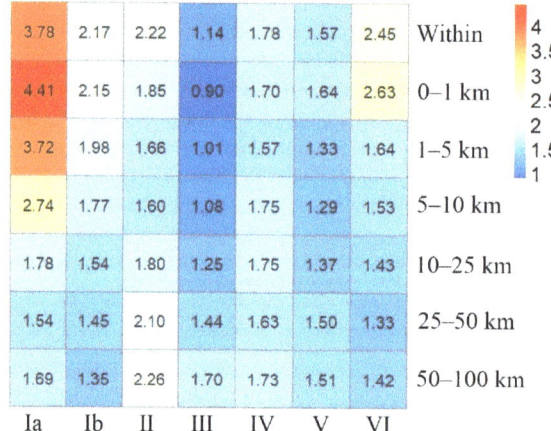

Figure 10. Growth rate of NTL within every type of PA and the buffer zones.

3.4. Trends in Different PAs and Buffer Zones

NTL DN values of all PAs and their buffers showed a significant increasing trend from 1992 to 2013 (Figure 11). Except for the type V PAs, the NTL trends within the other six types of PAs were lower than that in any other buffer zone at the 0–100 km. The change trend within type Ib PAs was the lowest of all the PAs, with a value close to 0. The change trend of type V (0.13 DN/year) was the highest, which was greater than the trend in 50–100 km buffer zone. The average trend of type V PAs buffers was the lowest among all types of PAs.

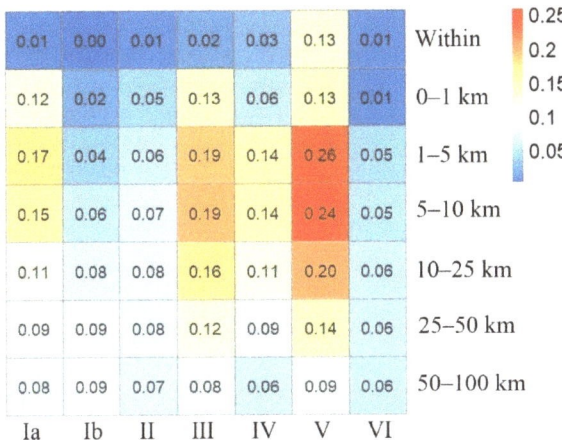

Figure 11. Trends within every type of PAs and of buffers.

The trend for type Ia PAs was significantly higher than that of type Ib, indicating that Ia was affected more by human development. The change trends of different types of PAs showed characteristics. The rank of mean values of change trends in buffer zones were V > III > Ia > IV > II > Ib > VI. The change trend of the type V PAs was the highest, e.g., 0.26/year in 1–5 km and 0.24/year in 5–10 km buffers.

The trends in type V, III, Ia, and IV PAs were concentrated in the 1–10 km buffer zone, indicating active urban development. The type VI, Ib, and II PAs had the lower change trends inside PAs, but higher change trends in the peripheral areas far from the PA boundary. The change trends in the area near the PA boundary between 0–10 km were relatively low.

4. Discussion

4.1. The NTL Distribution Pattern in Different Types of PAs

The primary objective of type Ia PAs is to conserve regionally, nationally, and globally outstanding ecosystems, species (occurrences or aggregations), and/or geodiversity features: these attributes will have been formed mostly or entirely by non-human forces [55]. However, the global NTL level inside type Ia PAs was not the lowest. The NTL value in type Ia PAs was 2.5 times higher than that of the type Ib PAs. Among seven types of PAs, the NTL growth rate inside type Ia was very high from 1992 to 2013, reaching 378%, which was far greater than the others. The NTL for 0–1 km and 1–5 km buffer zones had growth rates that were ranked highly, with values of 441% and 372%, respectively. The area with the highest NTL level around Ia consistently approached the boundary of the PAs. The brightest NTL areas between 1992 and 1999 was in the 25–50 km buffer zone, which shifted to the 5–10 km buffer zone between 2001 and 2003. According to the growth rate and trend of each buffer zone, the brightest NTL area would be encroaching in the 0–5 km area quickly. Human development in the 0–1 km outside PAs could have direct impacts [56,57]. Because the baselines of the nightlight level of different types of protected areas are very different, the actual change values corresponding to the same growth rate are very different even though the growth rates are the same. For example, the NTL growth rate of type Ia PAs exceeded the others, but the absolute increase of the NTL DN value inside and around the Ia PAs was still low in comparison to other types because its baseline of the nightlight level was far less than other types. It is noteworthy that, according to the guidelines, Ia will be degraded or destroyed when subjected to all but very limited human impact [55]. In addition, the area of category Ia was generally small, and human-induced impacts contributed more in small PAs than to stochastic processes [58].

For type Ib PAs, the average NTL DN value was 0.04, which was the lowest among all types. Also, the NTL levels in and out of the PAs boundaries was lower than type Ia PAs. The 50–100 km buffer zone was the area with the most concentrated human population around Ib PAs. The NTL distribution for type II PAs were similar to type VI and both were lower among the types of PAs. The NTL levels of types II and VI were not significantly higher than those of Ia and Ib.

The main function of type IV PAs is to protect certain species and habitats. The NTL levels inside type IV PAs and buffer zones were significantly higher than all other types except V. NTL levels within and outside type III PAs were high, second only to type V. NTL levels within and outside of type V PAs were much higher than other types.

The result suggested that the NTL levels were significantly lower within the PAs than that of the surrounding 0–100 km buffer zones. In particular, there was a big difference between inside and outside the boundary of types Ia and III PAs. For most PAs, the surrounding areas with the highest DN value were in the 1–25 km buffer zone.

4.2. Skyglow as a Biodiversity Threat

Skyglow occurs when artificial light is scattered by water, atmospheric molecules, or aerosols and returned to Earth. In this study, we used the NTL data to show the impact of urbanization on protected areas. However, we did not apply any light propagation models to integrate the skyglow effect due to indirect lights. Skyglow is of increasing concern since it is able to multiply and extend the light pollution effect and affect those areas with no direct light pollution. Because of skyglow, the biological impacts of artificial light are not limited to the vicinity of the light source and may spread over much larger extents via several mechanisms. Individual lights may be visible kilometers away from their source, and the addition of artificial skyglow can extinguish such lunar light cycles and permanently remove dark nights from a landscape [59,60]. Therefore, artificial lighting may have an effect on natural ecosystems even when the source of lighting lies kilometers away. Most organisms have evolved molecular circadian clocks controlled by natural day–night cycles. These clocks play key roles in metabolism, growth, and behavior [61]. As the world grows ever-more illuminated, many light-sensitive species will be lost, especially in or near highly illuminated urban areas [62]. Light pollution threatens biodiversity through changed night habits, including organismal movements [63,64]; foraging [65–68]; interspecific interactions [69]; communication [70,71]; reproduction [72] of insects, amphibians, fish, birds, bats, and other animals; and it can disrupt plants resulting in phenological changes by distorting their natural day–night cycle [63].

Therefore, measures are necessary to prevent or reduce the ecological impact of night-time light pollution. Maintaining and increasing natural unlit areas is likely to be the most effective option for reducing the ecological effects of lighting [73]. From the lighting differences inside and outside the PAs, it can be seen that in the process of human development, PAs have greatly reduced human interference, so PAs are undoubtedly the best place to maintain darkness. More stringent control measures should be implemented within and around PAs, such as limiting the duration of lighting, reducing the trespass of lighting, changing the intensity of lighting, and changing the spectrum of lighting [73,74].

Setting an area surrounding PAs is an ideal option to provide a buffer against the light pollution released by human development. However, it cannot be solved by using only remote sensing data. Different types of PAs are in different natural and socio-economic conditions, and as such, the appropriate buffer radius should also be different. The success of planners in reducing the ecological impacts of light pollution will ultimately depend on an assessment of the critical mechanisms and thresholds that determine those impacts in a particular environment [73]. One possible way is to combine remote sensing data with biodiversity data to explore how biodiversity in protected areas with similar natural conditions respond to different lighting patterns. This may be of great significance for solving light pollution near the PAs and may also contribute to determining how much of a buffer distance should be set around the PAs.

4.3. Limitations of Night-Time Lighting Data

The 1992–2013 NTL data used in this study were derived from six different sensors. Therefore, it is difficult to discriminate whether the small difference between the light data of different years was caused by the sensor or by changes in the field light. Moreover, we are not sure how much data error was caused by the sensor. In this study, we compared the nightlight levels between 1992 and 2013. The time span and the difference in the NTL data were sufficiently large, such that the data error caused by the different sensor could be ignored.

In addition, the light data could not detect negative light growth (i.e., reductions in light) in each PA. To reduce the error caused by the sensor when processing the light data, we assumed that all regions were developing positively, and the DN value was increasing; thus, we assigned the larger DN value of the previous year to the darker pixels corresponding to the following year. If the actual nightlight of a certain area was dark each year, the DN value would remain unchanged. On the global scale, the nightlight within and around each type of PA was increasing each year, and there was no negative growth in the brightness of the PA. However, when assessing individual PAs, there may be negative growth due to increasingly strict regulations and improved awareness of protection, e.g., lighting tools could be replaced by those with a lower brightness and lighting time could be reduced. Therefore, the lighting data may not be suitable for the study of PAs on a small spatial scale, especially in developed countries where the management of PAs is more stringent.

Author Contributions: All authors contributed to this manuscript. Specific contributions include conceptualization, L.F., J.Z., and Y.W.; methodology, J.Z. and L.F.; software, L.F.; validation, L.F. and J.Z.; formal analysis, L.F.; data curation, L.F. and J.Z.; writing—original draft preparation, L.F. and J.Z.; writing—review and editing, L.F., J.Z., Y.W., Z.R., H.Z., and X.G.; visualization, J.Z. and L.F.; supervision, J.Z.; funding acquisition, J.Z.

Funding: This research was funded by the National Natural Science Foundation of China (grant nos. 41771450, 41871330, and 41630749), the Fundamental Research Funds for the Central Universities (grant no. 2412019BJ001), the Foundation of the Education Department of Jilin Province in the 13th Five-Year project (grant no. JJKH20190282KJ), and the Science and Technology Development Project of Jilin Province (grant no. 20190802024ZG).

Conflicts of Interest: The authors declare no conflict of interest.

References

1. UNEP-WCMC; IUCN. *Protected Planet Report 2016*; UNEP-WCMC: Cambridge, UK; IUCN: Gland, Switzerland, 2016; ISBN 978-92-807-3587-1. Available online: https://www.iucn.org/theme/protected-areas/publications/protected-planet-report (accessed on 12 June 2019).
2. Wang, Y. *Remote Sensing of Protected Lands*; CRC Press: Boca Raton, FL, USA, 2011; ISBN 978-1-4398-4187-7.
3. Elvidge, C.D.; Sutton, P.C.; Wagner, T.W.; Ryzner, R.; Vogelmann, J.V.; Goetz, S.J.; Smith, A.J.; Jantz, C.; Seto, K.C.; Imhoff, M.L.; et al. (Eds.) *Land Change Science*; Kluwer Academic Publishers: Dordrecht, The Netherlands, 2004; pp. 315–328.
4. Crabtree, R.L.; Sheldon, J.W.; Sheldon, J.W. Monitoring and Modeling Environmental Change in Protected Areas: Integration of Focal Species Populations and Remote Sensing. In *Remote Sensing of Protected Lands*; CRC Press: Boca Raton, FL, USA, 2012; pp. 495–524.
5. IUCN. Shaping a sustainable future. In *The IUCN Programme 2009–2012*; IUCN: Gland, Switzerland, 2008.
6. Heinen, J.; Hite, K. Protected natural areas. In *Encyclopedia of Earth*; Cutler, J., Cleveland, C.J., Eds.; Environmental Information Coalition, National Council for Science and the Environment: Washington, DC, USA, 2007.
7. Hoffmann, M.; Hilton-Taylor, C.; Angulo, A.; Böhm, M.; Brooks, T.M.; Butchart, S.H.M.; Carpenter, K.E.; Chanson, J.; Collen, B.; Cox, N.A.; et al. The Impact of Conservation on the Status of the World's Vertebrates. *Science* **2010**, *330*, 1503. [CrossRef]
8. Pereira, H.M.; Leadley, P.W.; Proença, V.; Alkemade, R.; Scharlemann, J.P.W.; Fernandez-Manjarrés, J.F.; Araújo, M.B.; Balvanera, P.; Biggs, R.; Cheung, W.W.L.; et al. Scenarios for Global Biodiversity in the 21st Century. *Science* **2010**, *330*, 1496. [CrossRef]
9. Watson, J.E.M.; Venter, O.; Lee, J.; Jones, K.R.; Robinson, J.G.; Possingham, H.P.; Allan, J.R. Protect the last of the wild. *Nature* **2018**, *563*, 27. [CrossRef] [PubMed]

10. Prendergast, J.R.; Quinn, R.M.; Lawton, J.H.; Eversham, B.C.; Gibbons, D.W. Rare species, the coincidence of diversity hotspots and conservation strategies. *Nature* **1993**, *365*, 335. [CrossRef]
11. Finlayson, M.; Cruz, R.D.; Davidson, N.; Alder, J.; Cork, S.; Groot, R.S.D.; Leveque, C.; Milton, G.R.; Peterson, G.; Pritchard, D.; et al. *Millennium Ecosystem Assessment: Ecosystems and Human Well-Being: Wetlands and Water Synthesis*; World Resources Institute: Washington, DC, USA, 2005.
12. Gaston, K.J.; Jackson, S.F.; Cantúsalazar, L.; Cruzpiñón, G. The Ecological Performance of Protected Areas. *Annu. Rev. Ecol. Evol. Syst.* **2008**, *39*, 93–113. [CrossRef]
13. Walker, R.; Moore, N.J.; Arima, E.; Perz, S.; Simmons, C.; Caldas, M.; Vergara, D.; Bohrer, C. Protecting the Amazon with protected areas. *Proc. Natl. Acad. Sci. USA* **2009**, *106*, 10582–10586. [CrossRef] [PubMed]
14. George, W.; Paul, E.; Bean, W.T.; Burton, A.C.O.; Brashares, J.S. Accelerated human population growth at protected area edges. *Science* **2008**, *321*, 123–126.
15. Scherl, L.M.; Wilson, A.; Wild, R.; Blockhus, J.; Franks, P.; Mcneely, J.A.; Mcshane, T.O. Can protected areas contribute to poverty reduction? opportunities and limitations. *Biodiversity* **2010**, *11*, 5–7.
16. Zhou, Y.; Li, X.; Asrar, G.R.; Smith, S.J.; Imhoff, M. A global record of annual urban dynamics (1992–2013) from nighttime lights. *Remote Sens. Environ.* **2018**, *219*, 206–220. [CrossRef]
17. Li, X.; Zhou, Y. Urban mapping using DMSP/OLS stable night-time light: A review. *Int. J. Remote Sens.* **2017**, *38*, 6030–6046. [CrossRef]
18. Rodrigues, A.S.; Andelman, S.J.; Bakarr, M.I.; Boitani, L.; Brooks, T.M.; Cowling, R.M.; Fishpool, L.D.; Da, F.G.; Gaston, K.J.; Hoffmann, M. Effectiveness of the global protected area network in representing species diversity. *Nature* **2004**, *428*, 640–643. [CrossRef] [PubMed]
19. Bruner, A.G.; Gullison, R.E.; Rice, R.E.; Fonseca, G.A. Effectiveness of parks in protecting tropical biodiversity. *Science* **2001**, *291*, 125–128. [CrossRef] [PubMed]
20. Andam, K.S.; Ferraro, P.J.; Pfaff, A.; Sanchez-Azofeifa, G.A.; Robalino, J.A. Measuring the Effectiveness of Protected Area Networks in Reducing Deforestation. *Proc. Natl. Acad. Sci. USA* **2008**, *105*, 16089–16094. [CrossRef] [PubMed]
21. Brun, C.; Cook, A.R.; Lee, J.S.H.; Wich, S.A.; Lian, P.K.; Carrasco, L.R. Analysis of deforestation and protected area effectiveness in Indonesia: A comparison of Bayesian spatial models. *Glob. Environ. Chang.* **2015**, *31*, 285–295. [CrossRef]
22. Foley, J.A.; Ruth, D.; Asner, G.P.; Carol, B.; Gordon, B.; Carpenter, S.R.; F Stuart, C.; Coe, M.T.; Daily, G.C.; Gibbs, H.K. Global consequences of land use. *Science* **2005**, *309*, 570–574. [CrossRef]
23. Mcdonald, R.I.; Kareiva, P.; Forman, R.T.T. The implications of current and future urbanization for global protected areas and biodiversity conservation. *Biol. Conserv.* **2008**, *141*, 1695–1703. [CrossRef]
24. Radeloff, V.C.; Stewart, S.I.; Hawbaker, T.J.; Gimmi, U.; Pidgeon, A.M.; Flather, C.H.; Hammer, R.B.; Helmers, D.P. Housing growth in and near United States protected areas limits their conservation value. *Proc. Natl. Acad. Sci. USA* **2010**, *107*, 940–945. [CrossRef]
25. Li, X.; Zhou, Y.; Asrar, G.R.; Mao, J.; Li, X.; Li, W. Response of vegetation phenology to urbanization in the conterminous United States. *Glob. Chang. Biol.* **2017**, *23*, 2818–2830. [CrossRef]
26. Elvidge, C.D.; Sutton, P.C.; Baugh, K.E.; Ziskin, D.C.; Anderson, S. National Trends in Satellite Observed Lighting: 1992–2009. In Proceedings of the Agu Fall Meeting, San Francisco, CA, USA, 5–9 December 2011.
27. Hsu, F.C.; Baugh, K.E.; Ghosh, T.; Zhizhin, M.; Elvidge, C.D. DMSP-OLS Radiance Calibrated Nighttime Lights Time Series with Intercalibration. *Remote Sens.* **2015**, *7*, 1855–1876. [CrossRef]
28. Zhang, Q.; Pandey, B.; Seto, K.C. A Robust Method to Generate a Consistent Time Series from DMSP/OLS Nighttime Light Data. *IEEE Trans. Geosci. Remote Sens.* **2016**, *54*, 5821–5831. [CrossRef]
29. Gaston, K.J.; Bennie, J. Demographic effects of artificial nighttime lighting on animal populations. *Doss. Environ.* **2014**, *22*, 323–330. [CrossRef]
30. Gaston, K.J.; Duffy, J.P.; Bennie, J. Quantifying the erosion of natural darkness in the global protected area system. *Conserv. Biol.* **2015**, *29*, 1132–1141. [CrossRef] [PubMed]
31. Longcore, T.; Rich, C. Ecological Light Pollution. *Front. Ecol. Environ.* **2004**, *2*, 191–198. [CrossRef]
32. Bennie, J.; Duffy, J.P.; Davies, T.W.; Correa-Cano, M.E.; Gaston, K.J. Global Trends in Exposure to Light Pollution in Natural Terrestrial Ecosystems. *Remote Sens.* **2015**, *7*, 2715–2730. [CrossRef]
33. Freitas, J.R.; Bennie, J.; Mantovani, W.; Gaston, K.J. Exposure of tropical ecosystems to artificial light at night: Brazil as a case study. *PLoS ONE* **2017**, *12*, e0171655. [CrossRef]

34. Koen, E.L.; Minnaar, C.; Roever, C.L.; Boyles, J.G. Emerging threat of the 21st century lightscape to global biodiversity. *Glob. Chang. Biol.* **2018**, *24*, 2315–2324. [CrossRef]
35. Zhao, M.; Zhou, Y.; Li, X.; Cao, W.; He, C.; Yu, B.; Li, X.; Elvidge, C.D.; Cheng, W.; Zhou, C. Applications of satellite remote sensing of nighttime light observations: Advances, challenges, and perspectives. *Remote Sens.* **2019**, *11*, 1971. [CrossRef]
36. Wei, J.; He, G.; Leng, W.; Long, T.; Guo, H. Characterizing Light Pollution Trends across Protected Areas in China Using Nighttime Light Remote Sensing Data. *ISPRS Int. J. Geo-Inf.* **2018**, *7*, 243.
37. Xiang, W.; Tan, M. Changes in Light Pollution and the Causing Factors in China's Protected Areas, 1992–2012. *Remote Sens.* **2017**, *9*, 1026. [CrossRef]
38. Davies, T.W.; Duffy, J.P.; Bennie, J.; Gaston, K.J. Stemming the Tide of Light Pollution Encroaching into Marine Protected Areas. *Conserv. Lett.* **2015**, *9*, 164–171. [CrossRef]
39. Raap, T.; Pinxten, R.; Eens, M. Artificial light at night causes an unexpected increase in oxalate in developing male songbirds. *Conserv. Physiol.* **2018**, *6*, coy005. [CrossRef] [PubMed]
40. Geldmann, J.; Joppa, L.N.; Burgess, N.D. Mapping change in human pressure globally on land and within protected areas. *Conserv. Biol.* **2015**, *28*, 1604–1616. [CrossRef] [PubMed]
41. Theobald, D.M.; Miller, J.R.; Hobbs, N.T. Estimating the cumulative effects of development on wildlife habitat. *Landsc. Urban Plan.* **1997**, *39*, 25–36. [CrossRef]
42. He, C.; Li, J.; Chen, J.; Shi, P.; Chen, J.; Pan, Y.; Li, J.; Zhuo, L.; Toshiaki, I. The urbanization process of Bohai Rim in the 1990s by using DMSP/OLS data. *J. Geogr. Sci.* **2006**, *16*, 174–182. [CrossRef]
43. Liu, Z.; He, C.; Zhang, Q.; Huang, Q.; Yang, Y. Extracting the dynamics of urban expansion in China using DMSP-OLS nighttime light data from 1992 to 2008. *Landsc. Urban Plan.* **2012**, *106*, 62–72. [CrossRef]
44. Chen, X.; Nordhaus, W.D. Using luminosity data as a proxy for economic statistics. *Proc. Natl. Acad. Sci. USA* **2011**, *108*, 8589. [CrossRef]
45. Pandey, B.; Joshi, P.K.; Seto, K.C. Monitoring urbanization dynamics in India using DMSP/OLS night time lights and SPOT-VGT data. *Int. J. Appl. Earth Obs. Geoinf.* **2013**, *23*, 49–61. [CrossRef]
46. Zhang, Q.; Seto, K.C. Mapping urbanization dynamics at regional and global scales using multi-temporal DMSP/OLS nighttime light data. *Remote Sens. Environ.* **2011**, *115*, 2320–2329. [CrossRef]
47. Elvidge, C.; Imhoff, M.; Baugh, K.; Hobson, V.; Nelson, I.; Safran, J.; Dietz, J.; Tuttle, B. Night-time lights of the world: 1994–1995. *ISPRS J. Photogramm. Remote Sens.* **2001**, *56*, 81–99. [CrossRef]
48. Seto, K.C.; Zhang, Q. Can Night-Time Light Data Identify Typologies of Urbanization? A Global Assessment of Successes and Failures. *Remote Sens.* **2013**, *5*, 3476–3494.
49. Lyytimäki, J.; Tapio, P.; Assmuth, T. Unawareness in environmental protection: The case of light pollution from traffic. *Land Use Policy* **2012**, *29*, 598–604. [CrossRef]
50. de Miguel, A.S.; Zamorano, J.; Castano, J.G.; Pascual, S. Evolution of the energy consumed by street lighting in Spain estimated with DMSP-OLS data. *J. Quant. Spectrosc. Radiat. Transf.* **2014**, *139*, 109–117. [CrossRef]
51. He, C.; Ma, Q.; Li, T.; Yang, Y.; Liu, Z. Spatiotemporal dynamics of electric power consumption in Chinese Mainland from 1995 to 2008 modeled using DMSP/OLS stable nighttime lights data. *J. Geogr. Sci.* **2012**, *22*, 125–136. [CrossRef]
52. Wu, J.; He, S.; Peng, J.; Li, W.; Zhong, X. Intercalibration of DMSP-OLS night-time light data by the invariant region method. *Int. J. Remote Sens.* **2013**, *34*, 7356–7368. [CrossRef]
53. Li, X.; Zhou, Y. A Stepwise Calibration of Global DMSP/OLS Stable Nighttime Light Data (1992–2013). *Remote Sens.* **2017**, *9*, 637.
54. Yue, S.; Pilon, P.; Cavadias, G. Power of the Mann–Kendall and Spearman's rho tests for detecting monotonic trends in hydrological series. *J. Hydrol.* **2002**, *259*, 254–271. [CrossRef]
55. Dudley, N. *Guidelines for Applying Protected Area Management Categories*; IUCN: Gland, Switzerland, 2008; ISBN 978-2-8317-1086-0.
56. Radeloff, V.C.; Hammer, R.B.; Stewart, S.I. Rural and Suburban Sprawl in the U.S. Midwest from 1940 to 2000 and Its Relation to Forest Fragmentation. *Conserv. Biol.* **2005**, *19*, 793–805. [CrossRef]
57. Crooks, K.R.; Soulé, M.E. Mesopredator release and avifaunal extinctions in a fragmented system. *Nature* **1999**, *400*, 563–566. [CrossRef]
58. Woodroffe, R. Edge Effects and the Extinction of Populations Inside Protected Areas. *Science* **1998**, *280*, 2126–2128. [CrossRef]

59. Davies, T.; Bennie, J.; Inger, R.; Gaston, K. Artificial light alters regimes of natural sky brightness. *Sci. Rep.* **2013**, *3*, 1722. [CrossRef]
60. Puschnig, J.; Posch, T.; Uttenthaler, S. Night sky photometry and spectroscopy performed at the Vienna University Observatory. *J. Quant. Spectrosc. Radiat. Transf.* **2014**, *139*, 64–75. [CrossRef]
61. Dunlap, J. Molecular Bases for Circadian Clocks. *Cell* **1999**, *96*, 271–290. [CrossRef]
62. Hölker, F.; Wolter, C.; Perkin, E. Light Pollution as a Biodiversity Threat. *Trends Ecol. Evol.* **2010**, *25*, 681–682. [CrossRef] [PubMed]
63. Peters, A.; Verhoeven, K.J.F. Impact of Artificial Lighting on the Seaward Orientation of Hatchling Loggerhead Turtles. *J. Herpetol.* **1994**, *28*, 112–114. [CrossRef]
64. Emma Louise, S.; Gareth, J.; Stephen, H. Street lighting disturbs commuting bats. *Curr. Biol. CB* **2009**, *19*, 1123–1127.
65. Rydell, J. Seasonal use of illuminated areas by foraging northern bats Eptesicus nilssoni. *Ecography* **1991**, *14*, 203–207. [CrossRef]
66. Buchanan, B.W. Effects of enhanced lighting on the behaviour of nocturnal frogs. *Anim. Behav.* **1993**, *45*, 893–899. [CrossRef]
67. Bird, B.L.; Branch, L.C.; Miller, D.L. Effects of Coastal Lighting on Foraging Behaviorof Beach Mice. *Conserv. Biol.* **2004**, *18*, 1435–1439. [CrossRef]
68. Santos, C.D.; Miranda, A.C.; Granadeiro, J.P.; Lourenco, P.M.; Saraiva, S.; Palmeirim, J.M. Effects of artificial illumination on the nocturnal foraging of waders. *Acta Oecol.* **2010**, *36*, 166–172. [CrossRef]
69. Svensson, A.M.; Rydell, J. Mercury vapour lamps interfere with the bat defence of tympanate moths (*Operophtera* spp.; Geometridae). *Anim. Behav.* **1998**, *55*, 223–226. [CrossRef]
70. Baker, B.J.; Richardson, J. The effect of artificial light on male breeding-season behaviour in green frogs, Rana clamitans melanota. *Can. J. Zool.* **2006**, *84*, 1528–1532. [CrossRef]
71. Miller, M.W. Apparent Effects of Light Pollution on Singing Behavior of American Robins. *Condor* **2006**, *108*, 130. [CrossRef]
72. Boldogh, S.; Dobrosi, D.; Samu, P. The effects of the illumination of buildings on house-dwelling bats and its conservation consequences. *Acta Chiropterol.* **2007**, *9*, 527–534. [CrossRef]
73. Gaston, K.J.; Davies, T.W.; Bennie, J.; Hopkins, J.; Fernandez-Juricic, E. Review: Reducing the ecological consequences of night-time light pollution: Options and developments. *J. Appl. Ecol.* **2012**, *49*, 1256–1266. [CrossRef]
74. Kyba, C.C.M.; Hölker, F. Do artificially illuminated skies affect biodiversity in nocturnal landscapes? *Landsc. Ecol.* **2013**, *28*, 1637–1640. [CrossRef]

© 2019 by the authors. Licensee MDPI, Basel, Switzerland. This article is an open access article distributed under the terms and conditions of the Creative Commons Attribution (CC BY) license (http://creativecommons.org/licenses/by/4.0/).

Article

Long-Term Spatiotemporal Dynamics of Terrestrial Biophysical Variables in the Three-River Headwaters Region of China from Satellite and Meteorological Datasets

Xiangyi Bei [1], Yunjun Yao [1,*], Lilin Zhang [2], Tongren Xu [3], Kun Jia [1], Xiaotong Zhang [1], Ke Shang [1], Jia Xu [1] and Xiaowei Chen [1]

[1] State Key Laboratory of Remote Sensing Science, Faculty of Geographical Science, Beijing Normal University, Beijing 100875, China
[2] Faculty of Geo-Information and Earth Observation (ITC), University of Twente, Enschede 7500 AE, The Netherlands
[3] State Key Laboratory of Earth Surface Processes and Resource Ecology, Faculty of Geographical Science, Beijing Normal University, Beijing 100875, China
* Correspondence: yaoyunjun@bnu.edu.cn; Tel.: +86-10-5880-3002

Received: 19 June 2019; Accepted: 8 July 2019; Published: 10 July 2019

Abstract: Terrestrial biophysical variables play an essential role in quantifying the amount of energy budget, water cycle, and carbon sink over the Three-River Headwaters Region of China (TRHR). However, direct field observations are missing in this region, and few studies have focused on the long-term spatiotemporal variations of terrestrial biophysical variables. In this study, we evaluated the spatiotemporal dynamics of biophysical variables including meteorological variables, vegetation, and evapotranspiration (ET) over the TRHR, and analyzed the response of vegetation and ET to climate change in the period from 1982 to 2015. The main input gridded datasets included meteorological reanalysis data, a satellite-based vegetation index dataset, and the ET product developed by a process-based Priestley–Taylor algorithm. Our results illustrate that: (1) The air temperature and precipitation over the TRHR increased by 0.597 °C and 41.1 mm per decade, respectively, while the relative humidity and surface downward shortwave radiation declined at a rate of 0.9% and 1.8 W/m^2 per decade during the period 1982–2015, respectively. We also found that a 'dryer warming' tendency and a 'wetter warming' tendency existed in different areas of the TRHR. (2) Due to the predominant 'wetter warming' tendency characterized by the increasing temperature and precipitation, more than 56.8% of areas in the TRHR presented a significant increment in vegetation (0.0051/decade, $p < 0.05$), particularly in the northern and western meadow areas. When energy was the limiting factor for vegetation growth, temperature was a considerably more important driving factor than precipitation. (3) The annual ET of the TRHR increased by 3.34 mm/decade ($p < 0.05$) with an annual mean of 230.23 mm/year. More importantly, our analysis noted that ET was governed by terrestrial water supply, e.g., soil moisture and precipitation in the arid region of the western TRHR. By contrast, atmospheric evaporative demand derived by temperature and relative humidity was the primary controlling factor over the humid region of the southeastern TRHR. It was noted that land management activities, e.g., irrigation, also had a nonnegligible impact on the temporal and spatial variation of ET.

Keywords: terrestrial biophysical variables; Three-River Headwaters Region of China; spatiotemporal dynamics; climate change; vegetation index; evapotranspiration

1. Introduction

The variability of terrestrial biophysical variables influences the function of ecosystem components, which is likely to alter terrestrial ecological processes [1]. As one of the largest Chinese nature reserves, the Three-River Headwaters Region (TRHR) has a relatively high altitude and severe climate conditions, which makes its ecosystem extremely sensitive and vulnerable [2]. In the last few decades, due to intensified climate change and uncontrolled development activities, several ecological issues, including the recession of glaciers and tundra, wetland shrinkage, and grassland desertification, have emerged over the TRHR, resulting in complex biophysical interactions and an irreversible effect on the ecosystem [3]. Noticing the importance and urgency of environment protection, the Chinese government has implemented a series of environmental protection policies over the TRHR since the early 21st century [4]. The Sanjiangyuan National Nature Reserve (SNNR) [5] as well as the Ecological Protection and Restoration Program (EPRP) [6] were established to conserve and rehabilitate the ecological environment, including retiring livestock, restoring degraded grassland, and ecological migration. Although these projects have greatly improved the resilience of the ecosystems, there are still large uncertainties in the spatiotemporal dynamics of the terrestrial biophysical variables. Therefore, comprehensive assessment of the terrestrial biophysical variation is a prerequisite for studying the interaction among ecological environment dynamics and provides instructive information about the hydrology, geographical ecology, and water resource management.

The air temperature (Ta) of the TRHR is undergoing significant warming, and has done over the last few decades [7,8]. Previous studies have shown that the rising trend of temperature over the TRHR is obviously larger than that in other regions in China [9,10]. The obvious warming trend, coupled with the accelerated carbon cycle between the land and atmosphere, has a significant impact on the biophysical processes, including the water cycle and energy exchange [11]. Recently, several studies based on ground observations found that the TRHR experienced a sustained warming and wetting trend over the past few decades [12]. For instance, Chong et al. [13] revealed that both Ta and precipitation (P) showed a significant upward trend (0.31 °C and 10.6 mm per decade, respectively) based on ground measurements from 21 meteorological sites distributed in the TRHR during 1956–2012. Significant warming and intensified P were also detected by Tong et al. [14], who suggested that Ta and P had increased by 0.9 °C and 102 mm in the past 20 years, respectively. The reduction of terrestrial relative humidity (RH) and solar radiation (Rs) were also captured during observations of the Tibetan Plateau, which correlate with rapid climate warming. However, in situ observations have their stubborn limitations as their representativeness of regional-scale climatic parameters remains problematic due to the terrestrial heterogeneity [15]. Fortunately, data assimilation techniques can provide optimal integrated information from site measurements, weather forecast products, and remote sensing data [16]. With the continuous accumulation of emerging forcing datasets produced by the data assimilation technique, it has become meaningful to further evaluate the long-term spatiotemporal information regarding climate change over the TRHR.

The pronounced climate warming along with the redistribution of precipitation patterns significantly influences the vegetation through a series of biophysical processes [17]. In this context, the remotely sensed normalized difference vegetation index (NDVI) has been widely used to detect the temporal variation of vegetation in the TRHR at multiple scales [18]. In past decades, the TRHR was under pressure to sustain increasing livestock grazing and suffered from an alpine grassland degradation problem. Liu et al. [19] reported that continuous and obvious grassland degradation had occured since the 1970s, experiencing fragmentation, desertification, and degradation to "black soil beach" [20]. In order to protect the grassland resource, a series of national nature reserve projects and ecological policies were established within the TRHR during the 21st century [21]. Recent studies have indicated that the slight increment in vegetation density (0.047/decade) is mainly attributed to the implementation of ecological restoration programs over the TRHR during 2001–2010 [22]. These findings were also demonstrated by Liu et al. [23], who found that the NDVI of the TRHR increased by 0.012/decade over the past 12 years (2000–2011), which is consistent with the ongoing "warm and moist" trend. Understanding the variation in vegetation is often limited by the relatively brief dataset sequences, resulting in inconsistent accepted conclusions about the definite tendency of vegetation coverage in the TRHR. Therefore, it is critical to analyze the detailed variation of vegetation cover and the response of vegetation to climate change.

The fluctuation of climate and vegetation also has significant impacts on the surface water budget, particularly for evapotranspiration (ET), a crucial component of the terrestrial hydrological cycle [24]. ET is the sum of the evaporation from the land surface and the transpiration from plants into the atmosphere, and links the water budget, carbon sink, and energy exchange [25,26]. Therefore, the long-term variation of regional ET is of significance to monitor the biophysical processes and climate change. However, accurate simulations of the long-term ET of the TRHR remain a major challenge due to the lack of adequate and robust ground observations to determine regional ET over the TRHR. Moreover, datasets, such as the MOD16 product, from some global ET datasets are missing over the TRHR due to their existing gaps [27]. Recently, several satellite-based models and approaches have been developed to estimate the spatiotemporal ET in the TRHR over the last few decades [28]. For instance, based on a revised semi-empirical algorithm, Yao et al. [29] illustrated that there was no statistically significant trend in ET over the TRHR during the period 1982–2010. Xu et al. [30] found that ET showed a slight decreasing trend at the rate of 3.3 mm/decade from 2000 through 2014 in the TRHR by using an enhanced surface energy balance system (SEBS) algorithm. The simulated results were limited by the relatively short time span of the dataset and the uncertainties of model parameterization [31,32]. There are still large uncertainties about the spatiotemporal dynamics of ET over the complicated topography and heterogeneous surface of the TRHR. Thus, a robust assessment of the long-term variation of ET at a regional scale over the TRHR is in great demand for understanding the water cycle under an environment of rapid climate change.

As one of the most sensitive areas for climate change with complex terrain and high altitude, the TRHR is an ideal natural experimental area for investigating the response of terrestrial processes to climate change. Numerous studies have attempted to evaluate the interaction of the terrestrial biophysical variables (including climate, vegetation indices, and ET) by using different algorithms and datasets at multiple scales. For example, Zhang et al. [33] estimated the net primary productivity (NPP) of the TRHR using the Carnegie-Ames-Stanford approach (CASA) model, and found that the vegetation had a general increasing trend from 1982 to 2012, and pointed out that solar radiation was the primary factor controlling the increment of vegetation, with an average contribution of 0.73. Based on Gravity Recovery and Climate Experiment (GRACE) satellite data and Moderate Resolution Imaging Spectroradiometer (MODIS) NDVI data, Xu et al. [30] suggested that soil moisture and total water storage were major determining drivers in vegetation greening. However, large discrepancies still exist in the spatiotemporal variation of terrestrial biophysical variables over the TRHR due to the differences in temporal series, spatial scale, algorithm, and data sources, which have hampered attempts to accurately evaluate long-term biophysical variation. Moreover, the spatial–temporal dynamics of climate change, vegetation growth, and water cycling have seldom been simultaneously discussed over the TRHR. As a result, little is accurately known about the spatiotemporal characterization of the response of terrestrial biophysical variables over the TRHR to climate change on large spatial scales and over long time periods.

In this study, we analyzed the spatiotemporal dynamics of terrestrial biophysical variables over the TRHR using a meteorological dataset, satellite-based vegetation index dataset, and a satellite-derived ET product from 1982 through to 2015, and investigated the main influencing factors accounting for biophysical variation. We had three major objectives. First, we analyzed the spatial patterns and trends of climate factors including Ta, P, RH, and Rs from 1982 through to 2015 over the TRHR of China. Second, we analyzed the spatiotemporal variation in the NDVI and ET from 1982 to 2015. Finally, we detected the response of vegetation and ET to climate change.

2. Materials and Methods

2.1. Study Area

The Three-River Headwaters Region (31°38′–36°20′N, 89°31′–102°14′E) is located in southern Qinghai Province, the hinterland of the Tibetan Plateau (TP) (Figure 1). This region is the headstream of three major Asian rivers, including the Yellow, Yangtze, and Lantsang Rivers, and is known as the "Chinese water tower", supporting approximately 40% of the world's population [34]. The TRHR covers an area of 350,000 km^2, which supplies 49% of the total water of the Yellow River, 25% of the total water of the Yangtze River, and 15% of the total water of the Lantsang River. The TRHR is a central part of the highest and largest plateau in the world, and constitutes mountainous landforms with an average elevation of more than 4000 m. Due to its unique location and complex topography, the TRHR is characterized by a typical plateau climate with a low air temperature, high daily temperature range, and strong solar radiation. The climate of the TRHR is wet and moist in summers, and cool and dry in winters, with distinct wet and dry seasons. The average mean temperature ranges from −5.6 to 3.8 °C and the annual rainfall ranges from 262.2 to 772.8 mm with a notably decreasing trend from southeast to northwest [35]. The TRHR has the richest biodiversity, and contains the largest Chinese alpine wetlands ecosystem. The main ecosystem type of the TRHR is grassland including alpine meadow and alpine steppe, accounting for approximately 76% and 23% of the grasslands, respectively [36].

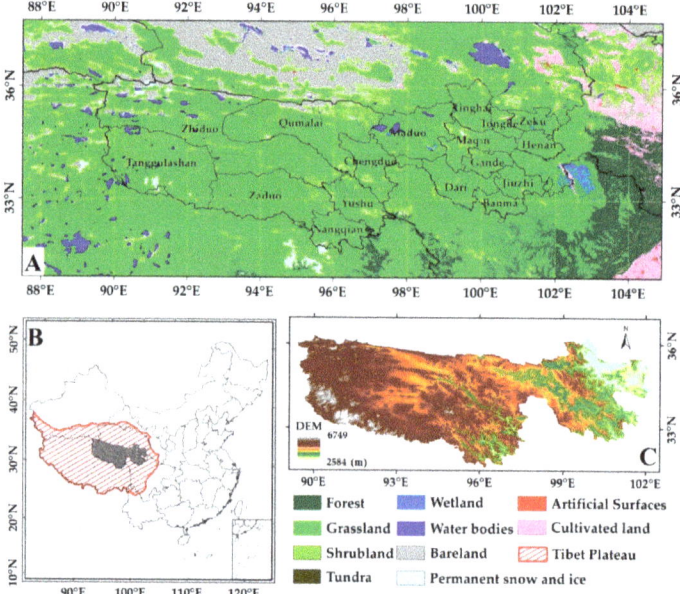

Figure 1. Maps showing the location of the study sites. (**A**) The distribution of land cover types. (**B**) The Tibetan Plateau (TP) and Three-River Headwaters Region (TRHR). (**C**) The digital elevation model (DEM) data of the TRHR with a spatial resolution of 250 m. The study area consists of 17 counties and cities delineated by the Ecological Protection and Restoration Program, including Zeku, Tongde, Henan, Xinghai, Maqin, Gande, Jiuzhi, Dari, Banma, Maduo Chengduo, Yushu, Nangqian, Qumalai, Zaduo, Zhiduo, and Tanggulashan.

2.2. Climate Data

In this study, the China Meteorological Forcing Dataset (CMFD) was developed by the Institute of Tibetan Plateau Research, Chinese Academy of Sciences [37,38]. This product covers the period of 1982–2015, with a spatial and temporal resolution of 0.1° every 3 h. The instantaneous 2 m air temperature, surface pressure, specific humidity, and wind speed were produced by merging 740 meteorological observations and environmental data sources. The precipitation dataset was developed by merging three precipitation datasets including the in situ observations dataset, the Tropical Rainfall Measuring Mission (TRMM) 3B42 precipitation products, and the Asian Precipitation—Highly Resolved Observational Data Integration Towards Evaluation of the Water Resources project. The CMFD downward shortwave radiation dataset was constructed by the Global Energy and Global Energy and Water cycle Experiment-Surface Radiation Budget (GEWEX-SRB) radiation data and meteorological station measurements [39].

To obtain the surface net radiation (Rn), we calculated Rn using the method proposed by Wang et al. [40]. This method accurately simulates the Rn value combined shortwave radiation measurement with meteorological observations, which is suitable for various land cover types with a correlation coefficient of 0.99. We also used the model-derived soil moisture (SM) dataset provided by the National Centers for Environmental Prediction (NCEP), which contains monthly SM from 1982 to 2015, with a spatial resolution of $0.5° \times 0.5°$. The specific dataset sources and detailed information on the datasets are provided in Table 1.

Table 1. Datasets for the meteorological reanalysis data and satellite data used in this study. NDVI, normalized difference vegetation index; ET, evapotranspiration.

Data	Name	Spatial Resolution	Temporal Resolution	Unit	Period
Climate Data	Precipitation	$0.1° \times 0.1°$	3 h	mm/h	1982–2015
	Pressure	$0.1° \times 0.1°$	3 h	Pa	1982–2015
	Specific Humidity	$0.1° \times 0.1°$	3 h	kg/kg	1982–2015
	Wind Speed	$0.1° \times 0.1°$	3 h	m/s	1982–2015
	Shortwave radiation	$0.1° \times 0.1°$	3 h	W/m^2	1982–2015
	Longwave radiation	$0.1° \times 0.1°$	3 h	W/m^2	1982–2015
	Temperature	$0.1° \times 0.1°$	3 h	K	1982–2015
	Soil moisture	$0.5° \times 0.5°$	1 month	-	1982–2015
Satellite Data	NDVI	8 km	16 day	-	1982–2015
	ET	$0.1° \times 0.1°$	daily	mm/day	1982–2015
	DEM	90 m	yearly	m	2003
	Land Cover	30 m	yearly	-	2010

2.3. Satellite Data

2.3.1. GIMMS NDVI Product

To quantify the variation of vegetation dynamics at regional scales, we used the Global Inventory Modeling and Mapping Studies (GIMMS) NDVI3g product derived from the Advanced Very High Resolution Radiometer (AVHRR) sensor National Oceanic and Atmospheric Administration (NOAA) polar satellite series with a spatial resolution of 8 km and a 15-day interval [41,42]. The GIMMS NDVI product has already been corrected to minimize the effects of clouds and aerosols using the maximum value composite (MVC) method. Previous studies have demonstrated that this dataset can reflect the real response of vegetation to climate change and provides more accuracy when evaluating the long-term trends of vegetation activity [43]. In this study, we extracted the subset of coverage in the TRHR from the global bimonthly NDVI for the period 1982–2015 and resampled the bimonthly NDVI of the study area to a daily value with a resolution of $0.1° \times 0.1°$.

2.3.2. ET Product

Considering that the MOD16 ET product is missing in the TRHR, we used the ET product produced by the modified satellite-based Priestley–Taylor algorithm (Appendix A) driven by net radiation (Rn), air temperature (Ta), diurnal temperature range (DT), and the NDVI [44]. This product has been validated at 16 eddy covariance (EC) flux tower sites, and performed better than MODIS ET products at a regional scale, with a higher squared correlation coefficient (R^2) and a lower root mean square error (RMSE) [45]. The modified satellite-based Priestley–Taylor (MS-PT) product has provided more reliable and long-term spatiotemporal variations of the ET estimations of China [46].

2.3.3. DEM Data

We used the global digital elevation model (DEM) data with a spatial resolution of 250 m acquired from 90 m Shuttle Radar Topography Mission (SRTM) images (version 004) (http://srtm.csi.cgiar.org/) in Geo-TIFF format.

2.3.4. Land Cover Data

The GlobeLand30 product developed by the National Geomatics Center of China (NGCC) provides detailed land cover information about a global coverage of high-resolution imagery at 30 m for the years 2000 and 2010 [47]. It is generated from the Thematic Mapper (TM), Enhanced Thematic Mapper plus (ETM+) of America Land Resources Satellite (Landsat) and the multispectral images of the China Environmental Disaster Alleviation Satellite (HJ-1) developed by integrating the pixel-object knowledge-based approach with other auxiliary datasets. This dataset is freely available and consists of 10 land cover types, including forest, grassland, shrubland, wetland, water bodies, tundra, bare land, artificial surfaces, cultivated land, permanent snow, and ice, with an overall accuracy of 80.33% [48].

2.4. Data Analysis

The Mann–Kendall test, as a nonparametric method for testing trends, and is also satisfactory for examining the significance of trends in a time series [49]. The statistics of variance can be described as follows:

$$S = \sum_{n=1}^{i-1} \sum_{m=n+1}^{i} sgn(x_m - x_n), \tag{1}$$

$$sgn(x_m - x_n) = \begin{cases} 1 & x_m - x_n > 0 \\ 0 & x_m - x_n = 0 \\ -0 & x_m - x_n < 0 \end{cases}, \tag{2}$$

$$Z = \begin{cases} \frac{S-1}{\sqrt{Var(S)}} & S > 0 \\ 0 & S = 0 \\ \frac{S+1}{\sqrt{Var(S)}} & S < 0 \end{cases}, \tag{3}$$

$$Var(S) = \frac{1}{18}[i(i+1)(2i+5) - \sum_{i=1}^{n} t_i(t_i-1)(2t_i+5)], \tag{4}$$

where i is the number of data points in the sequence, and t_i is the number of data values. Statistic Z, as a standard normal variable, was used to evaluate the statistical significance. The Mann–Kendall test is applied on a time series for all biophysical variables, and if the Z value is less than or equal to the significance level ($\alpha = 0.05$), a significant trend of the variable will be detected. In this study, the Mann–Kendall test for trends and linear regression analysis was used to detect and estimate the annual and seasonal trend of biophysical variables, with significance defined as $p < 0.05$.

Pearson's Correlation Coefficient was used to evaluate the correlation between the climate variables and vegetation index as well as between the climate variables and ET to determine the response of vegetation and ET to climate change.

$$r = \frac{\sum_{i=1}^{n}(X_i - \overline{X})(Y_i - \overline{Y})}{\sqrt{\sum_{i=1}^{n}(X_i - \overline{X})^2}\sqrt{\sum_{i=1}^{n}(Y_i - \overline{Y})^2}},\tag{5}$$

where X represents the climate variables, Y represents the vegetation index or ET, and n is the number of samples.

3. Results

3.1. Spatial and Seasonal Patterns of Terrestrial Biophysical Variables in the Three-River Headwaters Region

3.1.1. Climate Variables

Figure 2 shows the spatial distribution of the climate variables (Ta, P, RH, and Rs) over the TRHR at annual and seasonal scales during 1982–2015. Influenced by the typical plateau continental climate, the climate variables have distinctly different spatial patterns. On an annual basis, the annual mean Ta (Figure 2a) of the TRHR ranged from −12 to 6 °C, with an average of −4.2 °C. The multiyear average P (Figure 2b) varied from 162 to 781 mm, with an average of 424 mm. Both Ta and P presented obvious decreasing trends from the southeast to the northwest, which corresponded to the water and energy gradients of the TRHR. The 400 mm contour lines of annual precipitation roughly divide the TRHR into semi-arid and semi-humid climates from northwest to southeast. As a major part of the Tibetan Plateau, the climate of the TRHR is also influenced by atmospheric circulation and topographical features [50]. Figure 2c shows the spatial distribution of annual mean RH over the TRHR, with an average value of 52.3%. The decreasing trend of RH is noticeable from southeast to northwest, which is consistent with the distribution pattern of cloud cover [51]. By contrast, the annual mean Rs spatially decreased from west to east, ranging from 196 to 232 W/m². There were abundant solar energy resources in the TRHR due to the high altitude, thin atmosphere, and few anthropogenic activities [52].

On a seasonal basis, the climate of the TRHR is characterized by cold and dry winters, and cool and rainy summers. The spatial distributions of seasonal Ta and P were similar to the multiyear patterns of Ta and P as averaged during 1982–2015. The mean Ta was below −5 °C, and the P was less than 10 mm/month in winter (DJF, December, January, and February), whereas in summer (JJA, June, July, and August), the average P accounted for more than 80% of the total annual P, and the average Ta was about 5 °C. Moreover, we also found an obvious and clear distinction between dry and wet season over the TRHR. The southeastern area of TRHR remained the most humid region in other seasons with the exception of winter. During the Asian summer monsoon period, the mean surface RH of JJA was relatively higher than that of other seasons. The seasonal mean Rs of MAM (March, April, and May) and JJA were approximately 230–290 W/m², which were much higher than the values of SON (September, October, and November) and DJF.

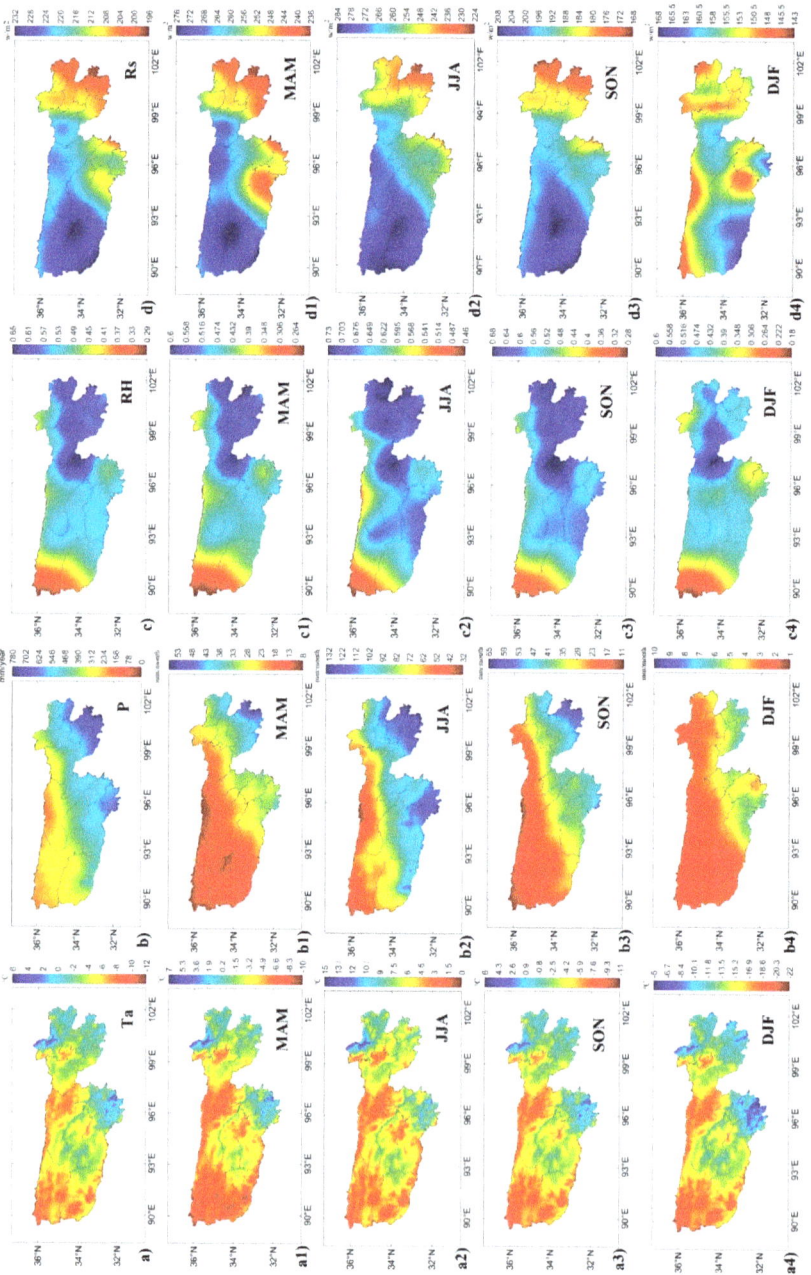

Figure 2. Multiyear average and seasonal spatial patterns of the climate variables: (**a**) temperature, (**b**) precipitation, (**c**) relative humidity, (**d**) downward shortwave radiation. (**1**) MAM (March, April, and May); (**2**) JJA (June, July, and August); (**3**) SON (September, October, and November); (**4**) DJF (December, January, and February). Precipitation is in units of mm/month. Radiation is in units of W/m².

3.1.2. Normalized Difference Vegetation Index

Figure 3a illustrates the annual mean NDVI, which presents an increasing trend from northwest to southeast over the TRHR during the period 1982–2015. Higher NDVI values were mainly distributed in the southeastern part of the TRHR, where the main land use type is forest and temperate grassland. Meanwhile, this region has sufficient precipitation and warmer temperatures that are suitable for vegetation growth. By contrast, the northwestern part of the TRHR has a relatively cold and dry climate, resulting in a lower NDVI value. Our findings are consistent with those of Zhong et al. [53], who also found that the spatial distribution of the NDVI was influenced by the Asian monsoon over the Tibetan Plateau.

As shown in Figure 3, seasonal NDVI has a decreasing trend from southeast to northwest across all four seasons. However, there were distinctive differences in the seasonal average of the NDVI value. During winter (DJF), the NDVI value was below 0.24 in the majority of the region under the dormancy condition of vegetation and lower air temperature. The NDVI reached a maximum value of 0.8, accompanied by increasing precipitation and rising temperatures in summer (JJA). When the rainy season had passed, the NDVI value began to decrease in autumn (SON) and winter (DJF), with an average of 0.15 and 0.09, respectively.

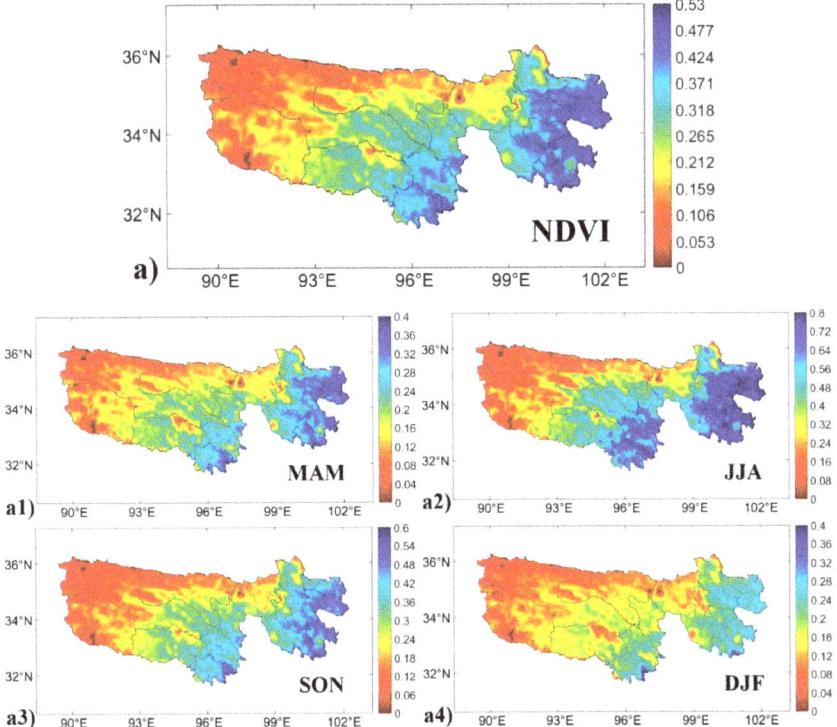

Figure 3. (**a**) Spatial patterns of the multiyear average NDVI in the TRHR. (**a1**) MAM (March, April, and May); (**a2**) JJA (June, July, and August); (**a3**) SON (September, October, and November); (**a4**) DJF (December, January, and February).

3.1.3. Evapotranspiration

Figure 4a shows the spatial distribution of the multiyear (1982–2015) average ET over the TRHR, and the annual mean ET was approximately 230.23 mm/year. As shown in Figure 4a, the multiyear average ET decreased from southeast to northwest, which was similar to the spatial patterns of Ta and P. Higher ET mainly occurred in the moister and warmer regions, including the eastern and southern parts of the TRHR, whereas the northwestern part of the TRHR with less P and lower Ta had the lowest ET value. Furthermore, the spatial pattern of ET was also affected by land management, such as agricultural irrigation, that caused a positive trend of ET in the cropland areas.

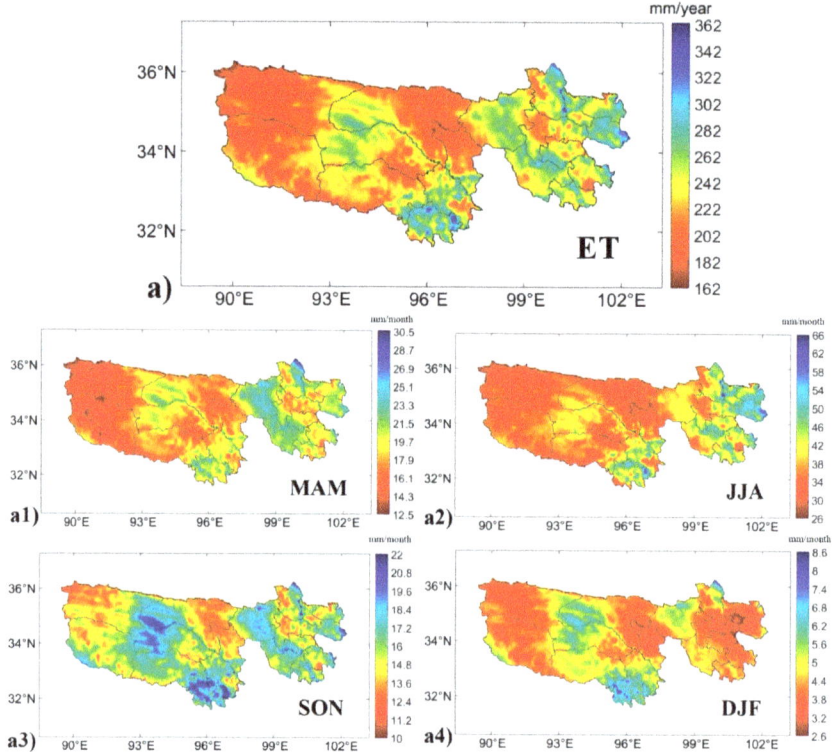

Figure 4. (**a**) Spatial patterns of the multiyear ET of the TRHR. Multiyear seasonal patterns of ET: (**a1**) MAM (March, April, and May); (**a2**) JJA (June, July, and August); (**a3**) SON (September, October, and November); (**a4**) DJF (December, January, and February). ET is in units of mm/month.

As shown in Figure 4, the multiyear average seasonal patterns of ET exhibited obvious seasonality with reasonable seasonal cycles (higher ET in the summer wet season and lower ET in the winter dry season). Distinct fluctuations of ET throughout the four seasons corresponded to the plateau mountain climate system. In spring (MAM) and autumn (SON), the ET was less than 26 mm/month due to the lack of available energy and temperature. The seasonal ET reached the largest value (26–66 mm/month) in summer (JJA), accompanied by the maximum Ta and P in the whole year. By contrast, ET dropped to its lowest value in winter (DJF), which is when vegetation turns to dormancy, and the temperature declines.

3.2. Interannual and Seasonal Variation of Terrestrial Biophysical Variables in the Three-River Headwaters Region

3.2.1. Climate Variables

Figure 5 shows the variation trend of the meteorological variables (Ta, P, RH, and Rs) over the TRHR during 1982–2015. An increasing trend in Ta appeared over the TRHR with an average value of 0.597 °C/decade, which is much higher than the global warming average of 0.12 °C/decade [54], and 96.3% of the pixels showed a significant increasing trend ($p < 0.05$). The trend of Ta over Maduo, the north of Chengduo, and the east of Qumalai was relatively higher than in other areas and the maximum reached 1.47 °C/decade. As shown in Figure 5b, P also experienced a positive trend during the period of 1982–2015 over the TRHR. We found that the P substantially increased in arid areas, with a linear tendency of 41.1 mm/decade ($p < 0.05$). By contrast, a significant decreasing trend of RH over the majority of the region was evident, which corresponded to the warming tendency over the TRHR. A negative Rs trend occurred in the southeastern region, with an average of 3.05 W/m^2 per decade ($p < 0.05$). Some scientists suggested that the decline in Rs is consistent with solar dimming over the TP due to an increase in the amount of water vapor and the atmospheric concentrations of aerosols [55].

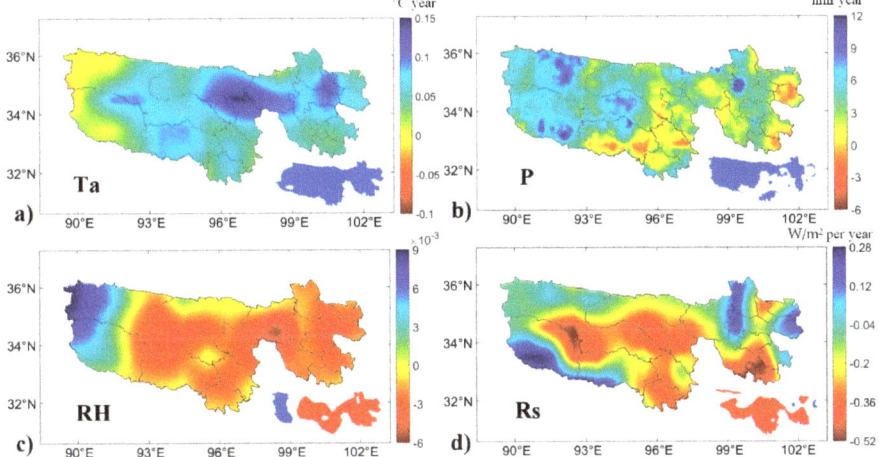

Figure 5. Spatial distributions of climate variable trends in the TRHR over the period 1982–2015: (**a**) temperature; (**b**) precipitation; (**c**) relative humidity; (**d**) downward shortwave radiation. The inset panels show the area where the climate variables trends were statistically significant ($p < 0.05$). Blue represents a significant increase and red represents a significant decrease.

Figure 6 shows the interannual and seasonal climate variables (Ta, P, RH, and Rs) of the TRHR during 1982–2015. Both annual and seasonal mean Ta and P showed a significant positive trend, with a linear trend of 0.6 °C/decade and 41.2 mm/decade, respectively. Figure 6a illustrates that a significantly increasing Ta has occurred since 1998, coincident with the last major El Nino event in 1998. During this period, the most significant increase in Ta occurred in the winter (0.901 °C/decade, $p < 0.01$), followed by autumn (0.57 °C/decade, $p < 0.01$), summer (0.475 °C/decade, $p < 0.01$), and spring (0.445 °C/decade, $p < 0.01$). Similarly, P also showed a significant positive trend in all four seasons ($p < 0.01$), with the largest P increases in summer (6.67 mm/decade, $p < 0.01$), and the rates for spring, autumn, and winter were 3.54, 2.74, and 0.806 mm per decade ($p < 0.01$), respectively. In addition, a severe drought was also detected in the summer of 2006, and the annual P decreased to 370 mm/year due to the abnormally high Ta and low P [14]. By contrast, the interannual RH and Rs of the TRHR showed a decreasing trend over the whole period. The largest decline in the regional mean surface RH occurred in winter (DJF) at 2.3%/decade ($p < 0.01$), which corresponded with the temperature rising in winter. A significant decrease of Rs occurred in summer (4.57 W/m^2 per decade, $p < 0.01$), while in other seasons, Rs presented a slight negative trend with no statistical significance.

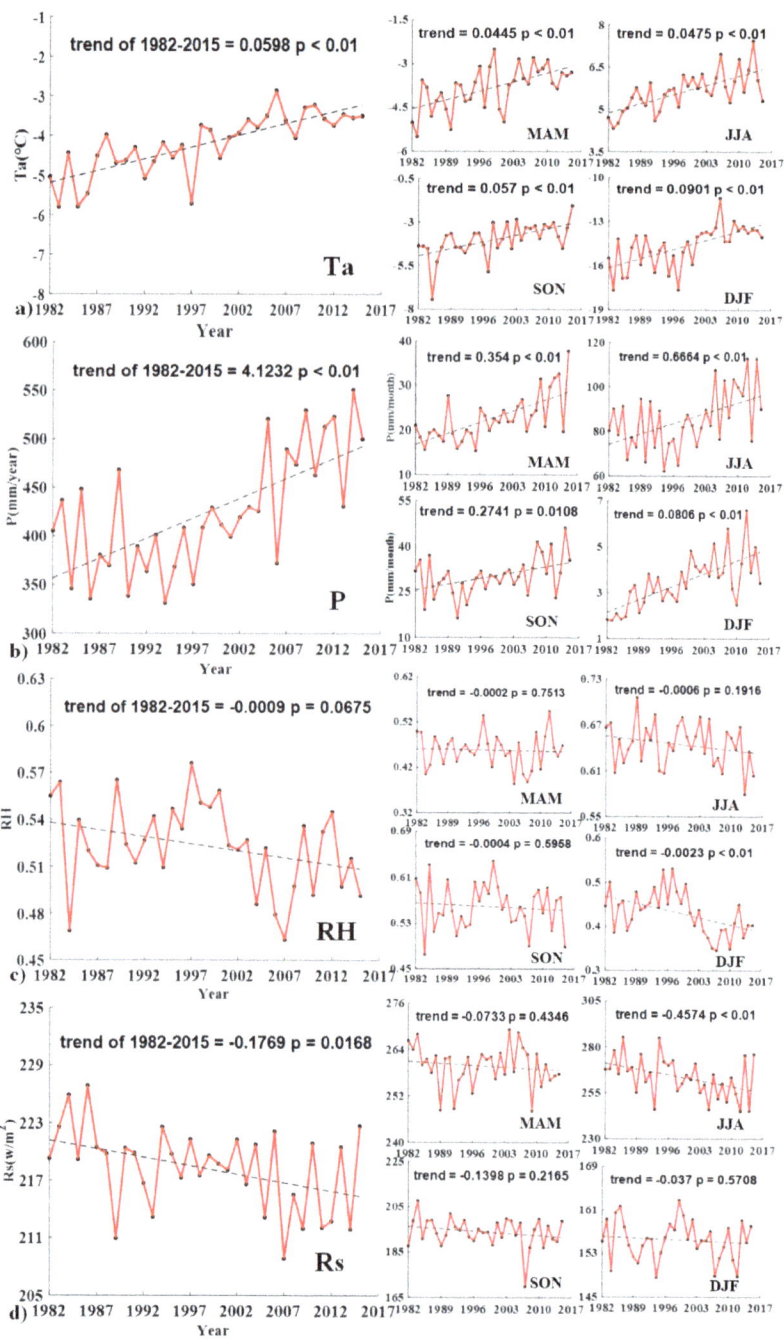

Figure 6. Interannual and seasonal variability of climate variables trends over TRHR in the period 1982–2015 (**a**) temperature; (**b**) precipitation; (**c**) relative humidity; (**d**) downward shortwave radiation.

3.2.2. Normalized Difference Vegetation Index

Figure 7a shows the spatial distribution of the NDVI trend over the TRHR during 1982–2015. Our results showed that the vegetation in the TRHR experienced slight greening and over 77.6% of the area showed a slight increasing trend, of which 56.8% significantly increased at a rate of 0.0051/decade ($p < 0.05$). In particular, a significant increase in the annual NDVI occurred in the northern and western part of the TRHR, where the main land use type is alpine and subalpine meadows. Only a tiny portion of the region had a significant decreasing trend, which was mainly distributed in Chengduo and Yushu counties, and the majority of the midland region did not exhibit significant changes in vegetation cover. The increasing trend of NDVI was similar to the findings of Xu et al. [56], who found that the vegetation coverage of the TRHR showed a consistent and slight increase in the period of 1982–2006.

The annual and seasonal NDVI also presented a slightly enhanced trend, particularly after the implementation of the TRHR project (2005 to 2012) [57]. This indicated that the implementation of ecological projects also promotes vegetation growth and gradually reverses the degradation of grassland ecosystems. Specifically, the largest significant increase in the NDVI occurred in spring at the rate of 0.003/decade ($p < 0.01$), which contributed most to the interannual NDVI increase trend.

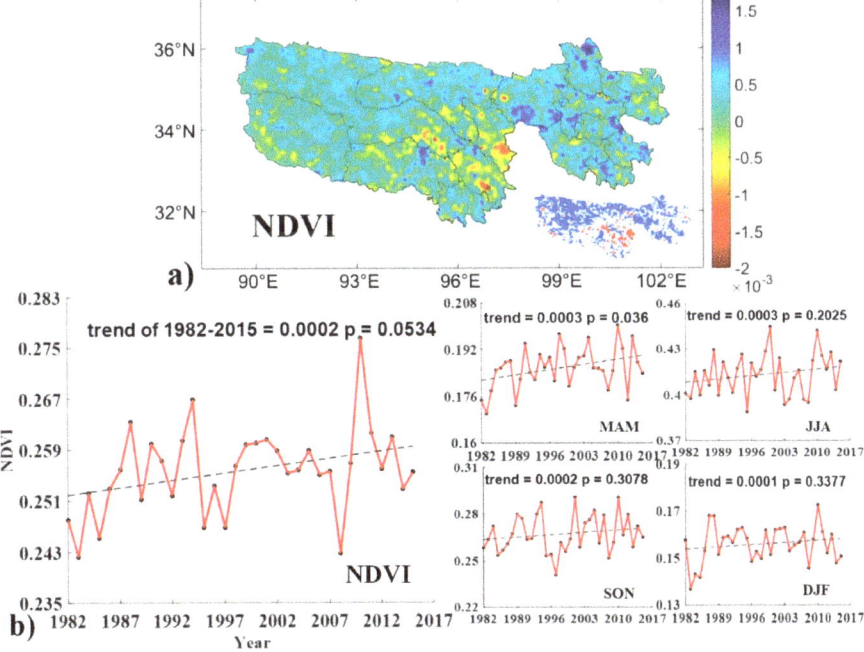

Figure 7. (**a**) Spatial distributions of the NDVI trends. (**b**) Interannual and seasonal variability of the NDVI trends. The inset panel shows the area where the NDVI trend was statistically significant ($p < 0.05$). Blue represents a significant increase and Red represents a significant decrease.

3.2.3. Evapotranspiration

Spatial patterns of the ET trend over the TRHR were detected from 1982 through 2015. There were significant differences in the ET between the southeastern and northwestern parts of the region. Figure 8a shows that the ET has increased, on average, by 3.34 mm/decade over the TRHR, which corresponded to the expected acceleration associated with rising air temperature. About 26.5% of the pixels showed a significant increasing trend over the TRHR, while only 3.81% of the pixels showed a significantly decreasing trend ($p < 0.05$). A significant positive ET trend was mainly distributed in the core of the Sanjiangyuan National Nature Reserve, namely, the east and west regions of the TRHR, with a linear tendency of 1.2 mm/year per decade, while the Dari and Banma counties showed a negative ET trend.

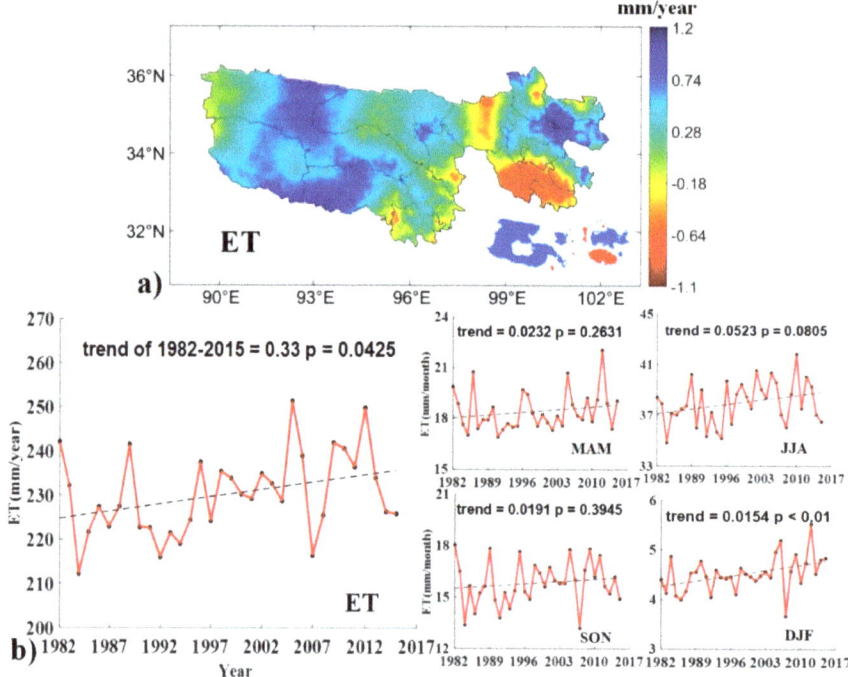

Figure 8. (a) Spatial distributions of the ET trends; (b) Interannual and seasonal variability of the ET trends. The inset panel shows the area where the ET trend was statistically significant ($p < 0.05$). Red represents a significant increase and blue represents a significant decrease.

As shown in Figure 8b, the spatially averaged ET has increased, on average, by 3.3 W/m^2 per decade ($p < 0.05$) over the entire TRHR during 1982–2015. The value of ET has obviously fluctuated since 2000, indicative of strong regional variations controlled by the monsoon climate system and the arid climate system. Considering the seasonal difference in climatic conditions, we further calculated the trend of ET across four seasons. Figure 8c illustrates that the ET trend in winter had a significant increase with a linear tendency of 0.154 ($p < 0.05$), while in other seasons, the ET presented a slight increasing trend with no statistical significance ($p > 0.05$). It is evident that the temperature warming in winter had a significant positive effect on water cycling. The trend magnitudes of the annual and seasonal terrestrial biophysical variables are summarized in Table 2.

Table 2. The Mann–Kendall test results for the terrestrial biophysical variable trends.

Biophysical Variable	Season	Z	β	R/A
	MAM	3.3207	0.0441	A
	JJA	4.0915	0.0475	A
Ta	SON	3.3800	0.0476	A
	DJF	4.2101	0.0844	A
	Year	5.2275	0.0568	A

Table 2. Cont.

Biophysical Variable	Season	Z	β	R/A
P	MAM	4.0322	0.3319	A
	JJA	2.5498	0.6663	A
	SON	2.4609	0.2943	A
	DJF	4.5066	0.0820	A
	Year	4.0915	0.3755	A
RH	MAM	−0.5336	−0.0002	R
	JJA	−1.1415	−0.0005	R
	SON	−0.5633	−0.0004	R
	DJF	−2.4905	−0.0022	R
	Year	−1.69	−0.0009	R
Rs	MAM	−0.8302	−0.0750	R
	JJA	−2.6091	−0.5360	A
	SON	−0.770	−0.0715	R
	DJF	−0.5929	−0.0300	R
	Year	−1.9865	−0.1776	A
NDVI	MAM	1.6603	0.0003	R
	JJA	1.3046	0.0003	R
	SON	1.1415	0.0002	R
	DJF	0.42991	0.0000	R
	Year	1.7345	0.0002	R
ET	MAM	1.2453	0.0233	R
	JJA	1.5121	0.0528	R
	SON	1.0081	0.0219	R
	DJF	2.5795	0.0146	A
	Year	1.9272	0.3501	R

R: reject hypothesis H0; A: accept hypothesis H0.

3.3. Vegetation Greening and ET Variation Response to Climate Change

Correlation analysis was used to investigate the relationship between each climate factor (Ta, P, RH, Rs) and the NDVI over the TRHR during 1982–2015. We found that over 57.54% of the area of the TRHR had a moderate positive correlation between the NDVI and Ta, and the maximum coefficient was about 0.89 (Figure 9a). When water was the limiting factor for vegetation growth in the western part of the TRHR, a strong correlation existed between the NDVI and P with a maximum coefficient of 0.74 (Figure 9b). The relationship between the NDVI and P was much weaker than that between the NDVI and Ta, which indicated that increasing temperature appeared to be the driving factor for vegetation greening, and better at explaining this phenomenon in comparison to P. Compared with Ta and P, no strong coherent spatial patterns were found in the relationship between the NDVI and annual Rs and annual RH, with a negative correlation coefficient of 0.3 (Figure 10).

We further conducted a correlation analysis between the ET and each energy- or water-limiting factor (Ta, P, RH, Rs, NDVI, potential ET (PET), and soil moisture (SM)) (Figure 11). The results showed that SM was the primary factor in controlling ET change in the western TRHR during the period 1982–2015. Given the fact that this area is located at arid and semi-arid climatic zones, the terrestrial moisture limitation is expected to be the most important driver of ET variation [58]. Similarly, over 55.21% of pixels showed a moderate positive correlation between precipitation and ET (Figure 11b), which can be attributed to the fact that ET corresponds well with surface moisture supply in a region with scarce water. The infrequent rainfall causes shortages in soil moisture and further feedbacks to the decreases in ET.

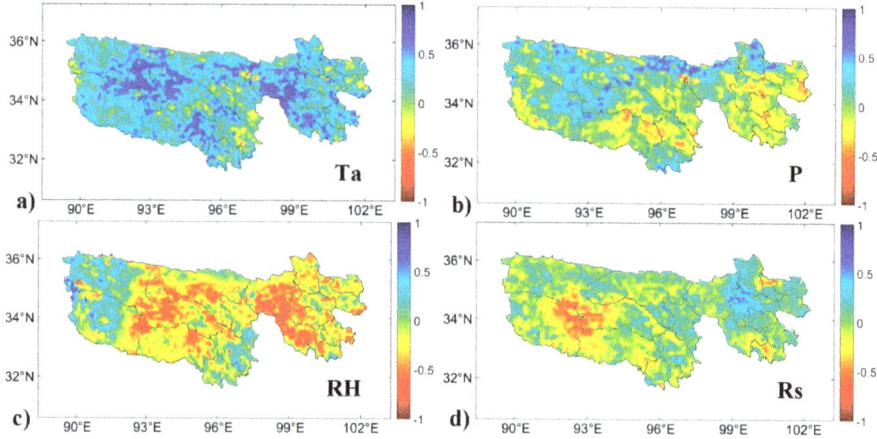

Figure 9. Maps of the relationship between the NDVI and climate variables: (**a**) temperature; (**b**) precipitation; (**c**) relative humidity; (**d**) downward shortwave radiation.

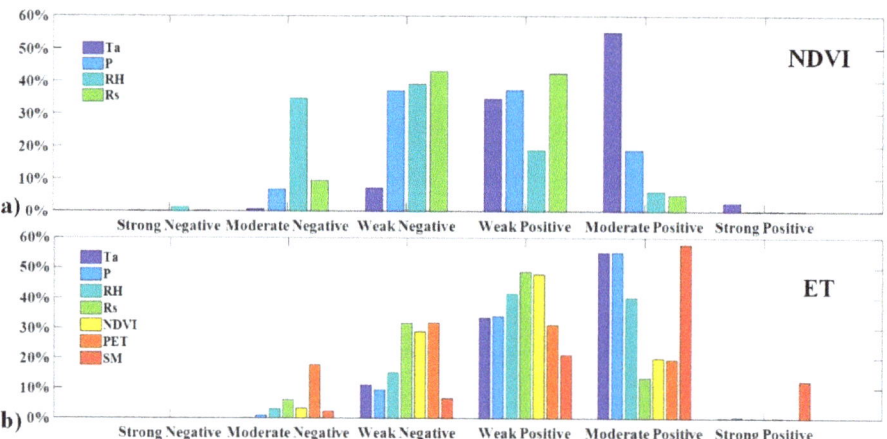

Figure 10. The frequency of correlation coefficient with (**a**) NDVI, (**b**) ET. The degree correlation was classified into six ranks: Strong Negative ($-1 < R < -0.7$); Moderate Negative ($-0.7 < R < -0.3$); Weak Negative ($-0.3 < R < 0$); Weak Positive ($0 < R < 0.3$); Moderate Positive ($0.3 < R < 0.7$); and Strong Positive ($0.7 < R < 1$).

In the relatively humid area of the TRHR, ET showed a positive correlation with Ta, accounting for approximately 55.3%. Ta was the primary indicator governing ET variation in the unrestricted water region, where ET corresponded well to atmospheric energy demand. The NDVI was also an important dominant factor in controlling the increasing ET in the southern part of the TRHR. The relatively higher plant transpiration and canopy conductance contributed to the increment of ET [59]. As shown in Figure 10c, the decline of RH has continuously contributed to the decrease in ET over the southeastern part of the TRHR. The rising temperature was expected to feedback to the atmosphere and consequently decreased the RH and ET, which implied that this area is projected to be drier. Previous studies have proposed that there is a complementary relationship in the ET and potential ET (PET) [60]. Zhang et al. [61] point out that vapor transfer power was suppressed due to the low Ta and vapor pressure deficit (VPD) in the TRHR. The negative correlation between the ET and PET revealed by this study supports their findings.

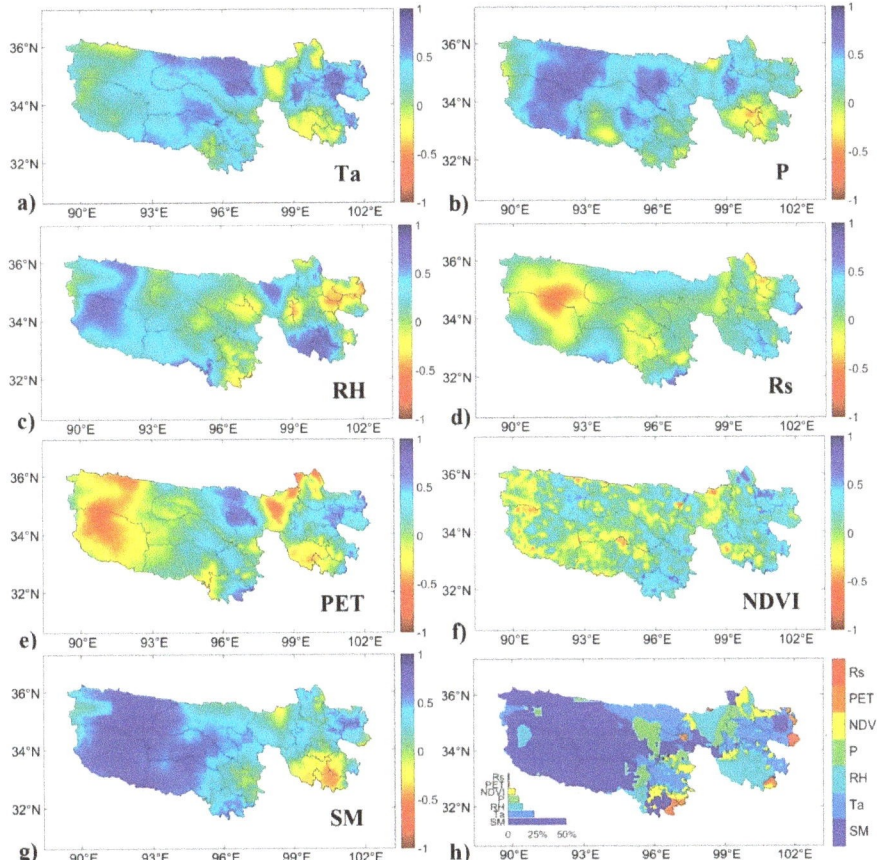

Figure 11. Maps of the spatial distribution of correlation coefficient (r) between annual ET: (**a**) temperature, (**b**) precipitation, (**c**) relative humidity, (**d**) downward shortwave radiation, (**e**) potential evapotranspiration, (**f**) NDVI, (**g**) soil moisture, (**h**) spatial distribution of most related driving variables for annual ET during 1982–2015 over the TRHR.

Land use and land cover change can also have substantial influences on the biophysical variables in hydrologic processes and terrestrial energy exchange by affecting the patterns of ET. We further investigated responses in the distribution of the multiyear average ET to the difference of land cover and use type. As shown in Figure 12, cropland had the highest ET values. The lowest annual ET occurred in the artificial surface and bare land. For each vegetation type, forest had the highest ET, followed by grassland and shrubland. This can be explained by forest ecosystems having relatively higher total root biomass and deeper effective rooting depth, thereby having the potential to create positive transpiration forcing [62]. The ET value of cropland was generally higher than that of forest, where it was noted that artificial management, e.g., agriculture irrigation, has a nonnegligible impact on the variation of ET.

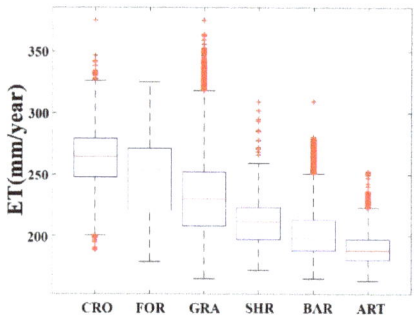

Figure 12. Box plots of per pixel annual average ET (mm/year) for each land cover type from 1982 to 2015 over the TRHR. CRO: cropland; FOR: forest; GRA: grassland; SHR: shrubland; BAR: bare land; ART: artificial surfaces.

4. Discussion

This study analyzed the long-term spatiotemporal dynamics of terrestrial climate variables from an interannual and seasonal perspective over the TRHR of China. The rising Ta and P and decline in the RH and Rs over the TRHR were similar to the trends observed over the Tibetan Plateau (TP), where annual Ta and P increased by 0.447 °C and 9.97 mm per decade at the 27 meteorological stations during 1961–2001, as reported by Xu et al. [50]. Using observations from 78 China Meteorological Administration (CMA) stations, Yang et al. [63] also demonstrated that the TP experienced a rapid warming and wetting tendency in the period of 1984–2006. The overall rapid climate warming tendency over the TP has been demonstrated in numerous studies by observing stations [51,64,65], by oxygen isotope analysis of ice cores [66], and by satellite remote sensing products [67,68], with the observed warming rate ranging from 0.16 to 0.67 °C per decade during the past few decades [69]. In comparison with previous studies regarding climate change, our findings improve the spatial information over heterogeneous landscapes and present long-term distribution patterns of annual and seasonal climate variables at a regional scale and provide a new understanding of the climate change in the TRHR in recent years.

Climate change has an inevitable and significant impact on vegetation dynamics, particularly in the extremely sensitive ecosystem of the TRHR. The response of vegetation to climate change has been discussed by many studies, where the results differed according to the different vegetation types, the plant physiological processes, and environmental factors. Previous studies have pointed out that CO_2 fertilization effects explain 70% of the observed greening trend in the tropics, whereas climate change contributes most to the vegetation greening of the TP [70]. Du et al. [71] proposed that solar radiation was the key factor governing the vegetation greening on the TP. This result is reasonable because sufficient solar radiation can promote the photosynthesis and respiration of vegetation, which is beneficial to plant growth [72]. According to Xu et al. [30], the averaged NDVI of the growing season was positively correlated with the summer Ta in the high-cold region, which indicated that the response of vegetation to temperature was likely to be more intense under climate warming. This conclusion supported our results to a certain extent. We found that when energy was the limiting factor for vegetation growth, Ta was a considerably more important driving factor than water. However, the effect of temperature on vegetation was obviously less than that of moisture in the water limiting area. These results can be explained by the fact that the climate condition in the TRHR is characterized by relatively abundant P during the growing season and lower temperature across the whole year [73]. We can conclude that increases in either Ta or P are predicted to have a positive influence on vegetation greening. These findings are in line with emerging evidence that the potential benefits from the climate "warming and moisture" trend are closely related to the increment of vegetation through alteration of vegetation phenology and prolonged growing season length in the TRHR [74].

The upward trend of ET we reported is consistent with long-term trend analysis, which indicated that the ET has significantly increased since the 1960s, especially in the central area of the TP [61,69,75]. The rising trend of ET corresponded to the significant increase in precipitation, the reduction of RH, and sunshine duration during the same period over the TP [76]. Yin et al. [77] suggested that the upward trend of ET was mainly constrained by the soil water supply, and linked with increased P, which is consistent with our results. In the arid and semi-arid regions, the increased P promoted the water availability for ET, resulting in the increment of ET. The pattern of increasing ET was matched by an increasing P in the western part of the TRHR, which was also confirmed by Yao et al. [29], who reported that P was the primary contributor to increasing ET during 1982–2010. However, in well-watered regions, climate (Ta, RH) and vegetation factors were considered to be related more to the ET dynamic. This result also agrees well with the study by Song et al. [78] on the TP, where dependencies of ET on leaf area index (LAI) and Ta appeared to be largely independent of moisture supply. Atmospheric demand was recognized as an important controlling factor on the long-term variations of ET. This inconsistent result can be explained by the different atmospheric energy demand or surface moisture supply in different regions [79]. In addition, the land use and land cover change (LUCC), and anthropogenic activities, such as agriculture irrigation and afforestation projects, also have a substantial influence on the variation of ET [80].

A long-term spatiotemporal biophysical dynamic provides more accurate estimates of climate change, vegetation greening, and ET variation in the TRHR. Although several products have been extensively validated and confirmed in different scales and regions, considerable uncertainties still exist. Regarding the climate forcing dataset, the accuracy of the reanalysis may be limited by the scarce measurements in the TRHR. Yang [81] et al. compared the shortwave radiation data of the CMFD product against the 579 in situ observations in China and found that the CMFD provided the closest match with ground measurements, with a 0.02 relative bias and a 5.6 root-mean-square error (RMSE) during 2008–2010. However, the precipitation data were detected to have an abnormal underestimation after August 2014. The inaccuracy of the precipitation data was also evaluated by Wang et al. [82], who found that the precipitation was overestimated at 90 stations over the TP. The biases of the CMFD dataset led to substantial errors in climate variation. Aside from the climate dataset, the uncertainties were also associated with the GIMMS NDVI data series driven by AVHRR. Kern et al. [83] suggested that there was a significant disagreement relationship between AVHRR NDVI3g and the MODIS NDVI dataset. Moreover, the influence of the canopy and soil background, aerosol effects, and cloud contamination were not completely eliminated due to the limitations of the AVHRR instruments [43,84]. The modified satellite-based Priestley–Taylor (MS-PT) algorithm produced a more accurate product as daily ET estimates exhibited a higher R^2 (0.87) and lower RMSE (12.5 W/m^2) than the original PT algorithm in regional ET simulations. However, there are still large uncertainties due to the different parameterization schemes of evaporation constraint. One limitation of the MS-PT product is that it shows large differences in daily ET estimates among the different ecosystem types [85]. Yao et al. [86] evaluated the performance of PT products at different biomes and demonstrated that the MS-PT model performed better in forest and village sites, with a higher R^2 of 0.93 and lower RMSE of 11.9 W/m^2, whereas in grassland sites, the algorithms may not capture the soil moisture constraint, resulting in underestimating the ET value, which makes the simulated ET value uncertain in the alpine grassland ecosystems of the TRHR.

5. Conclusions

In this study, we applied satellite data products in combination with meteorological reanalysis datasets to evaluate the interannual and seasonal dynamics of terrestrial biophysical variables, including the meteorological variables, vegetation, and evapotranspiration (ET) over the Three-River Headwaters Region (TRHR). We then further investigated the response of vegetation and ET to climate change during the period 1982–2015. Our results showed that the Ta and P increased by 0.597 °C and 41.1 mm per decade, while the RH and Rs declined at a rate of 0.9% and 1.8 W/m^2 per decade, respectively. The largest upward movement of Ta associated with the decline in RH occurred in winter (0.901/decade and 0.6%/decade, respectively), and the increment of P and the reduction of Rs were largest in summer (6.66 mm/decade and 4.57 W/m^2, respectively). A 'dryer warming' tendency and a 'wetter warming' tendency exist in different areas of the TRHR. Generally, most areas of the TRHR became warmer and moister, except for some areas in the southern TRHR, with a trend of being dryer and warmer.

Our findings illustrate that the NDVI of the TRHR fluctuated in the period 1982–2015, with a slight increase (0.0051/decade) particularly in the northern and western meadow areas. The NDVI significantly increased over 56.8% of the TRHR, and the largest increment occurred in spring, followed by summer. In well-watered regions, Ta was the primary driver of vegetation greening, while in the water limiting areas, vegetation growth was mainly governed by the variation of P. Our results suggest that the warming and wetting tendencies of the climate characterized by increasing Ta and P contribute most to the increment of vegetation in the TRHR.

The annual mean terrestrial ET was about 230.23 mm/year and varied 162 mm/year to 362 mm/year from the northwest to southeast over the TRHR in the period from 1982 to 2015. The ET of the TRHR showed a significant increasing trend at a rate of 3.34 mm/decade, particularly in winter (0.154 mm/decade), which corresponded to the expected acceleration associated with climate warming. In the arid region of western TRHR, ET was limited by the terrestrial water supply, which includes soil moisture (SM) and P. By contrast, atmospheric evaporative demand derived from Ta and relative humidity (RH) were the main controlling factors over the relatively humid region of southeastern TRHR. In addition, the intensification of agriculture irrigation is also responsible for the temporal and spatial variation of ET. Moreover, the impacts of carbon flux and anthropogenic disturbance on the biophysical variables need further exploration.

Author Contributions: Conceptualization, Y.Y.; Formal analysis, K.J.; Investigation, X.C.; Methodology, K.S.; Software, L.Z.; Supervision, T.X.; Validation, J.X.; Visualization, X.Z.; Writing—original draft, X.B.

Funding: This work was also partially supported by the National Key Research and Development Program of China (No. 2016YFB0501404) and the Natural Science Fund of China (41671331).

Acknowledgments: The authors thank Xianhong Xie and Bo Jiang from Beijing Normal University for their helpful suggestions. The authors thank Jie He and Kun Yang from the Institute of Tibetan Plateau Research, Chinese Academy of Sciences (http://westdc.westgis.ac.cn/) for providing the China Meteorological Forcing Dataset. The GIMMS NDVI product was obtained from NOAA (http://islscp2.sesda.com/ISLSCP21/data), and the land cover type product of GlobeLand30 was obtained online (http://www.globallandcover.com). The Climate Prediction Center soil moisture dataset was obtained online (http://www.esrl.noaa.gov/psd/).

Conflicts of Interest: The authors declare no conflict of interest.

Appendix A Algorithms

The MS-PT algorithm can be described as

$$ET = ET_s + ET_c + ET_{ic} + ET_{ws}, \tag{A1}$$

$$ET_s = (1 - f_{wet})f_{sm}\alpha\frac{\Delta}{\Delta + \gamma}(R_{ns} - G), \tag{A2}$$

$$ET_c = (1 - f_{wet})f_c f_T \alpha\frac{\Delta}{\Delta + \gamma}R_{nv}, \tag{A3}$$

$$ET_{ic} = f_{wet}\alpha\frac{\Delta}{\Delta + \gamma}R_{nv}, \tag{A4}$$

$$ET_{ws} = f_{wet}\alpha\frac{\Delta}{\Delta + \gamma}(R_{ns} - G), \tag{A5}$$

where ET_c is the canopy transpiration, ET_s is the unsaturated soil evaporation, ET_{ic} is the canopy interception evaporation, and ET_{ws} is the saturated wet soil surface evaporation. Moreover, f_{wet} is the relative surface wetness (f_{sm}^4), in which f_{sm} refers to soil moisture constraint and can be derived from ATI ($ATI = (\frac{1}{DT})^{DT/DT_{max}}$, $DT_{max} = 40\ °C$), f_T represents plant temperature constraint $\left(\exp(-(T_{max} - T_{opt})/T_{opt})^2\right)$, T_{opt} is an optimum temperature (25 °C), R_{ns} is the surface net radiation to the soil ($R_{ns} = R_n(1 - f_c)$), G is soil heat flux ($\mu R_n(1 - f_c)$, $\mu = 0.18$), R_{nv} represents the surface net radiation to the vegetation ($R_{nv} = R_n f_c$), f_c is the vegetation cover fraction ($f_c = (NDVI - NDVI_{min}/(NDVI_{max} - NDVI_{min}))$, and $NDVI_{min}$ and $NDVI_{max}$ are the minimum and maximum NDVI, respectively. Δ is the slope of the saturate vapor pressure curve, and γ is the psychrometric constant (0.066 kPa/ °C).

References

1. Duveiller, G.; Hooker, J.; Cescatti, A. The mark of vegetation change on Earth's surface energy balance. *Nat. Commun.* **2018**, *9*, 679. [CrossRef]
2. Jiang, C.; Zhang, L. Ecosystem change assessment in the Three-river Headwater Region, China: Patterns, causes, and implications. *Ecol. Eng.* **2016**, *93*, 24–36. [CrossRef]
3. Li, W.H.; Zhao, X.Q.; Zhang, X.Z.; Shi, P.L.; Wang, X.D.; Zhao, L. Change mechanism in main ecosystems and its effect of carbon source/sink function on the Qinghai-Tibetan Plateau. *Chin. J. Nat.* **2013**, *35*, 172–178.
4. Wang, Z.; Song, K.; Hu, L. China's Largest Scale Ecological Migration in the Three-River Headwater Region. *Ambio* **2010**, *39*, 443–446. [CrossRef] [PubMed]
5. Shao, Q.; Liu, J.; Huang, L.; Fan, J.; Xinliang, X.U.; Wang, J. Integrated assessment on the effectiveness of ecological conservation in Sanjiangyuan National Nature Reserve. *Geogr. Res.* **2013**, *32*, 1645–1656.
6. Li, X.-L.; Brierley, G.; Shi, D.-J.; Xie, Y.-L.; Sun, H.-Q. Ecological Protection and Restoration in Sanjiangyuan National Nature Reserve, Qinghai Province, China. In *Perspectives on Environmental Management and Technology in Asian River Basins*; Springer: Berlin/Heidelberg, Germany, 2012; pp. 93–120.
7. Zhang, Y.; Zhang, S.; Zhai, X.; Xia, J. Runoff variation and its response to climate change in the Three Rivers Source Region. *J. Geogr. Sci.* **2012**, *22*, 781–794. [CrossRef]
8. Liu, X.; Zhu, X.; Pan, Y.; Zhu, W.; Zhang, J.; Zhang, D. Thermal growing season and response of alpine grassland to climate variability across the Three-Rivers Headwater Region, China. *Agric. For. Meteorol.* **2016**, *220*, 30–37. [CrossRef]
9. Chen, H.; Zhu, Q.; Peng, C.; Wu, N.; Wang, Y.; Fang, X.; Gao, Y.; Zhu, D.; Yang, G.; Tian, J. The impacts of climate change and human activities on biogeochemical cycles on the Qinghai-Tibetan Plateau. *Glob. Chang. Biol.* **2013**, *19*, 2940–2955. [CrossRef]
10. Jiang, C.; Li, D.; Gao, Y.; Liu, W.; Zhang, L. Impact of climate variability and anthropogenic activity on streamflow in the Three Rivers Headwater Region, Tibetan Plateau, China. *Theor. Appl. Climatol.* **2017**, *129*, 667–681. [CrossRef]
11. Yao, Y.; Liang, S.; Li, X.; Chen, J.; Wang, K.; Jia, K.; Cheng, J.; Jiang, B.; Fisher, J.B.; Mu, Q. A satellite-based hybrid algorithm to determine the Priestley–Taylor parameter for global terrestrial latent heat flux estimation across multiple biomes. *Remote Sens. Environ.* **2015**, *165*, 216–233. [CrossRef]
12. Jiang, C.; Zhang, L. Climate Change and Its Impact on the Eco-Environment of the Three-Rivers Headwater Region on the Tibetan Plateau, China. *Int. J. Environ. Res. Public Health* **2015**, *12*, 12057–12081. [CrossRef] [PubMed]
13. Chong, J.; Li, D.; Gao, Y.; Liu, X.; Liu, W.; Zhang, L. Spatiotemporal variability of streamflow and attribution in the Three-Rivers Headwater Region, northwest China. *J. Water Clim. Chang.* **2016**, *7*, 637–649.
14. Tong, L.; Xu, X.; Fu, Y.; Li, S. Wetland Changes and Their Responses to Climate Change in the "Three-River Headwaters" Region of China since the 1990s. *Energies* **2014**, *7*, 2515–2534. [CrossRef]
15. Liu, S.; Li, X.; Xu, Z.; Che, T.; Xiao, Q.; Ma, M.; Liu, Q.; Jin, R.; Guo, J.; Wang, L. The Heihe Integrated Observatory Network: A basin-scale land surface processes observatory in China. *Vadose Zone J.* **2018**, *17*. [CrossRef]
16. Zhao, T.; Congbin, F.U.; Zongjian, K.E.; Guo, W. Global Atmosphere Reanalysis Datasets: Current Status and Recent Advances. *Adv. Earth Sci.* **2010**, *25*, 242–254.
17. Gao, Q.; Guo, Y.; Xu, H.; Ganjurjav, H.; Li, Y.; Wan, Y.; Qin, X.; Ma, X.; Liu, S. Climate change and its impacts on vegetation distribution and net primary productivity of the alpine ecosystem in the Qinghai-Tibetan Plateau. *Sci. Total Environ.* **2016**, *554*, 34–41. [CrossRef] [PubMed]

18. Liang, T.; Yang, S.; Feng, Q.; Liu, B.; Zhang, R.; Huang, X.; Xie, H. Multi-factor modeling of above-ground biomass in alpine grassland: A case study in the Three-River Headwaters Region, China. *Remote Sens. Environ.* **2016**, *186*, 164–172. [CrossRef]
19. Liu, J.; Xu, X.; Shao, Q. The spatil and temporal characteristics of grassland degradation in the three-river headwaters region in Qinghai Province. *Acta Geogr. Sin.* **2008**, *63*, 364–376.
20. Chen, Q. Causes of Grassland Degradation in Dari County of Qinghai Province. *Acta Pratac.* **1998**, *7*, 44–48.
21. Cai, H.; Yang, X.; Xu, X. Human-induced grassland degradation/restoration in the central Tibetan Plateau: The effects of ecological protection and restoration projects. *Ecol. Eng.* **2015**, *83*, 112–119. [CrossRef]
22. Huixia, L.I.; Liu, G.; Bojie, F.U. Response of vegetation to climate change and human activity based on NDVI in the Three-River Headwaters region. *Acta Ecol. Sin.* **2011**, *31*, 5495–5504.
23. Liu, X.; Zhang, J.; Zhu, X.; Pan, Y.; Liu, Y.; Zhang, D.; Lin, Z. Spatiotemporal changes in vegetation coverage and its driving factors in the Three-River Headwaters Region during 2000–2011. *J. Geogr. Sci.* **2014**, *24*, 288–302. [CrossRef]
24. Yao, Y.; Liang, S.; Li, X.; Chen, J.; Liu, S.; Jia, K.; Zhang, X.; Xiao, Z.; Fisher, J.B.; Mu, Q. Improving global terrestrial evapotranspiration estimation using support vector machine by integrating three process-based algorithms. *Agric. For. Meteorol.* **2017**, *242*, 55–74. [CrossRef]
25. Wang, K.; Dickinson, R.E. A review of global terrestrial evapotranspiration: Observation, modeling, climatology, and climatic variability. *Rev. Geophys.* **2012**, *50*. [CrossRef]
26. Yao, Y.; Hong, Y.; Zhang, N.; Chen, J.; Cheng, J.; Zhao, S.; Zhang, X.; Jiang, B.; Sun, L.; Jia, K. Bayesian multimodel estimation of global terrestrial latent heat flux from eddy covariance, meteorological, and satellite observations. *J. Geophys. Res. Atmos.* **2014**, *119*, 4521–4545. [CrossRef]
27. Chang, Y.; Qin, D.; Ding, Y.; Zhao, Q.; Zhang, S. A modified MOD16 algorithm to estimate evapotranspiration over alpine meadow on the Tibetan Plateau, China. *J. Hydrol.* **2018**, *561*, 16–30. [CrossRef]
28. Liu, W. Evaluating remotely sensed monthly evapotranspiration against water balance estimates at basin scale in the Tibetan Plateau. *Hydrol. Res.* **2018**, *49*, 1977–1990. [CrossRef]
29. Yao, Y.; Zhao, S.; Wan, H.; Zhang, Y.; Jiang, B.O.; Jia, K.; Liu, M.; Jinhui, W.U. Satellite evidence for no change in terrestrial latent heat flux in the Three-River Headwaters region of China over the past three decades. *J. Earth Syst. Sci.* **2016**, *125*, 1245–1253. [CrossRef]
30. Xu, M.; Kang, S.; Chen, X.; Wu, H.; Wang, X.; Su, Z. Detection of hydrological variations and their impacts on vegetation from multiple satellite observations in the Three-River Source Region of the Tibetan Plateau. *Sci. Total Environ.* **2018**, *639*, 1220–1232. [CrossRef]
31. Peng, J.; Loew, A.; Chen, X.; Ma, Y.; Su, Z. Comparison of satellite based evapotranspiration estimates over the Tibetan Plateau. *Hydrol. Earth Syst. Sci.* **2016**, *20*, 3167–3182. [CrossRef]
32. Yao, Y.; Liang, S.; Yu, J.; Chen, J.; Liu, S.; Lin, Y.; Fisher, J.B.; McVicar, T.R.; Cheng, J.; Jia, K. A simple temperature domain two-source model for estimating agricultural field surface energy fluxes from Landsat images. *J. Geophys. Res. Atmos.* **2017**, *122*, 5211–5236. [CrossRef]
33. Zhang, Y.; Zhang, C.; Wang, Z.; Chen, Y.; Gang, C.; An, R.; Li, J. Vegetation dynamics and its driving forces from climate change and human activities in the Three-River Source Region, China from 1982 to 2012. *Sci. Total Environ.* **2016**, *563*, 210–220. [CrossRef] [PubMed]
34. Li, L.; Li, F.X.; Guo, A.H.; Zhu, X.D. Study on the Climate Change Trend and Its Catastrophe over "Sanjiangyuan" Region in Recent 43 Years. *J. Nat. Resour.* **2006**, *21*, 79–85.
35. Fang, Y. Managing the Three-Rivers Headwater Region, China: From Ecological Engineering to Social Engineering. *Ambio* **2013**, *42*, 566–576. [CrossRef] [PubMed]
36. Wu, J.; Feng, Y.; Zhang, J.; Zhang, X.; Song, C. Identifying the Relative Contributions of Climate and Grazing to Both Direction and Magnitude of Alpine Grassland Productivity Dynamic from 1993 to 2011 on the Northern Tibetan Plateau. In Proceedings of the EGU General Assembly Conference, Vienna, Austria, 23–28 April 2017; p. 136.
37. Chen, Y.; Yang, K.; Jie, H.; Qin, J.; Shi, J.; Du, J.; He, Q. Improving land surface temperature modeling for dry land of China. *J. Geophys. Res. Atmos.* **2011**, *116*. [CrossRef]
38. He, J.; Yang, K. *China Meteorological Forcing Dataset*; Cold and Arid Regions Science Data Center: Lanzhou, China, 2011.
39. Yang, K.; Koike, T.; Ye, B. Improving estimation of hourly, daily, and monthly solar radiation by importing global data sets. *Agric. For. Meteorol.* **2006**, *137*, 43–55. [CrossRef]

40. Wang, K.; Liang, S. Estimation of Surface Net Radiation from Solar Shortwave Radiation Measurements. In Proceedings of the IGARSS 2008—IEEE International Geoscience and Remote Sensing Symposium, Boston, MA, USA, 7–11 July 2008; pp. 483–486.
41. Tucker, C.J.; Newcomb, W.W.; Dregne, H.E. AVHRR data sets for determination of desert spatial extent. *Int. J. Remote Sens.* **1994**, *15*, 3547–3565. [CrossRef]
42. Tucker, C.J.; Pinzon, J.E.; Brown, M.E.; Slayback, D.A.; Pak, E.W.; Mahoney, R.; Vermote, E.F.; Saleous, N.E. An extended AVHRR 8 m NDVI dataset compatible with MODIS and SPOT vegetation NDVI data. *Int. J. Remote Sens.* **2005**, *26*, 4485–4498. [CrossRef]
43. Zeng, F.-W.; Collatz, G.; Pinzon, J.; Ivanoff, A. Evaluating and quantifying the climate-driven interannual variability in Global Inventory Modeling and Mapping Studies (GIMMS) Normalized Difference Vegetation Index (NDVI3g) at global scales. *Remote Sens.* **2013**, *5*, 3918–3950. [CrossRef]
44. Yao, Y.; Liang, S.; Cheng, J.; Liu, S.; Fisher, J.B.; Zhang, X.; Jia, K.; Zhao, X.; Qin, Q.; Zhao, B. MODIS-driven estimation of terrestrial latent heat flux in China based on a modified Priestley–Taylor algorithm. *Agric. For. Meteorol.* **2013**, *171–172*, 187–202. [CrossRef]
45. Yao, Y.; Liang, S.; Zhao, S.; Zhang, Y.; Qin, Q.; Cheng, J.; Jia, K.; Xie, X.; Zhang, N.; Liu, M. Validation and application of the modified satellite-based Priestley-Taylor algorithm for mapping terrestrial evapotranspiration. *Remote Sens.* **2014**, *6*, 880–904. [CrossRef]
46. Zhang, L.; Yao, Y.; Wang, Z.; Jia, K.; Zhang, X.; Zhang, Y.; Wang, X.; Xu, J.; Chen, X. Satellite-Derived Spatiotemporal Variations in Evapotranspiration over Northeast China during 1982–2010. *Remote Sens.* **2017**, *9*, 1140. [CrossRef]
47. Chen, J.; Chen, J.; Liao, A.; Cao, X.; Chen, L.; Chen, X.; He, C.; Han, G.; Peng, S.; Lu, M. Global land cover mapping at 30 m resolution: A POK-based operational approach. *ISPRS J. Photogr. Remote Sens.* **2015**, *103*, 7–27. [CrossRef]
48. Jun, C.; Ban, Y.; Li, S. China: Open access to Earth land-cover map. *Nature* **2014**, *514*, 434. [CrossRef] [PubMed]
49. Atta-ur-Rahman; Dawood, M. Spatio-statistical analysis of temperature fluctuation using Mann–Kendall and Sen's slope approach. *Clim. Dyn.* **2017**, *48*, 783–797. [CrossRef]
50. Xu, Z.; Gong, T.; Li, J. Decadal trend of climate in the Tibetan Plateau—Regional temperature and precipitation. *Hydrol. Process. Int. J.* **2008**, *22*, 3056–3065. [CrossRef]
51. You, Q.; Min, J.; Lin, H.; Pepin, N.; Kang, S. Observed climatology and trend in relative humidity in the central and eastern Tibetan Plateau. *J. Geophys. Res. Atmos.* **2015**, *120*, 3610–3621. [CrossRef]
52. Feng, Y.; Li, Y. Estimated spatiotemporal variability of total, direct and diffuse solar radiation across China during 1958–2016. *Int. J. Climatol.* **2018**, *38*, 4395–4404. [CrossRef]
53. Zhong, L.; Ma, Y.; Salama, M.S.; Su, Z. Assessment of vegetation dynamics and their response to variations in precipitation and temperature in the Tibetan Plateau. *Clim. Chang.* **2010**, *103*, 519–535. [CrossRef]
54. Ji, F.; Wu, Z.; Huang, J.; Chassignet, E.P. Evolution of land surface air temperature trend. *Nat. Clim. Chang.* **2014**, *4*, 462–466. [CrossRef]
55. Yang, K.; Ding, B.; Qin, J.; Tang, W.; Lu, N.; Lin, C. Can aerosol loading explain the solar dimming over the Tibetan Plateau? *Geophys. Res. Lett.* **2012**, *39*. [CrossRef]
56. Xu, G.; Zhang, H.; Chen, B.; Zhang, H.; Innes, J.; Wang, G.; Yan, J.; Zheng, Y.; Zhu, Z.; Myneni, R. Changes in vegetation growth dynamics and relations with climate over China's landmass from 1982 to 2011. *Remote Sens.* **2014**, *6*, 3263–3283. [CrossRef]
57. Shao, Q.; Fan, J.; Liu, J.; Lin, H.; Wei, C.; Xinliang, X.U.; Jinsong, G.E.; Dan, W.U.; Zhiqiang, L.I.; Gong, G. Assessment on the effects of the first-stage ecological conservation and restoration project in Sanjiangyuan region. *Acta Geogr. Sin.* **2016**, *71*, 3–20.
58. Zeng, Z.; Wang, T.; Zhou, F.; Ciais, P.; Mao, J.; Shi, X.; Piao, S. A worldwide analysis of spatiotemporal changes in water balance-based evapotranspiration from 1982 to 2009. *J. Geophys. Res. Atmos.* **2014**, *119*, 1186–1202. [CrossRef]
59. Liu, X.; Zhu, X.; Zhu, W.; Pan, Y.; Zhang, C.; Zhang, D. Changes in Spring Phenology in the Three-Rivers Headwater Region from 1999 to 2013. *Remote Sens.* **2014**, *6*, 9130–9144. [CrossRef]
60. Yao, Y.; Zhao, S.; Zhang, Y.; Jia, K.; Liu, M. Spatial and Decadal Variations in Potential Evapotranspiration of China Based on Reanalysis Datasets during 1982–2010. *Atmosphere* **2014**, *5*, 737–754. [CrossRef]

61. Zhang, Y.; Liu, C.; Tang, Y.; Yang, Y. Trends in pan evaporation and reference and actual evapotranspiration across the Tibetan Plateau. *J. Geophys. Res. Atmos.* **2007**, *112*. [CrossRef]
62. Yang, Y.; Donohue, R.J.; Mcvicar, T.R. Global estimation of effective plant rooting depth: Implications for hydrological modelling. *Water Resources Res.* **2016**, *52*, 8260–8276. [CrossRef]
63. Yang, K.; Zhou, D.; Wu, B.; Foken, T.; Qin, J.; Zhou, Z. Response of hydrological cycle to recent climate changes in the Tibetan Plateau. *Clim. Chang.* **2011**, *109*, 517–534. [CrossRef]
64. Guo, D.; Wang, H. The significant climate warming in the northern Tibetan Plateau and its possible causes. *Int. J. Climatol.* **2012**, *32*, 1775–1781. [CrossRef]
65. Li, L.; Yang, S.; Wang, Z.; Zhu, X.; Tang, H. Evidence of warming and wetting climate over the Qinghai-Tibet Plateau. *Arct. Antarct. Alp. Res.* **2010**, *42*, 449–457. [CrossRef]
66. An, W.; Hou, S.; Zhang, W.; Wang, Y.; Liu, Y.; Wu, S.; Pang, H. Significant recent warming over the northern Tibetan Plateau from ice core $\delta^{18}O$ records. *Clim. Past* **2016**, *11*, 2701–2728. [CrossRef]
67. Salama, M.S.; Van der Velde, R.; Zhong, L.; Ma, Y.; Ofwono, M.; Su, Z. Decadal variations of land surface temperature anomalies observed over the Tibetan Plateau by the Special Sensor Microwave Imager (SSM/I) from 1987 to 2008. *Clim. Chang.* **2012**, *114*, 769–781. [CrossRef]
68. Gao, Y.; Lan, C.; Zhang, Y. Changes in Moisture Flux over the Tibetan Plateau during 1979–2011 and Possible Mechanisms. *J. Clim.* **2014**, *27*, 1876–1893. [CrossRef]
69. Kuang, X.; Jiao, J.J. Review on climate change on the Tibetan Plateau during the last half century. *J. Geophys. Res. Atmos.* **2016**, *121*, 3979–4007. [CrossRef]
70. Zhu, Z.; Piao, S.; Myneni, R.B.; Huang, M.; Zeng, Z.; Canadell, J.G.; Ciais, P.; Sitch, S.; Friedlingstein, P.; Arneth, A. Greening of the Earth and its drivers. *Nat. Clim. Chang.* **2016**, *6*, 791–795. [CrossRef]
71. Du, J.; Zhao, C.; Shu, J.; Jiaerheng, A.; Yuan, X.; Yin, J.; Fang, S.; Ping, H. Spatiotemporal changes of vegetation on the Tibetan Plateau and relationship to climatic variables during multiyear periods from 1982–2012. *Environ. Earth Sci.* **2016**, *75*, 1–18. [CrossRef]
72. Gu, F.; Zhang, Y.; Huang, M.; Tao, B.; Guo, R.; Yan, C. Effects of climate warming on net primary productivity in China during 1961–2010. *Ecol. Evolut.* **2017**, *7*, 6736–6746. [CrossRef]
73. Xu, W.; Gu, S.; Zhao, X.; Xiao, J.; Tang, Y.; Fang, J.; Zhang, J.; Jiang, S. High positive correlation between soil temperature and NDVI from 1982 to 2006 in alpine meadow of the Three-River Source Region on the Qinghai-Tibetan Plateau. *Int. J. Appl. Earth Obs. Geoinf.* **2011**, *13*, 528–535. [CrossRef]
74. Han, Z.; Song, W.; Deng, X.; Xu, X. Grassland ecosystem responses to climate change and human activities within the Three-River Headwaters region of China. *Sci. Rep.* **2018**, *8*, 9079. [CrossRef]
75. Zhang, T.; Gebremichael, M.; Meng, X.; Wen, J.; Iqbal, M.; Jia, D.; Yu, Y.; Li, Z. Climate-related trends of actual evapotranspiration over the Tibetan Plateau (1961–2010). *Int. J. Climatol.* **2018**, *38*, e48–e56. [CrossRef]
76. Zhang, H.; Sun, J.; Xiong, J. Spatial-Temporal Patterns and Controls of Evapotranspiration across the Tibetan Plateau (2000–2012). *Adv. Meteorol.* **2017**, *2017*. [CrossRef]
77. Yin, Y.; Wu, S.; Zhao, D. Past and future spatiotemporal changes in evapotranspiration and effective moisture on the Tibetan Plateau. *J. Geophys. Res. Atmos.* **2013**, *118*. [CrossRef]
78. Song, L.; Zhuang, Q.; Yin, Y.; Zhu, X.; Wu, S. Spatio-temporal dynamics of evapotranspiration on the Tibetan Plateau from 2000 to 2010. *Environ. Res. Lett.* **2017**, *12*, 014011. [CrossRef]
79. Jung, M.; Reichstein, M.; Ciais, P.; Seneviratne, S.I.; Sheffield, J.; Goulden, M.L.; Bonan, G.; Cescatti, A.; Chen, J.; Jeu, R.D. Recent decline in the global land evapotranspiration trend due to limited moisture supply. *Nature* **2010**, *467*, 951–954. [CrossRef]
80. Xiao, J. Satellite evidence for significant biophysical consequences of the "Grain for Green" Program on the Loess Plateau in China. *J. Geophys. Res. Biogeosci.* **2015**, *119*, 2261–2275. [CrossRef]
81. Yang, F.; Lu, H.; Yang, K.; He, J.; Wang, W.; Wright, J.S.; Li, C.; Han, M.; Li, Y. Evaluation of multiple forcing data sets for precipitation and shortwave radiation over major land areas of China. *Hydrol. Earth Syst. Sci.* **2017**, *21*, 5805–5821. [CrossRef]
82. Wang, Y.; Nan, Z.; Chen, H.; Wu, X. Correction of Daily Precipitation Data of ITPCAS Dataset over the Qinghai-Tibetan Plateau with KNN Model. In Proceedings of the 2016 IEEE International Geoscience and Remote Sensing Symposium (IGARSS), Beijing, China, 10–15 July 2016; pp. 593–596.
83. Kern, A.; Marjanović, H.; Barcza, Z. Evaluation of the Quality of NDVI3g Dataset against Collection 6 MODIS NDVI in Central Europe between 2000 and 2013. *Remote Sens.* **2016**, *8*, 955. [CrossRef]

84. Pinzon, J.; Tucker, C. A non-stationary 1981–2012 AVHRR NDVI3g time series. *Remote Sens.* **2014**, *6*, 6929–6960. [CrossRef]
85. Yao, Y.; Liang, S.; Li, X.; Zhang, Y.; Chen, J.; Jia, K.; Zhang, X.; Fisher, J.B.; Wang, X.; Zhang, L. Estimation of high-resolution terrestrial evapotranspiration from Landsat data using a simple Taylor skill fusion method. *J. Hydrol.* **2017**, *553*, 508–526. [CrossRef]
86. Yao, Y.; Liang, S.; Jian, Y.; Zhao, S.; Yi, L.; Jia, K.; Zhang, X.; Jie, C.; Xie, X.; Liang, S. Differences in estimating terrestrial water flux from three satellite-based Priestley-Taylor algorithms. *Int. J. Appl. Earth Obs. Geoinf.* **2017**, *56*, 1–12. [CrossRef]

© 2019 by the authors. Licensee MDPI, Basel, Switzerland. This article is an open access article distributed under the terms and conditions of the Creative Commons Attribution (CC BY) license (http://creativecommons.org/licenses/by/4.0/).

Article

Monitoring and Assessment of Wetland Loss and Fragmentation in the Cross-Boundary Protected Area: A Case Study of Wusuli River Basin

Chunyan Lu [1,2,3], Chunying Ren [2,*], Zongming Wang [2,4], Bai Zhang [2], Weidong Man [5], Hao Yu [2], Yibin Gao [1,3] and Mingyue Liu [5]

1. College of Computer and Information Sciences, Fujian Agriculture and Forestry University, Fuzhou 350002, China; luchunyan@fafu.edu.cn (C.L.); Gao_YB@fafu.edu.cn (Y.G.)
2. Key Laboratory of Wetland Ecology and Environment, Northeast Institute of Geography and Agroecology, Chinese Academy of Sciences, Changchun 130102, China; zongmingwang@iga.ac.cn (Z.W.); zhangbai@iga.ac.cn (B.Z.); yuhao@iga.ac.cn (H.Y.)
3. Research Centre of Resource and Environment Spatial Information Statistics of Fujian Province, Fujian Agriculture and Forestry University, Fuzhou 350002, China
4. National Earth System Science Data Center, Beijing 100101, China
5. College of Mining Engineering, North China University of Science and Technology, Tangshan 063210, China; manwd@ncst.edu.cn (W.M.); liumy917@ncst.edu.cn (M.L.)
* Correspondence: renchy@iga.ac.cn; Tel.: +86-431-8554-2297

Received: 1 October 2019; Accepted: 30 October 2019; Published: 3 November 2019

Abstract: Comparative evaluation of cross-boundary wetland protected areas is essential to underpin knowledge-based bilateral conservation policies and funding decisions by governments and managers. In this paper, wetland change monitoring for the Wusuli River Basin in the cross-boundary zone of China and Russia from 1990 to 2015 was quantitatively analyzed using Landsat images. The spatial-temporal distribution of wetlands was identified using a rule-based object-oriented classification method. Wetland dynamics were determined by combining annual land change area (ALCA), annual land change rate (ALCR), landscape metrics and spatial analysis in a geographic information system (GIS). A Mann–Kendall test was used to evaluate changing climate trends. Results showed that natural wetlands in the Wusuli River Basin have declined by 5625.76 km^2 in the past 25 years, especially swamp/marsh, which decreased by 26.88%. Specifically, natural wetlands declined by 49.93% in the Chinese section but increased with an ALCA of 16.62 km^2/y in the Russian section during 1990–2015. Agricultural encroachment was the most important reason for the loss and degradation of natural wetlands in the Wusuli River Basin, especially in China. Different population change trends and conservation policies in China and Russia affected natural wetland dynamics. The research offers an efficient and effective method to evaluate cross-boundary wetland change. This study provides important scientific information necessary for developing future ecological conservation and management of cross-boundary wetlands.

Keywords: cross-boundary protected area; rule-based object-oriented classification; wetland dynamics; Wusuli River Basin; rate of change

1. Introduction

Wetlands are among the most productive ecosystems on Earth. They provide a wide variety of ecological functions and values, ranging from flood control to groundwater aquifer recharge and discharge, carbon sequestration, and water quality improvement, and they harbor a large part of the Earth's biodiversity [1–3]. They also supply many services for humans, such as food, water, recreation and space for living. In many countries, the local economy depends on wetlands for fisheries, reed harvesting, grazing, and tourism development [4–7].

Human activities and global climate change, including construction of canals and dams, agricultural cultivation, residential and industrial development, as well as droughts [8–10], currently affect most wetland ecosystems with ever-increasing intensity and scope [11]. Considerable evidence has shown that wetlands have experienced alarming rates of loss and degradation, with their ecological functions and biodiversity declining at local, regional and global scales. In the Sanjiang Plain, Northeast China, the wetland area had decreased by 53% as a result of farmland reclamation, and their ecosystem service values noticeably declined from 1980 to 2000 [6]. Greece lost approximately 70% of its wetlands between 1920 and 1991 [12]. Over the last century, depending on the region, 31%–95% of wetlands have been destroyed or strongly modified along the west coast of the Pacific [13]. According to an OECD/IUCN (Organization for Economic Co-operation and Development/International Union for Conservation of Nature) report [14], the world may have lost 50% of its wetlands since 1900, and land conversion into agriculture was the principal cause. Unfortunately, owing to the lack of a detailed wetland inventory and inconsistent wetland definitions, the wetland extent has not been precisely defined in several major regions of the world, such as Russia, South America, and Africa [15,16]. Moreover, the analysis of the underlying factors of wetland loss and fragmentation, such as population pressure, political institutions, economic development, and ecological conservation measures, is lacking currently [17,18]. To prevent further wetland loss and degradation as well as to identify valuable wetland protected areas (WPAs), it is essential to inventory and monitor wetlands and their adjacent uplands to analyze change factors, collect baseline data and support decision making in terms of long-term strategies for wetland conservation [19,20].

By their nature, wetland areas are relatively inaccessible and it is difficult to conduct traditional field surveys. However, remote sensing techniques make it possible to observe inaccessible zones or remote targets repeatedly, and thus allow for more effective monitoring of wetland change and distribution [5,21]. A geographic information system (GIS) is a valuable tool for studying the nature of wetlands and assessing their dynamics at different spatial scales [22,23]. Compared with conventional methods, remote sensing and GIS are often preferred tools for monitoring or mapping wetlands because they are relatively fast, time-saving and cost-effective.

Establishing WPAs is considered to be one of the most effective strategies for conserving and managing wetland resources worldwide [24]. Research monitoring WPAs has focused on the situation within a single country [25,26]. There are few studies aimed at the WPAs between countries [27,28]. Generally, the boundary among countries is a political one, which is inconsistent with the ecology and environment borderline, while species distribution and ecological processes do not designate or discriminate explicitly due to the existence of the national boundary [29]. Moreover, different countries in the world implement political institutions and livelihood strategies. There are even, in some cases, various contradictories in terms of land use policies. In particular, some neighboring countries or areas have similar climate and geographical conditions, but their political and socio-economic regimes are often very different. Under this circumstance, if the ecology system of one side changed, that of the other side would be affected to some degree, which accordingly causes the bilateral ecology system to undergo a fragile development process [30]. So far, a unanimous awareness, which it is difficult to achieve conservation effectiveness of cross-border WPAs only by one single country enforcing efforts, has been recognized in the worldwide [31]. Thus, it is informative and significant to conduct cross-border studies on wetland monitoring and assessment because such investigations will help determine how wetland dynamics are driven by differing socio-economic and political conditions, develop knowledge-based wetland conservation and management strategies on behalf of neighboring countries, as well as provide references on future cross-boundary wetland studies [32,33].

The Wusuli River Basin is located on the border of China and Russia and, by virtue of the suitable terrain, climate and natural conditions, is one of the most important WPAs in the Eurasian continent. The main objectives of this study were to (1) conduct a wetland mapping and inventory in the Wusuli

River Basin; (2) characterize the dynamics of wetlands from 1990 to 2015, and conversions between wetlands and other land cover types; (3) analyze the possible influences of anthropogenic activities and climate change on the spatio-temporal wetland dynamics; and (4) propose more feasible conservation and management measures from the perspective of bilateral cooperation. To fulfill these objectives, remotely sensed data were used to map land cover using rule-based object-oriented classification and visual interpretation. GIS was used to analyze the wetland dynamics.

2. Materials and Methods

2.1. Study Area

The Wusuli River Basin is situated in the cross-boundary zone of Northeast China and the Far East region of Russia. Based on the Shuttle Radar Topography Mission (SRTM) 90 m Digital Elevation Model (DEM) (http://srtm.csi.cgiar.org/index.asp), we defined the entire boundary of the Wusuli River Basin with spatial analyst tools in ArcGIS 10 [34] (Figure 1).

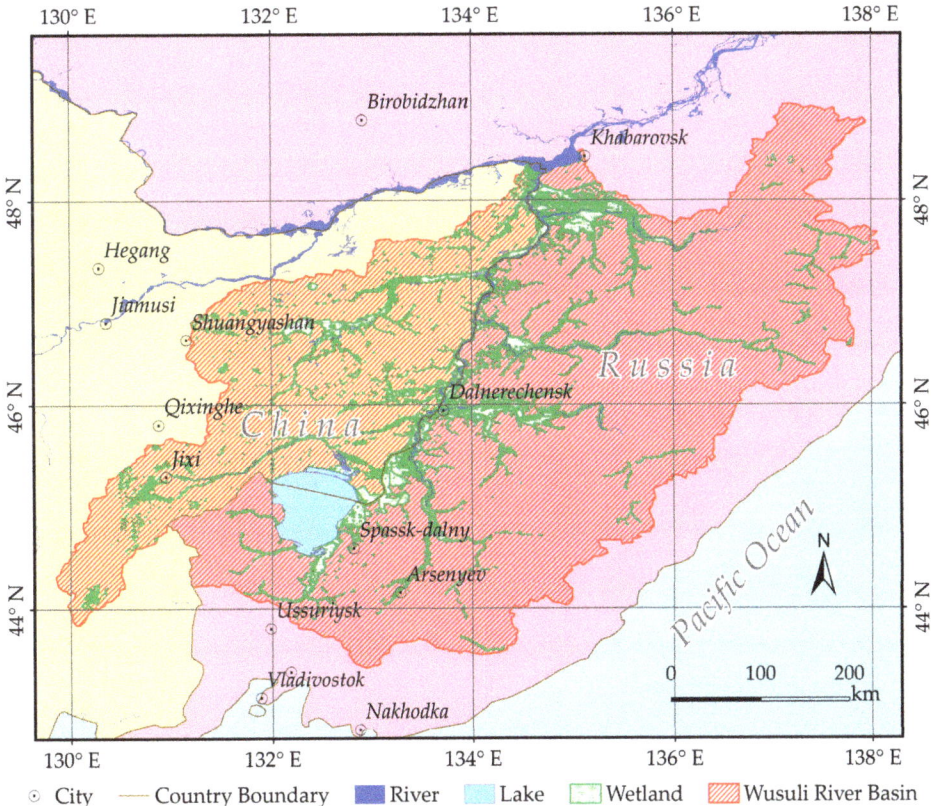

Figure 1. Location of the Wusuli River Basin.

The latitude of this area ranges from 43°25′N to 48°56′N and the longitude ranges from 129°50′E to 138°05′E. The watershed area is about 195.06 × 10³ km² in total, of which the Russian and the Chinese sections account for 69.06% and 30.94%, respectively. According to the administrative boundary, the basin runs from west to east across the Heilongjiang province of China and the Primorsky Krai province of Russia. In terms of the natural landscape, the Chinese section of the Wusuli River Basin is located on the East Sanjiang Plain, and the Russian section is part of the West Sikhot Mountains. The average elevation of this basin is 354 m, and the climate is characterized by a cold and dry winter and a warm and rainy summer.

The Wusuli River forms the border between China and Russia in the Wusuli River Basin, and the shared boundary stretches about 901.34 km. In 2015, the Chinese population in the basin was approximately 4,213,763 (http://data.stats.gov.cn/english/), and the Russian population was 878,007 (http://www.gks.ru/). Agriculture and coal mining are the main industries in the Chinese section, and more than 50% of the region is a plain (i.e., Sanjiang Plain), which is one of the most vital grain production bases in China. More than 70% of land is covered by forests in the Russian section, and timber and mining are the major industries—agricultural land covers less than 6% [35].

In the basin, wetlands serve as a stopover and nesting area for substantial migratory and waterfowl bird populations, such as *Grus japonensis*, *Ciconia ciconia*, *Larus ridibundus*, *Aix galericulata*, and *Tetrao tetrix* [36]. In addition, they play a vital role in stabilizing regional water supplies, ameliorating floods and drought and purifying polluted water. Furthermore, fish harvesting from wetlands is a significant economic resource for regional communities. Therefore, wetlands in the Wusuli River Basin are of great value for ecological balance, sustainable development and human well-being.

2.2. Data Preparation and Fieldwork

Cloud-free Landsat Thematic Mapper (TM) and Operational Land Imager (OLI) images (30 m spatial resolution) were chosen as the basic data sources to analyze the temporal and spatial wetland dynamics for the Wusuli River Basin in 1990, 2000 and 2015. These images were obtained in the growing season from May to September to minimize the effect of seasonal variations on the accuracy of land cover classification. Each image was acquired for the same month or the same vegetation growing period between 1990 and 2015 (Table 1). All images were geo-rectified with the registration error being less than half a pixel and atmospherically corrected using the Fast Line-of-sight Atmospheric Analysis of Spectral Hypercubes (FLAASH) [37].

Table 1. Specific description of each image used for land cover classification.

Path/Row	1990		2000		2015	
	Date	Sensor	Date	Sensor	Date	Sensor
111/27	17 Sept.	TM	12 Sept.	TM	17 Jul.	OLI
111/28	10 Aug.	TM	12 Sept.	TM	31 Aug.	OLI
111/29	16 Oct.	TM	12 Sept.	TM	15 Aug.	OLI
112/26	19 Jul.	TM	18 Aug.	TM	8 Jul.	OLI
112/27	19 Jul.	TM	3 Sept.	TM	8 Jul.	OLI
112/28	5 Sept.	TM	3 Sept.	TM	25 Aug.	OLI
112/29	1 Aug.	TM	13 Aug.	TM	25 Aug.	OLI
112/30	19 Jul.	TM	12 Jul.	TM	3 Jun.	OLI
113/26	9 Sept.	TM	21 Sept.	TM	17 Sept.	OLI
113/27	23 May	TM	21 Sept.	TM	9 May	OLI
113/28	9 Sept.	TM	21 Sept.	TM	14 Sept.	OLI
113/29	7 Jul.	TM	16 Jul.	TM	28 Jul.	OLI
113/30	12 Sept.	TM	7 Sept.	TM	28 Jul.	OLI
114/27	12 Jun.	TM	14 Sept.	TM	4 Jun.	OLI
114/28	19 Sept.	TM	15 Sept.	TM	8 Sept.	OLI
114/29	19 Sept.	TM	12 Sept.	TM	8 Sept.	OLI
114/30	19 Sept.	TM	29 Jun.	TM	4 Jun.	OLI
114/31	20 Sept.	TM	29 Jun.	TM	4 Jun.	OLI
115/27	25 Jun.	TM	17 Jun.	TM	15 Sept.	OLI
115/28	26 Jun.	TM	17 Jun.	TM	16 Sept.	OLI
115/29	26 Sept.	TM	17 Jun.	TM	15 Sept.	OLI

Data for the annual average temperature and annual precipitation from 1990 to 2015 were collected at 92 meteorological stations in and around the study area (Figure 2). This allowed us to analyze climate change factors driving wetland change. These meteorological data were interpolated to obtain a spatially continuous surface. The choice of spatial interpolation methods is referenced in Lu et al. [38].

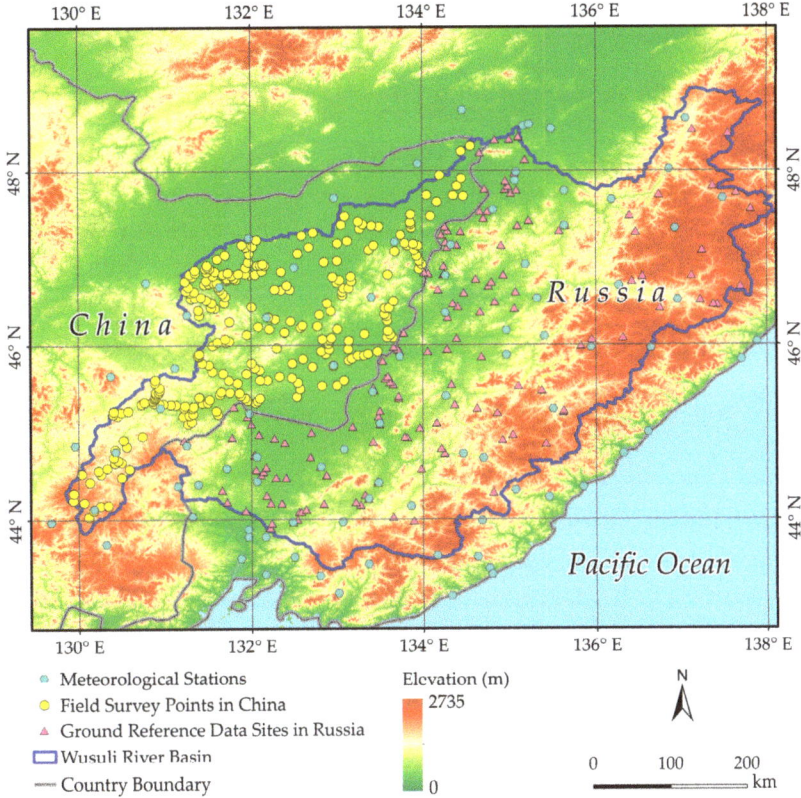

Figure 2. The location of meteorological stations, field survey points and ground reference data sites.

From 2012 to 2015, 315 ground survey points were collected in the Chinese section of the basin watershed. Owing to limited accessibility to the Russian section of the basin, visual inspection of high-resolution images from Google Earth, online photos and literature searches were carried out to collect land cover information during the period 2010–2015. This generated an additional 256 reference points. All field survey points and the reference data sites were used to evaluate the accuracy of the land cover classification results in 2015 (Figure 2). Owing to the lack of field survey data in 1990 and 2000, 600 independent points for each year were created by spatial analysis of create random points in ArcGIS 10 [34]. These random points were classified into different land cover types (described in Section 2.3) by consulting with experienced interpreters and experts, and were then used as validation points.

2.3. Land Cover Classification System

Considering systematically the wetland classification of Ramsar Wetland Convention, the purpose of our study, and the specific conditions of the land cover type in the study area, a landscape classification system was established for this study, including nine land cover types (i.e., swamp/marsh, natural open water, human-made wetland, woodland, grassland, paddy field, dry farmland, built-up land

and barren land). These were incorporated into seven categories (i.e., natural wetland, human-made wetland, woodland, grassland, cropland, built-up land and barren land). The detailed description of the classification system is given in Table 2.

Table 2. Description of the landscape classification system used by this study.

Landscape Category	Land Cover Type	Description
Natural wetland (NAW)	Swamp/Marsh (SAM)	Lands with a permanent mixture of water and herbaceous or woody vegetation that cover extensive areas
	Natural open water (NOW)	Rivers and lakes
Human-made wetland (HUW)	Human-made wetland (HUW)	Manufactured facilities for water, reservoirs, channels, ponds, and ditches, etc.
Woodland (WOL)	Woodland (WOL)	Broadleaved forest, mixed broadleaf-conifer forest, needle-leaved forest and shrubs
Grassland (GRL)	Grassland (GRL)	Natural areas with herbaceous vegetation
Cropland (CRL)	Paddy field (PAF)	Cropland that has enough water supply and irrigating facilities for planting paddy rice
	Dry farmland (DRF)	Cropland for planting dry farming crops without water supply and irrigation facilities
Built-up land (BUL)	Built-up land (BUL)	Lands used for urban and rural settlements, factories or transportation facilities
Barren land (BAL)	Barren land (BAL)	Sandy land and areas with less than 5% vegetation cover

2.4. Rule-Based Object-Oriented Classification Method

Compared with the frequent generation of 'salt-and-pepper' effects based on pixel-based classification methods [39], the object-oriented classification method can not only effectively avoid the 'salt-and-pepper' effects, but also reduce the 'within-class' spectral variation through segmenting an image into groups of contiguous and homogeneous pixels (image objects) as the mapping unit [40]. Moreover, besides of the spectral properties of the objects, their shape, texture and geometric features are also taken into account in the classification process of the object-oriented classification [8,41]. As a result, more effective and accurate performances are obtained than with pixel-based approaches [42,43].

To develop land cover maps for the study area in 1990, 2000 and 2015, a rule-based object-oriented classification method was applied to perform image segmentation and classify image objects into specific land cover types in the study. The layers which were selected to segment are Band 1 (0.45–0.52 μm), 2 (0.52–0.60 μm), 3 (0.63–0.69 μm), 4 (0.76–0.90 μm), 5 (1.55–1.75 μm), and 7 (2.08–2.35 μm) for Landsat TM, as well as Band 2 (0.450–0.515 μm), Band 3 (0.525–0.600 μm), Band 4 (0.630–0.680 μm), Band 5 (0.845–0.885 μm), Band 6 (1.560–1.660 μm), and Band 7 (2.100–2.300 μm) for Landsat OLI. The eCognition Developer 8.7.1 [44] was used to classify the images.

First, an optimal segmentation scale model referenced by Lu et al. [45] was used, in which a selected image scene was processed and grouped into homogeneous pixels (image objects) with an optimal segmentation scale. Each object resulting from this segmentation had minimal spectral variability [40,46] and the boundaries of these objects approximately followed the outline of individual land cover types. After segmentation, the segmented objects were categorized using a set of classification rules.

Considering the importance of vegetation growth and water content in wetland classification [47], the normalized difference vegetation index (NDVI) [Equation (1)] and land surface water index (LSWI) [Equation (2)] were used as rule layers to characterize vegetation and background soil, respectively.

$$NDVI = \frac{\rho_{nir} - \rho_{red}}{\rho_{nir} + \rho_{red}} \quad (1)$$

$$LSWI = \frac{\rho_{nir} - \rho_{swir}}{\rho_{nir} + \rho_{swir}} \quad (2)$$

where ρ_{red}, ρ_{nir} and ρ_{swir} are the reflectance values of Landsat TM Bands 3, 4 and 5 and Landsat OLI Bands 4, 5 and 6, respectively.

Several previous studies have reported that the hue of different band combinations can be a crucial factor for identifying different land cover types [48]. In this study, the hue was derived from a combination of Landsat TM Bands 5, 4, and 3 or Landsat OLI Bands 6, 5, and 4. The value range of the hue is between 0 and 1.

Different land cover types present distinct image textures, which is another variable necessary for land cover classification [49]. Based on the Haralick algorithm and gray level co-occurrence matrix (GLCM), the texture homogeneity ranging from 0 to 1 of each object was calculated [44]. The higher the value is, the higher the homogeneity.

The shape of an image object is also important to detect different land cover types. The shape index (SI) [Equation (3)] of an image object describes the smoothness of an image object border. The smoother the border of an image object is, the lower its shape index.

$$SI = \frac{b_v}{\sqrt[4]{P_v}} \tag{3}$$

where b_v is the border length of each image object, and P_v is the area of each image object.

After a series of pre-experiments, a classification rule set was developed (Figure 3). When the execution of classification rules was completed, the results were visually examined and modified for better precision. The overall accuracy, user accuracy and producer accuracy were used to assess the accuracy of the classification results.

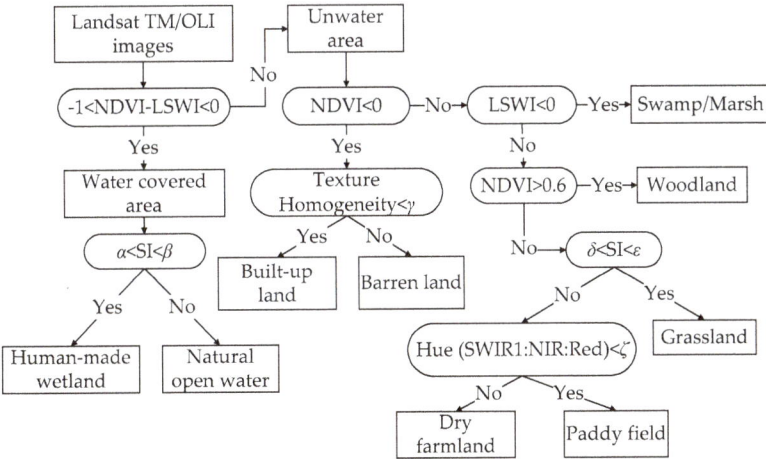

Figure 3. Rules for the land cove classification of the Wusuli River Basin (α, β, γ, δ, ε, and ζ, represent the selected classification parameters and each of them could vary for different images (Landsat TM/OLI: Landsat Thematic Mapper/Operational Land Imager; NDVI: Normalized difference vegetation index; LSWI: Land surface water index; SI: Shape index; SWIR1: Short wave infrared band1, corresponding to Landsat TM Band 5 and Landsat OLI Band 6, respectively; NIR: Near infrared band, corresponding to Landsat TM Band 4 and Landsat OLI Band 5, respectively; Red: Red band, corresponding to Landsat TM Band 3 and Landsat OLI Band 4, respectively; Unwater area: Land areas covered without water, including built-up area, barren land, woodland, grassland, swamp/marsh, dry farmland and paddy field).

2.5. Analysis of Land Cover Change

Two indices, annual land change area (ALCA) and annual land change rate (ALCR), were calculated to assess the dynamic degree of land cover types objectively. These are defined as follows:

$$ALCA = (U_b - U_a) \times \frac{1}{T} \qquad (4)$$

$$ALCR = \frac{U_b - U_a}{U_a} \times \frac{1}{T} \times 100\% \qquad (5)$$

where U_a and U_b represent the area of each land cover type at the beginning and the end of the study period, respectively, and T is the number of years. In the study, the time interval was divided into two stages: 1990–2000 and 2000–2015.

To analyze the spatial change characteristics of natural wetlands (i.e., swamp/marsh and natural open water) more explicitly, intersect overlay analysis in ArcGIS 10 [34] was used to create a conversion matrix between natural wetlands and other land cover types for the time periods 1990–2000 and 2000–2015. In addition, a Sankey diagram [50] was used to illustrate the conversion of all land cover types, as this can help visualize the temporal dynamics of all land cover types.

2.6. Calculation of Landscape Metrics

Landscape metrics can reflect the characteristics of changing landscape patterns, and allowed us to assess quantitatively the landscape change process. In the study, five landscape metrics were used to assess the change pattern of the natural wetland landscape, including the number of patches (NP), mean patch size (MPS), largest patch index (LPI), area-weighed mean shape index (AWMSI), and the interspersion and juxtaposition index (IJI).

NP is defined as the count of patches and is a simple measure of fragmentation of one landscape category. Although the NP of one landscape category may be important for ecological processes and landscape pattern, it cannot directly reflect information concerning the distribution, area and density of patches. MPS is defined as the average patch size, and LPI quantifies the percentage of the largest patch accounting for the total area of all patches belonging to a given landscape category. AWMSI is used to assess shape characteristics by calculating the sum of the area-weighted ratio between the perimeter and area of each patch. IJI represents interspersion and juxtaposition and can quantify the connectivity and distribution pattern between different patch types.

The detailed ecological significance and equations for selected landscape metrics are illustrated in Table 3. The calculation of landscape metrics was performed in Fragstats 4.2 [51].

2.7. Climate Change Analysis Based on Mann–Kendall Test

To measure the possible influence of climate change on the existence of wetlands, the changing climate trends were analyzed to determine whether climate change affected wetland dynamics. The statistical significance of the trends in annual average temperature and annual precipitation was measured using the Mann–Kendall test [52]. A trend is statistically significant if it is significant at the 5% level.

Table 3. Description of landscape metrics used by this study.

Landscape Metrics	Code	Calculation Equation *	Range and Units	Ecological Significance [53]
Number of patches	NP	n	≥ 1	NP can reflect the landscape spatial pattern and be used to describe the landscape heterogeneity. Moreover, there is also a positive correlation between the number of patches and landscape fragmentation, the value of NP is higher, the fragmentation more serious.
Mean patch size	MPS	$A/10000n$	>0 ha	MPS represents an average condition of landscape, which has two meanings in the analysis of landscape structure. On the one hand, MPS can describe the landscape scale, which indicates ecological characteristics to a certain extent. On the other hand, it can reflect the landscape fragmentation degree. It is demonstrated that one landscape with low MPS value has undergone more serious fragmentation than one landscape with high MPS value. Moreover, MPS is a kind of key index to quantify landscape heterogeneity.
Largest patch index	LPI	$\max_{j=1}^{n}(a_{ij})A \times 100\%$	From 0 to 100%	Largest patch area determines the ecological characteristics of dominant species, and influences the abundance, quantity and food chain of secondary species in the ecosystem, as well as the breeding of secondary species, etc. The value change of it can reflect the direction and intensity of human activities.
Area-weighed mean shape index	AWMSI	$\sum_{i=1}^{m}\sum_{j=1}^{n}\left[\left(\frac{0.25 P_{ij}}{\sqrt{a_{ij}}}\right)\left(\frac{a_{ij}}{A}\right)\right]$	>0	AWMSI is one of the most important indicators for measuring the complexity of landscape spatial pattern, and has impact on many ecological processes, such as the migration and foraging activities of animal species, cultivation and production efficiency of plant species, etc. Moreover, for the shape analysis of natural block or landscape, the edge effect which is attributed to the shape factor is a significant performance of ecological meaning.
Interspersion and juxtaposition index	IJI	$\dfrac{-\sum_{i=1}^{m}\sum_{k=i+1}^{m}\left[\left(\dfrac{e_{ik}}{\sum_{k=1}^{m} e_{ik}}\ln\left(\dfrac{e_{ik}}{\sum_{k=1}^{m} e_{ik}}\right)\right)\right]}{\ln(0.5|m(m-1)|)} \times 100$	From 0 to 100%	IJI is one of the most important indices for describing the landscape spatial pattern, and can reflect the restriction degree of natural and anthropic factors impacting on ecosystem distribution characteristics. The patches have the most intense aggregation if the value of IJI equals to 0. On the contrary, the patches distribute homogeneously within the landscape when the value of it is 100%. The value of IJI is higher, the influence degree of extra factors stronger.

* n represents the number of patches for one given land cover type; m is the number of land cover types in the landscape; A is the total area of patches belonging to a given land cover type; a_{ij} is the area of the ijth patch; P_{ij} is the perimeter of the ijth patch; e_{ij} is the total length of edges between the ith and jth land cover type.

3. Results

3.1. Accuracy Assessment of Land Cover Maps

Table 4 presents the accuracy assessment for the land cover types in each study year. The overall accuracies of all the classification results were more than 0.93, which means that our classification results were consistent with those obtained from the validation points.

Table 4. Summary of land cover classification accuracies in 1990, 2000 and 2015.

Land Cover Type	1990		2000		2015	
	Pro	Use	Pro	Use	Pro	Use
SAM	0.91 ± 0.01	0.95 ± 0.03	0.90 ± 0.01	0.96 ± 0.03	0.88 ± 0.01	0.94 ± 0.04
NOW	0.93 ± 0.01	0.92 ± 0.08	0.92 ± 0.02	0.90 ± 0.08	0.98 ± 0.01	0.93 ± 0.08
HUW	0.88 ± 0.02	0.90 ± 0.11	0.89 ± 0.01	0.91 ± 0.10	0.93 ± 0.02	0.92 ± 0.10
WOL	0.97 ± 0.02	0.94 ± 0.05	0.94 ± 0.04	0.96 ± 0.04	0.97 ± 0.04	0.94 ± 0.05
GRL	0.89 ± 0.03	0.90 ± 0.08	0.93 ± 0.05	0.92 ± 0.07	0.90 ± 0.05	0.93 ± 0.07
PAF	0.90 ± 0.03	0.94 ± 0.06	0.90 ± 0.03	0.95 ± 0.05	0.91 ± 0.00	0.94 ± 0.06
DRF	0.89 ± 0.01	0.93 ± 0.06	0.94 ± 0.01	0.94 ± 0.05	0.89 ± 0.01	0.91 ± 0.07
BUL	0.95 ± 0.01	0.97 ± 0.04	0.93 ± 0.01	0.98 ± 0.03	0.94 ± 0.01	0.97 ± 0.04
BAL	0.93 ± 0.03	0.94 ± 0.07	0.95 ± 0.02	0.92 ± 0.06	0.94 ± 0.03	0.91 ± 0.14
Overall accuracy	0.93 ± 0.04		0.94 ± 0.02		0.94 ± 0.03	

Note: Pro denotes producer accuracy; Use denotes user accuracy; the value after the symbol "±" represents the margin of error at confidence level 95%.

3.2. Temporal and Spatial Changes of Land Cover Types

Figure 4 illustrates the spatio-temporal distribution of each land cover type in the study area from 1990 to 2015. The comparisons of the percentage of each land cover type are depicted in Figure 5. Table 5 shows the ALCA and ALCR of each land cover type. The results indicate that the natural wetlands (i.e., swamp/marsh and natural open water) of the Wusuli River Basin experienced a gradual decrease from 13.79% of the total area (26,892.99 km^2) in 1990 to 10.91% of the total area (21,267.23 km^2) in 2015, with an ALCA of −225.03 km^2/y and an ALCR of −0.96%/y. Swamp/marsh decreased by 2.96% during the period 1990–2015, while the area of natural open water increased by 0.08%. From 1990 to 2000, swamp/marsh decreased dramatically, with a change rate of −42.81 km^2/y. During 2000–2015, the reduction rate of the swamp/marsh area slowed over time, with an average rate of loss of 90.45 km^2/y. Despite an annual reduction area of 85.87 km^2, woodland remained the dominant landscape type during the period 1990–2015, with the vast majority distributed in the Russian part of the catchment. Cropland expanded markedly with an average rate of gain of 236.01 km^2/y, especially in 1990–2000 with an ALCR of 417.91 km^2/y. Specifically, the distribution range of paddy field expanded from sporadic patches in 1990 to large-scale continuous areas in the Chinese section of the basin in 2015, and the area quadrupled. In contrast, dry farmland decreased overall from 15.33% of the total area in 1990 to 12.06% of the total area in 2015, mainly because of a rapid rate of decrease from 2000 to 2015 (ALCR was −545.27 km^2/y). Human-made wetland, grassland, built-up land, and barren land had a slight areal increase during the period 1990–2015 with an ALCA of 4.58 km^2/y, 18.61 km^2/y, 13.72 km^2/y, and 38.03 km^2/y, respectively.

In terms of the different countries, in the Russian section of the Wusuli River Basin, woodland still accounted for more than three-fourths of the total landscape area from 1990 to 2015, despite an annual decline of 55.09 km^2. For croplands, both dry farmland and paddy field decreased by a small proportion. However, swamp/marsh and natural open water increased with an ALCA of 11.68 km^2/y and 4.91 km^2/y, respectively.

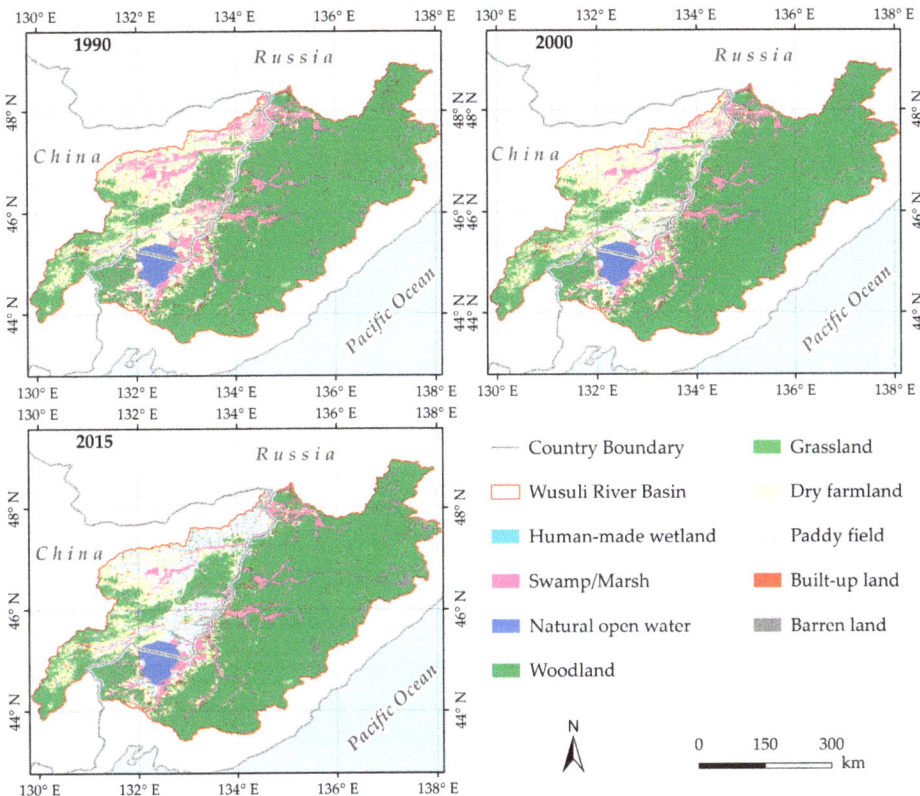

Figure 4. Land cover maps of the Wusuli River Basin in 1990, 2000 and 2015.

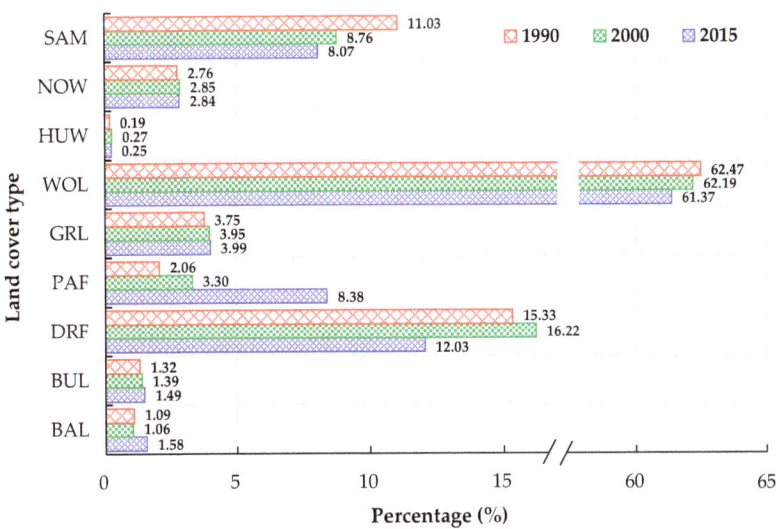

Figure 5. Area percentage of each land cover type for the Wusuli River Basin in 1990, 2000 and 2015.

Table 5. Annual land change area (ALCA) and annual land change rate (ALCR) of land cover types in the study area.

Landscape Category	Land Cover Type	China Portion						Russia Portion						Wusuli River Basin					
		ALCR (%/y)			ALCA (km²/y)			ALCR (%/y)			ALCA (km²/y)			ALCR (%/y)			ALCA (km²/y)		
		a	b	c	a	b	c	a	b	c	a	b	c	a	b	c	a	b	c
NAW	SAM	-4.49	-1.54	-2.30	-473.52	-89.45	-243.08	0.28	-0.01	0.11	30.71	-1.00	11.68	-2.06	-0.53	-1.08	-442.81	-90.45	-231.40
	NOW	0.54	-0.19	0.09	8.34	-3.13	1.46	0.26	0.04	0.13	9.98	1.52	4.91	0.34	-0.03	0.12	18.32	-1.61	6.36
HUW	HUW	5.41	-0.57	1.63	14.89	-2.43	4.49	0.98	-0.45	0.10	0.91	-0.46	0.09	4.30	-0.55	1.25	15.79	-2.89	4.58
WOL	WOL	-0.24	-0.11	-0.16	-46.37	-20.39	-30.78	-0.01	-0.08	-0.05	-9.12	-85.73	-55.09	-0.05	-0.09	-0.07	-55.48	-106.12	-85.87
GRL	GRL	-3.32	0.76	-1.02	-13.17	2.02	-4.06	0.74	0.05	0.33	51.58	3.38	22.66	0.52	0.07	0.25	38.41	5.40	18.61
CRL	PAF	8.62	12.22	17.10	249.98	660.14	496.08	-0.65	-0.01	-0.27	-7.23	-0.12	-2.97	6.05	10.24	12.28	242.75	660.02	493.11
	DRF	1.02	-2.09	-0.97	244.86	-551.57	-232.99	-1.17	0.12	-0.40	-69.71	6.30	-24.10	0.59	-1.72	-0.86	175.16	-545.27	-257.10
BUL	BUL	1.41	0.39	0.83	15.09	4.81	8.92	-0.03	0.56	0.32	-0.51	8.34	4.80	0.57	0.48	0.53	14.57	13.15	13.72
BAL	BAL	-2.70	-0.14	-1.14	-0.09	0.00	-0.04	-0.31	3.29	1.79	-6.49	67.77	38.06	-0.31	3.29	1.79	-6.58	67.77	38.03

Note: a, b, and c denote the stage of 1990–2000, 2000–2015 and 1990–2015, respectively.

In the Chinese section, croplands were the largest land cover type from 1990 to 2015. Paddy field and dry farmland showed the opposite areal change trends. During 1990–2015, paddy fields increased more than five-fold in area, while dry farmlands decreased by 9.65%. The area of natural wetland almost halved, and their areal proportion reduced from 20.05% of the total area of the Chinese section to 10.04%, with an ALCA of 241.62 km^2/y and an ALCR of −2.21%/y. The areal reduction of swamp/marsh accounted for the overwhelming majority of change in natural wetland with a loss rate of 243.08 km^2/y, whereas no significant change occurred in the area of natural open water.

3.3. Conversion between Natural Wetland and Other Land Cover Types

Figure 6 and Table 6 illustrate the conversions between natural wetland and other land cover types in terms of spatial distribution and area. Most of the natural wetland conversion occurred in the Chinese section of the Basin, while only a small proportion took place in the Russian section. Across the entire Wusuli River Basin, most of the natural wetland recession occurred from 1990 to 2000, attributed to the conversion of a large area to dry farmland and paddy field, which accounted for 78.51% and 15.16% of the natural wetland reduction in this stage, respectively. During the period 2000–2015, the percentage of conversion to dry farmland and paddy field was 43.78% and 50.87%, respectively. Between 1990 and 2000, 125.88 km^2 of natural wetlands were converted into human-made wetlands which reduced the area of natural wetlands. In terms of the transformation of other land cover types into natural wetlands, in both stages (1990–2000 and 2000–2015), the proportion of dry farmland converted into natural wetlands was the highest, accounting for 78.76% and 54.94%, respectively. Paddy field also contributed to the increase of natural wetlands, accounting for 14.22% and 16.36% of natural wetland area recovery in the stages 1990–2000 and 2000–2015, respectively. There was little reciprocal conversion among natural wetlands and woodland, grassland, built-up land or barren land during the period 1990–2015. These results suggest that the change in natural wetlands area can be attributed mainly to cropland reclamation and natural restoration from cropland.

For the Russian portion of the basin, the reclamation of natural wetlands covered a smaller area than their expansion during the two periods. Especially in the stage of 1990–2000, a total of 457.20 km^2 of natural wetlands were restored from cropland. Nevertheless, the reduced area of natural wetlands in the Chinese part of the basin was much larger than that of natural wetland restoration, suggesting a serious areal loss process.

Table 6. Conversion comparison between natural wetlands and other land cover types in the study area.

Change Type	Conversion Area/km^2					
	China Part		Russia Part		Wusuli River Basin	
	1990–2000	2000–2015	1990–2000	2000–2015	1990–2000	2000–2015
NAW→HUW	123.77	41.20	2.11	3.43	125.88	44.63
NAW→WOL	91.21	26.27	7.28	0.45	98.49	26.72
NAW→GRL	38.17	9.62	6.94	2.28	45.11	11.90
NAW→PAF	712.26	1043.13	11.97	0.00	724.23	1043.13
NAW→DRF	3720.86	897.74	29.89	0.12	3750.75	897.86
NAW→BUL	20.10	16.75	8.62	7.15	28.72	23.90
NAW→BAL	0.00	0.02	4.13	2.51	4.13	2.53
HUW→NAW	14.54	75.8	1.88	2.48	16.42	78.28
WOL→NAW	9.71	11.07	3.94	3.02	13.65	14.09
GRL→NAW	8.10	27.25	5.77	6.37	13.87	33.62
PAF→NAW	53.24	76.73	56.58	0.00	109.82	76.73
DRF→NAW	207.53	248.97	400.62	8.79	608.15	257.76
BUL→NAW	0.83	6.24	7.86	0.19	8.69	6.43
BAL→NAW	0.54	0.00	0.97	2.24	1.51	2.24

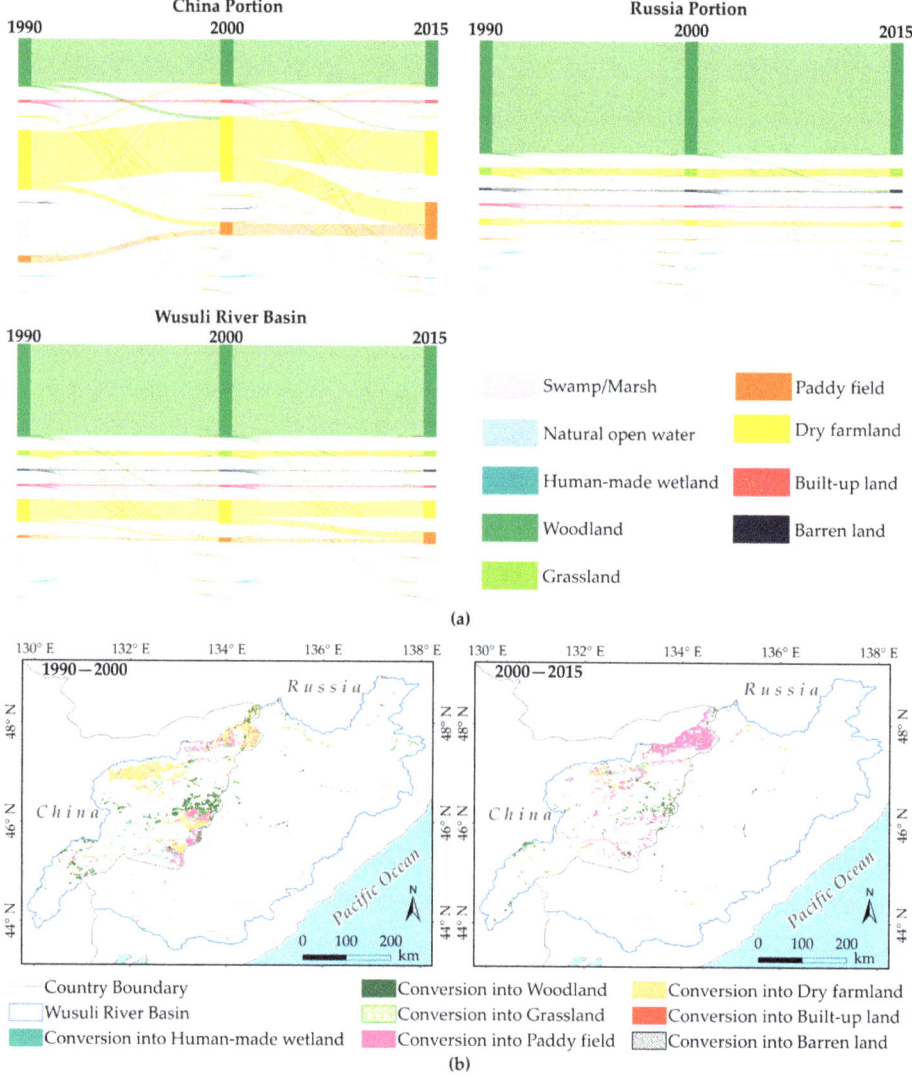

Figure 6. Dynamic conversion comparisons between natural wetlands and other land cover types. (**a**) Sankey diagram for comparison of total land cover dynamics from 1990 to 2015; (**b**) Spatial distribution of conversion between natural wetlands and other land cover types in two stages, 1990–2000 and 2000–2015.

3.4. Fragmentation and Improvement of Natural Wetland and Trend of Climate Change

Table 7 presents a comparison of the landscape metrics for natural wetlands in the Chinese and Russian sections of the Wusuli River Basin in 1990, 2000 and 2015. The landscape pattern of natural wetlands changed significantly in the Chinese part of the basin, while there was no significant change from 1990 to 2015 in the Russian part. During 1990–2015, despite a small increase in average patch area (MPS), the NP, LPI and AWMSI of natural wetlands decreased significantly in the Chinese portion, indicating that natural wetlands had undergone a loss and fragmentation process. In the Russian section, the NP, MPS, LPI and AWMSI of natural wetlands increased slightly, supporting the improvement which the natural wetlands had experienced in the Russian portion of the basin. There were more obvious changes in the IJI of natural wetlands in the Chinese part than in the Russian part of the basin, suggesting that the existence of natural wetland was subjected to more outside interference in the Chinese portion of the Wusuli River Basin.

Table 7. Landscape metrics comparison of the natural wetlands for the China portion and Russia portion of the Wusuli River Basin in1990, 2000 and 2015.

	Year	NP	MPS	LPI	AWMSI	IJI
China part	1990	3387	357.18	16.81	57.07	54.58
	2000	3647	204.16	5.92	32.13	59.78
	2015	1384	437.66	4.67	27.05	68.58
Russia part	1990	2158	685.60	5.00	24.34	73.75
	2000	2236	679.88	5.24	24.55	73.38
	2015	2217	686.06	5.24	24.39	74.02

Based on the Mann–Kendall test, there were no significant changes, at the 5% significant level, in the trend for the annual average temperature and annual precipitation in the Wusuli River Basin during the study period.

4. Discussion

4.1. Conservation and Threats of Anthropogenic Activities on Natural Wetland

Agriculture, considered as the primary foundation of a country's development, is important for ensuring national security and people's livelihoods [54]. In the past, reclaiming natural wetlands for croplands was seen as the best way to increase the cultivated land area to meet the need for grain production. Our results suggest that conversion into croplands was the primary contributor to natural wetland losses, especially in the Chinese section of the Wusuli River Basin (Table 5, Figure 6, and Table 6). Previous studies concerning the dynamics of natural wetlands in the Sanjiang Plain of China, have also shown that the loss and shrinkage of natural wetlands were generally caused by agricultural encroachment [55,56]. Indeed, there are a series of national and regional policies designed to stimulate natural wetland conversion into croplands in the Chinese part of the Wusuli River Basin [57]. In the 1990s, grain trade and crop cultivation were promoted by the establishment of a market-based economic system and comprehensive enforcement of a household responsibility system. The introduction of modern agricultural machinery also made agricultural encroachment more feasible [58,59]. To enhance grain security, the Heilongjiang province government has executed the project 'Land Regulation and Reclamation' since 2001. Owing to suitable geographical conditions, natural wetlands were seen as the most desirable land cover type for crop cultivation, especially in the Sanjiang Plain [60]. In 2004, the 'Reform of Rural Taxes and Administrative Charges' policy was first carried out in Heilongjiang Province [61], by which the agricultural tax was rescinded and subsides were granted to farmers according to their cultivated area. Because of the increase in farming profit in the context of the policy, significant areas of illegal cropland were developed in the Sanjiang Plain. During the past decade, a food security plan was launched in China aiming to increase an additional 50 million tons of grain production, which gave rise to a further wave of encroachment of croplands into natural wetlands [62].

As referenced by Mao et al. [54] and Lu et al. [63], agricultural plantation structures and hydraulic engineering construction are directly or indirectly responsible for the loss and degradation of natural wetland. From 1990 to 2015, the ratio of paddy fields to dry farmlands changed from 1:7.44 in 1990 to 1:1.44 in 2015, with a rapid expansion of paddy field (Figures 4 and 5). On one hand, large areas of new expanded paddy field were converted from natural wetlands. Our findings showed that the area of natural wetlands converted into paddy field was 724.23 km^2 and 1043.13 km^2 in the stages 1990–2000 and 2000–2015, respectively (Figure 6 and Table 6). On the other hand, irrigation in paddy fields consumed a vast amount of groundwater and surface water, which affected hydrological processes and threatened the water replenishment source for natural wetland [64]. Furthermore, with the increase in human-made wetland, substantial water sources in natural wetland are extracted for irrigation to meet the water demand for cultivated land, which undoubtedly aggravates threats to the existence of natural wetland. A previous study stated that the implementation of the 'Two Rivers and One Lake' project in Heilongjiang province, which was aimed at redirecting surface water to complement farmland irrigation, resulted in insufficient water resources for natural wetlands [65]. This is consistent with our research.

Population change is another common underlying force in natural wetland dynamics. Over the period 1990–2015, the population of Heilongjiang province in China increased, with rapid growth in the stage 1990–2000 (Figure 7a). Greater demand for grain was triggered by increases in the population, which promoted crop cultivation and stimulated the alteration of natural wetlands [66]. In contrast, there was a declining population trend in the Primorsky Krai province of Russia (Figure 7b). Due to this depopulation trend, reclaimed lands were abandoned and gradually turned into natural wetlands. Nearly 466 km^2 of croplands reverted to natural wetlands in the Russian section of the Wusuli River Bain from 1990 to 2015 (Table 6).

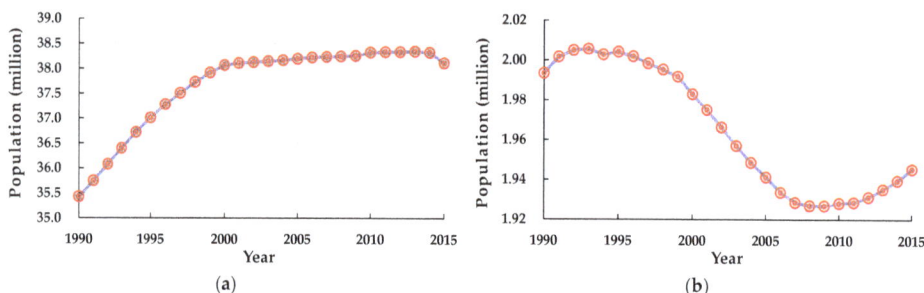

Figure 7. Population change for the Heilongjiang province of China (**a**) and the Primorsky Krai province of Russia (**b**) from 1990 to 2015.

As the eco-environmental values of natural wetlands have become more widely recognized, some ecological projects and conservation policies have been introduced to protect and restore them. Figure 8 illustrates pivotal ecological projects and conservation policies for natural wetland in China and Russia. It was found that the implementation of wetland protection measures in Russia occurred more than 60 years earlier than in China. Russia's accession to the Ramsar Wetland Convention was 17 years earlier than that of China. Compared with China, earlier environmental protection laws, which included the conservation and rational use of natural wetlands, were promulgated in Russia. All of these differences allow Russia has more prerequisites for wetland protection than China. This probably explains why the natural wetlands in the Russian region of the Wusuli River Basin have gone through a process of gradual restoration and improvement. In the Chinese part of the basin, although natural wetlands have experienced a loss and fragmentation process, the rate of area reduction has decreased over time (Table 5). Since the "Chinese wetland protection action plan" was initiated in 2000, many feasible and effective wetland protection and restoration measures have been implemented successively, which play a substantial role in natural wetland restoration and conservation [67]. Our results show that, on the one hand, the area of natural wetland reduced by cropland encroachment in 2000–2015 was only 25.56% of that in 1990–2000, on the other hand, the area of natural wetland restored from croplands in 2000–2015 was 1.25 times that in 1990–2000 in Chinese part of Wusuli River Basin (Table 5). Therefore, it can be inferred that, due to the implementation of conservation policies and measures, the destruction and disturbance caused by human activities to natural wetland has been mitigated to some degree in the Chinese region of the basin.

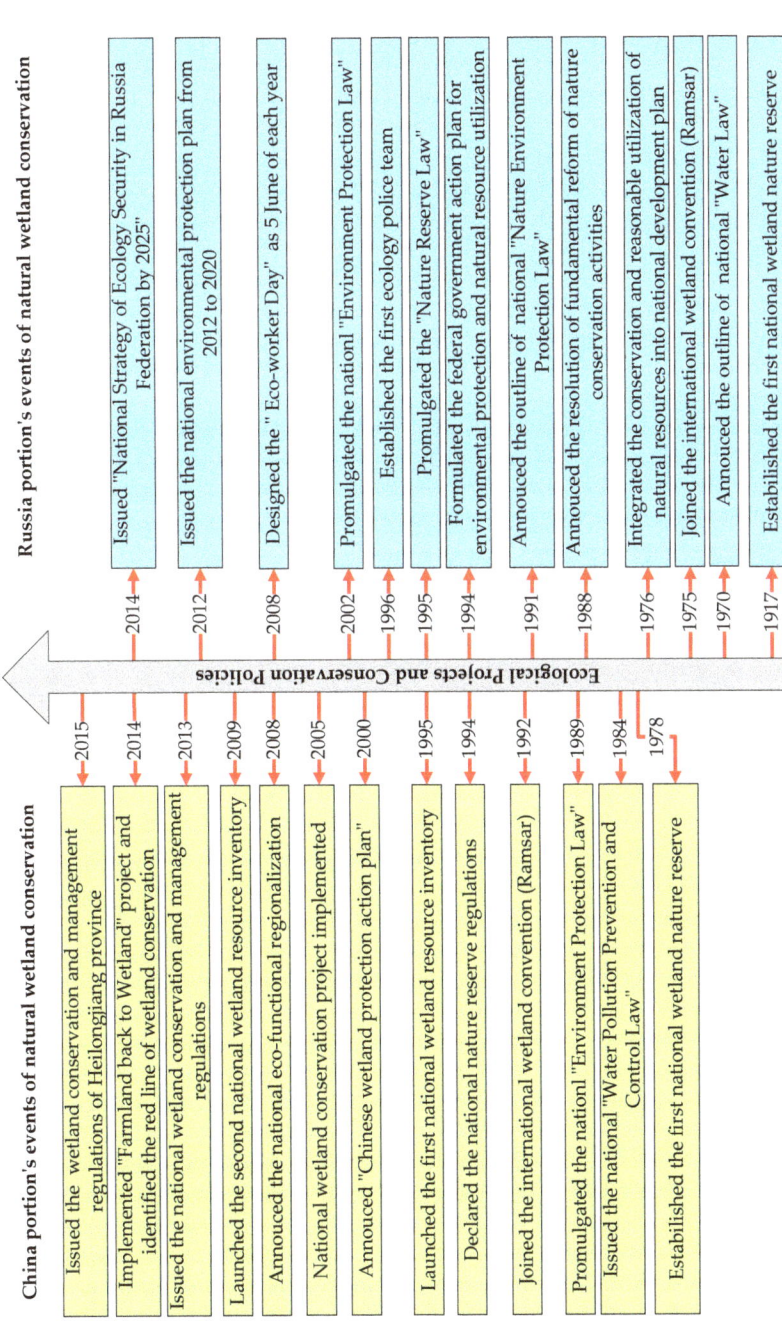

Figure 8. Comparison of ecological projects and conservation policies between China and Russia in Wusuli River Basin.

As mentioned above, agricultural activity has been the most important reason for the loss of natural wetlands in Wusuli River Basin, especially in the Chinese portion. Different demographic change trends and wetland protection levels in China and Russia region have also had opposing effects on wetland existence and restoration.

4.2. Further Studies Required on Remaining Natural Wetlands

The approach used in this paper provides a practical option for understanding the driving factors for cross-boundary areas. Combining remotely sensed data with spatial analysis is a tractable, effective and labor-saving method to determine wetland dynamics and their driving forces in neighboring countries. However, several further studies also should be carried out for effective conservation and management remaining natural wetlands.

On the one hand, more precise wetland monitoring data are needed. The resolution of Landsat TM/OLI images limits the smallest unit of wetland and land cover identifiable from the satellite images to 0.09 ha. Therefore, the existence and loss of wetlands smaller than 0.09 ha would not be captured in the study though small wetland patches are more likely to be influenced by human activities and climate change [36,45].

On the other hand, it should be noted that, although no significant climate change was observed during the study periods, the impacts of climate change on natural wetland dynamics should receive on-going attention in the context of global warming. At present, such assessment is in the qualitative stage. Therefore, more objective and quantitative approaches should be developed, especially for a long-term sequential research project [68].

4.3. Conservation Suggestions of Cross-Boundary WPA in Wusuli River Basin

Comparative studies across administrative borders or along transects are a promising alternative for understanding the driving forces associated different national development strategies and eco-environmental policies on wetland effects, which can help develop effective conservation measures at a regional or even a global scale. The results from this study on the spatial and temporal change characteristics, and landscape pattern comparison of natural wetlands for the Chinese and Russian sections of the Wusuli River Basin (Figures 5 and 6, and Table 7) can be taken as a guide for the formulation and implementation of conservation measure for wetlands in the Wusuli River Basin.

First, the Chinese and Russian governments should establish a bilateral cooperation mechanism to reinforce wetland ecosystem protection and maintain biodiversity. The managers, conservationists, and scientists of Russia and China should develop more feasible and effective plans to protect wetland ecosystems and to limit environmentally damaging human activities.

Second, regarding the Chinese government, more rigorous regulations and laws should be passed to prohibit people from converting natural wetlands into croplands [69,70]. For the areas in which natural wetlands have degraded, feasible wetland restoration projects should be implemented.

Third, establishing a wetland monitoring system is indispensable and allows for effective feedback on all aspects of wetlands. Moreover, adequate attention should be paid to the investigation and assessment of wetland biodiversity, which is related to the identification of key protected areas.

5. Conclusions

The monitoring and assessment for cross-boundary WPAs is essential to define the wetland dynamics as well as underpin knowledge-based conservation policies and funding decisions by bilateral government and managers. In the study, combining a rule-based object-oriented classification method, landscape metrics, spatial analysis and a Mann–Kendall test, we identified dynamic changes in natural wetlands and their influencing factors in the Wusuli River Basin from 1990 to 2015. Our results showed that the natural wetlands, as a whole, experienced a loss and fragmentation process, particularly in the Chinese section. Agricultural encroachment was the primary contributor to natural wetland degradation. In addition, differences in population trends and wetland conservation policies in the Chinese and Russian regions had differing effects on their natural wetland dynamics. The methods and results from this study can help our understanding of natural wetland changes and their driving forces in a cross-boundary study setting. These conclusions can be used as a guide for the bilateral government policies to effectively protect and manage natural wetlands.

Author Contributions: C.L. conceived and designed the research, process the Landsat image data, and wrote the manuscript draft. C.R., Z.W., and B.Z. helped to conceive the research and reviewed the manuscript. W.M. and H.Y. conducted the fieldwork and analyzed the land cover data. Y.G. analyzed the climate change trend. M.L. contributed materials.

Funding: This study was jointly funded by the National Key Research and Development Project (No. 2016YFA0602301), the Key Deployment Project of Chinese Academy of Sciences (No. KZZD-EW-08-02), the National Natural Science Foundation of China (No. 41730643), the funding from Youth Innovation Promotion Association of Chinese Academy of Sciences (2017277, 2012178), and the Fujian Natural Science Foundation General Program (2017J01457).

Acknowledgments: We appreciate the satellite and geospatial data provide by the Northeast Branch of National Earth System Science Data Center of China, as well as thank Leonie Seabrook, PhD, from Liwen Bianji, Edanz Group China (www.liwenbianji.cn/ac), for editing the English text of a draft of this manuscript.

Conflicts of Interest: The authors declare no conflict of interest.

References

1. Mitsch, W.J.; Gosselink, J.G. *Wetlands*, 2nd ed.; Van Nostrand Reinhold: New York, NY, USA, 1993; p. 722.
2. Kirwan, M.L.; Megonigal, J.P. Tidal wetland stability in the face of human impacts and sea-level rise. *Nature* **2013**, *504*, 53–60. [CrossRef]
3. Hu, S.J.; Niu, Z.G.; Chen, Y.F.; Li, L.F.; Zhang, H.Y. Global wetlands: Potential distribution, wetland loss, and status. *Sci. Total Environ.* **2017**, *586*, 319–327. [CrossRef]
4. Barbier, E.B.; Acreman, M.; Knowler, D. *Economic Valuation of Wetlands: A Guide for Policy Makers and Planners*, 1st ed.; Ramsar Convention Bureau: Gland, Switzerland, 1997; pp. 14–16.
5. Özesmi, S.L.; Bauer, M.E. Satellite remote sensing of wetlands. *Wetl. Ecol. Manag.* **2002**, *10*, 381–402. [CrossRef]
6. Wang, Z.M.; Zhang, B.; Zhang, S.Q.; Li, X.Y.; Liu, D.W.; Song, K.S.; Li, J.P.; Li, F.; Duan, H.T. Changes of land use and of ecosystem service values in Sanjiang Plain, Northeast China. *Environ. Monit. Assess.* **2006**, *112*, 69–91. [CrossRef] [PubMed]
7. Chaikumbung, M.; Doucouliagos, C.; Scarborough, H. The economic value of wetlands in developing countries: A meta-regression analysis. *Ecol. Econ.* **2016**, *124*, 164–174. [CrossRef]
8. Jia, M.M.; Wang, Z.M.; Zhang, Y.Z.; Mao, D.H.; Wang, C. Monitoring loss and recovery of mangrove forests during 42 years: The achievements of mangrove conservation in China. *Int. J. Appl. Earth Obs. Geoinf.* **2018**, *73*, 535–545. [CrossRef]
9. Mao, D.H.; He, X.Y.; Wang, Z.M.; Tian, Y.L.; Xiang, H.X.; Yu, H.; Man, W.D.; Jia, M.M.; Ren, C.Y.; Zheng, H.F. Diverse policies leading to contrasting impacts on land cover and ecosystem services in Northeast China. *J. Clean. Prod.* **2019**, *240*, 117961. [CrossRef]
10. Reis, V.; Hermoso, V.; Hamilton, S.K.; Ward, D.; Fluet-Chouinard, E.; Lehner, B.; Linke, S. A global assessment of inland wetland conservation status. *Bioscience* **2017**, *67*, 523–533. [CrossRef]
11. He, J.; Moffette, F.; Fournier, R.; Revéret, J.P.; Théau, J.; Dupras, J.; Boyer, J.P.; Varin, M. Meta-analysis for the transfer of economic benefits of ecosystem services provided by wetlands within two watersheds in Quebec, Canada. *Wetl. Ecol. Manag.* **2015**, *23*, 707–725. [CrossRef]
12. Gerakis, P.A. Conservation and management of Greek wetlands. In Proceedings of the Greek Wetlands Workshop, Thessaloniki, Greece, 17–21 April 1989; IUCN: Gland, Switzerland, 1992; p. 493.
13. Taft, O.W.; Haig, S.M.; Kiilsgaard, C. Use of radar remote sensing (RADARSAT) to map winter wetland habitat for shorebirds in an agricultural landscape. *Environ. Manag.* **2004**, *33*, 750–763. [CrossRef]
14. OECD/IUCN. *Guidelines for Aid Agencies for Improved Conservation and Sustainable Use of Tropical and Sub-Tropical Wetlands*, 1st ed.; OECD: Paris, France, 1996; pp. 7–8.
15. Rebelo, L.M.; Finlayson, C.M.; Nagabhatla, N. Remote sensing and GIS for wetland inventory, mapping and change analysis. *J. Environ. Manag.* **2009**, *90*, 2144–2153. [CrossRef] [PubMed]

16. Amler, E.; Schmidt, M.; Menz, G. Definitions and mapping of east African wetlands: A review. *Remote Sens.* **2015**, *7*, 5256–5282. [CrossRef]
17. Van Asselen, S.; Verburg, P.H.; Vermaat, J.E.; Janse, J.H. Drivers of wetland conversion: A global meta-analysis. *PLoS ONE* **2013**, *8*, e81292. [CrossRef] [PubMed]
18. Davidson, N.C. How much wetland has the world lost? Long-term and recent trends in global wetland area. *Mar. Freshw. Res.* **2014**, *65*, 934–941. [CrossRef]
19. Baker, C.; Lawrence, R.; Montagne, C.; Patten, D. Mapping wetlands and riparian areas using Landsat ETM+ imagery and decision-tree-based models. *Wetlands* **2006**, *26*, 465–474. [CrossRef]
20. Sica, Y.V.; Quintana, R.D.; Radeloff, V.C.; Gavier-Pizarro, G.I. Wetland loss due to land use change in the Lower Paraná River Delta, Argentina. *Sci. Total Environ.* **2016**, *568*, 967–978. [CrossRef]
21. Chen, L.F.; Jin, Z.Y.; Michishita, R.; Cai, J.; Yue, T.X.; Chen, B.; Xu, B. Dynamic monitoring of wetland cover changes using time-series remote sensing imagery. *Ecol. Inf.* **2014**, *24*, 17–26. [CrossRef]
22. Gottgens, J.F.; Swartz, B.P.; Kroll, R.W.; Eboch, M. Long-term GIS-based records of habitat changes in a Lake Erie coastal marsh. *Wetl. Ecol. Manag.* **1998**, *6*, 5–17. [CrossRef]
23. Garg, J.K. Wetland assessment, monitoring and management in India using geospatial techniques. *J. Environ. Manag.* **2015**, *148*, 112–123. [CrossRef]
24. Bürgi, M.; Hersperger, A.M.; Schneeberger, N. Driving forces of landscape change-current and new directions. *Landsc. Ecol.* **2005**, *19*, 857–868. [CrossRef]
25. Tuvi, E.L.; Vellak, A.; Reier, Ü.; Szava-kovats, R.; Pärtel, M. Establishment of protected areas in different ecoregions, ecosystems, and diversity hotspots under successive political systems. *Biol. Conserv.* **2011**, *5*, 1726–1732. [CrossRef]
26. Caddell, R. Nature Conservation in Estonia: From Soviet Union to European Union. *J. Balt. Stud.* **2009**, *3*, 307–332. [CrossRef]
27. Szantoi, Z.; Brink, A.; Buchanan, G.; Bastin, L.; Lupi, A.; Simonetti, D.; Mayaux, P.; Peedell, S.; Davy, J. A simple remote sensing based information system for monitoring sites of conservation importance. *Remote Sens. Ecol. Conserv.* **2016**, *2*, 16–24. [CrossRef]
28. Wang, X.P.; Guo, K. Basic implication of transboundary reserve and park for peace and their application. *Guihaia* **2004**, *3*, 220–223.
29. Zbicz, D.C. The "nature" of transboundary cooperation. *Environ. Sci. Policy Sustain. Dev.* **1999**, *3*, 15–16. [CrossRef]
30. Shi, L.Y.; Li, D.; Chen, L.; Zhao, Y. Transboundary protected areas as a means to biodiversity conservation. *Acta Ecol. Sin.* **2012**, *21*, 6892–6900.
31. Dahl, T.E.; Watmough, M.D. Current approaches to wetland status and trends monitoring in prairie Canada and the continental United States of America. *Can. J. Remote Sens.* **2007**, *33*, S17–S27. [CrossRef]
32. Xu, W.H.; Pimm, S.L.; Du, A.; Su, Y.; Fan, X.Y.; An, L.; Liu, J.G.; Ouyang, Z.Y. Transforming protected area management in China. *Trends Ecol. Evol.* **2019**, *34*, 762–766. [CrossRef]
33. Fent, A.; Bardou, R.; Carney, J.; Cavanaugh, K. Transborder political ecology of mangroves in Senegal and The Gambia. *Glob. Environ. Chang.* **2019**, *54*, 214–226. [CrossRef]
34. ESRI Inc. *ArcGIS Desktop: Release 10*; Environmental Systems Research Institute: Redlands, CA, USA, 2011.
35. Liu, H.Y.; Lu, X.G.; Wang, C.K. Study on the sustainable development of wetland resources in the Ussuri/Wusuli River basin. *Chin. Geogr. Sci.* **2000**, *10*, 270–275. [CrossRef]
36. Zhang, J.Y.; Ma, K.M.; Fu, B.J. Wetland loss under the impact of agricultural development in the Sanjiang Plain, NE China. *Environ. Monit. Assess.* **2010**, *166*, 139–148. [CrossRef] [PubMed]
37. EXELIS Inc. *ENVI®5.1 Fast Line-of-Sight Atmospheric Analysis of Hypercubes (FLAASH)*; EXELIS Inc.: Boulder, CO, USA, 2013.
38. Lu, C.Y.; Gu, W.; Dai, A.H.; Wei, H.Y. Assessing habitat suitability based on geographic information system (GIS) and fuzzy: A case study of *Schisandra sphenanthera* Rehd. et Wils. in Qinling Mountains, China. *Ecol. Model.* **2012**, *242*, 105–115. [CrossRef]
39. Xia, Q.; Qin, C.Z.; Li, H.; Huang, C.; Su, F.Z. Mapping mangrove forests based on multi-tidal high-resolution satellite imagery. *Remote Sens.* **2018**, *10*, 1343. [CrossRef]
40. Baatz, M.; Schäpe, A. Multiresolution segmentation: An optimization approach for high quality multi-scale image segmentation. In Proceedings of the 12th Angewandte Geographische Informations Verarbeitung, Heidelberg, Germany, 12 December 2000.

41. Baatz, M.; Schäpe, A. Multiresolution segmentation: An optimization approach for high quality multi-scale imagesegmentation. *Angew. Geogr. Inf.* **2000**, *12*, 12–23.
42. Liu, M.Y.; Li, H.Y.; Li, L.; Man, W.D.; Jia, M.M.; Wang, Z.M.; Lu, C.Y. Monitoring the invasion of *Spartina alterniflora* using multi-source high-resolution imagery in the Zhangjiang Estuary, China. *Remote Sens.* **2017**, *9*, 539. [CrossRef]
43. Conchedda, G.; Durieux, L.; Mayaux, P. An object-based method for mapping and change analysis in mangrove ecosystems. *ISPS J. Photogramm.* **2008**, *63*, 578–589. [CrossRef]
44. Definiens Imaging. In *eCognition Developer Software: 8.7.1. Reference Book*; Trimble GmbH: Raunheim, Germany, 2012.
45. Lu, C.Y.; Liu, J.F.; Jia, M.M.; Liu, M.Y.; Man, W.D.; Fu, W.W.; Zhong, L.X.; Lin, X.Q.; Su, Y.; Gao, Y.B. Dynamic analysis of mangrove forests based on an optimal segmentation scale model and multi-seasonal images in Quanzhou Bay, China. *Remote Sens.* **2018**, *10*, 2020. [CrossRef]
46. Blaschke, T. Object based image analysis for remote sensing. *ISPRS J. Photogram.* **2010**, *65*, 2–16. [CrossRef]
47. Adam, E.; Mutanga, O.; Rugege, D. Multispectral and hyperspectral remote sensing for identification and mapping of wetland vegetation: A review. *Wetl. Ecol. Manag.* **2010**, *18*, 281–296. [CrossRef]
48. Zhou, X.R.; Liu, J.; Liu, S.G.; Cao, L.; Zhou, Q.M.; Huang, H.W. A GIHS-based spectral preservation fusion method for remote sensing images using edge restored spectral modulation. *ISPRS J. Photogram.* **2014**, *88*, 16–27. [CrossRef]
49. Duque, J.C.; Patino, J.E.; Ruiz, L.A.; Pardo-Pascual, J.E. Measuring intra-urban poverty using land cover and texture metrics derived from remote sensing data. *Landsc. Urban Plan.* **2015**, *135*, 11–21. [CrossRef]
50. Cuba, N. Research note: Sankey diagrams for visualizing land cover dynamics. *Landsc. Urban Plan.* **2015**, *139*, 163–167. [CrossRef]
51. Fragstats Help Version 4.2. Available online: http://www.umass.edu/landeco/research/fragstats/documents/fragstats.help.4.2.pdf (accessed on 21 April 2015).
52. Wang, H.J.; Chen, Y.N.; Chen, Z.S.; Li, W.H. Changes in annual and seasonal temperature extremes in the arid region of china, 1960–2010. *Nat. Hazards* **2013**, *65*, 1913–1930. [CrossRef]
53. McGarigal, K.; Ene, E. *FRAGSTATS v4.2: A Spatial Pattern Analysis Program for Categorical Maps*; University of Massachusetts: Amherst, MA, USA, 2012.
54. Mao, D.H.; Luo, L.; Wang, Z.M.; Wilson, M.C.; Zeng, Y.; Wu, B.F.; Wu, J.G. Conversions between natural wetlands and farmland in China: A multiscale geospatial analysis. *Sci. Total Environ.* **2018**, *634*, 550–560. [CrossRef] [PubMed]
55. Song, K.S.; Wang, Z.M.; Li, L.; Tedesco, L.; Li, F.; Jin, C.; Du, J. Wetlands shrinkage, fragmentation and their links to agriculture in the Muleng–Xingkai Plain, China. *J. Environ. Manag.* **2012**, *111*, 120–132. [CrossRef] [PubMed]
56. Wang, Z.M.; Song, K.S.; Ma, W.H.; Ren, C.Y.; Zhang, B.; Liu, D.W.; Chen, J.M.; Song, C.C. Loss and fragmentation of marshes in the Sanjiang Plain, Northeast China, 1954–2005. *Wetlands* **2011**, *31*, 945–954. [CrossRef]
57. Wang, Z.M.; Wu, J.G.; Madden, M.; Mao, D.H. China's wetlands: Conservation plans and policy impacts. *AMBIO* **2012**, *41*, 782–786. [CrossRef]
58. Wang, G.G.; Liu, Y.S.; Li, Y.R.; Chen, Y.F. Dynamic trends and driving forces of land use intensification of cultivated land in China. *J. Geogr. Sci.* **2015**, *25*, 45–57. [CrossRef]
59. Song, K.S.; Wang, Z.M.; Du, J.; Liu, L.; Zeng, L.H.; Ren, C.Y. Wetland degradation: Its driving forces and environmental impacts in the Sanjiang Plain, China. *Environ. Manag.* **2014**, *54*, 255–271. [CrossRef]
60. Wang, H.X.; Zhang, M.H.; Cai, Y. Problems, challenges, and strategic options of grain security in China. *Adv. Agron.* **2009**, *103*, 101–147.
61. Tao, R.; Qin, P. How has rural tax reform affected farmers and local governance in China? *China World Econ.* **2007**, *15*, 19–32. [CrossRef]
62. Man, W.D.; Yu, H.; Li, L.; Liu, M.Y.; Mao, D.H.; Ren, C.Y.; Wang, Z.M.; Jia, M.M.; Miao, Z.H.; Lu, C.Y.; et al. Spatial expansion and soil organic carbon storage changes of croplands in the Sanjiang plain, China. *Sustainability* **2017**, *9*, 563. [CrossRef]
63. Lu, C.Y.; Wang, Z.M.; Li, L.; Wu, P.Z.; Mao, D.H.; Jia, M.M.; Dong, Z.Y. Assessing the conservation effectiveness of wetland protected areas in Northeast China. *Wetl. Ecol. Manag.* **2016**, *24*, 381–398. [CrossRef]

64. Chen, H.J.; Wang, G.P.; Lu, X.G.; Jiang, M.; Mendelssohn, I.A. Balancing the needs of China's wetland conservation and rice production. *Environ. Sci. Technol.* **2015**, *49*, 6385–6393. [CrossRef] [PubMed]
65. Man, W.D.; Wang, Z.M.; Liu, M.Y.; Lu, C.Y.; Jia, M.M.; Mao, D.H.; Ren, C.Y. Spatio-temporal dynamics analysis of cropland in Northeast China during 1990–2013 based on remote sensing. *Trans. Chin. Soc. Agric. Eng.* **2016**, *32*, 1–10. (In Chinese)
66. Grumbine, R.E. Assessing environmental security in China. *Front. Ecol. Environ.* **2014**, *12*, 403–411. [CrossRef]
67. Mao, D.H.; Wang, Z.M.; Wu, J.G.; Wu, B.F.; Zeng, Y.; Song, K.S.; Yi, K.P.; Luo, L. China's wetlands loss to urban expansion. *Land Degrad. Dev.* **2018**, *29*, 2644–2657. [CrossRef]
68. Junk, W.J.; An, S.Q.; Finlayson, C.M.; Gopal, B.; Květ, J.; Mitchell, S.A.; Mitsch, W.J.; Robarts, R.D. Current state of knowledge regarding the world's wetlands and their future under global climate change: A synthesis. *Aquat. Sci.* **2013**, *75*, 151–167. [CrossRef]
69. Stem, C.; Margoluis, R.; Salafsky, N.; Brown, M. Monitoring and evaluation in conservation: A review of trends and approaches. *Conserv. Biol.* **2005**, *19*, 295–309. [CrossRef]
70. Knight, A.T.; Driver, A.; Cowling, R.M.; Maze, K.; Desmet, P.G.; Lombard, A.T.; Rouget, M.; Botha, M.A.; Boshoff, A.F.; Castley, G.J.; et al. Designing systematic conservation assessments that promote effective implementation: Best practice from South Africa. *Conserv. Biol.* **2006**, *20*, 739–750. [CrossRef]

© 2019 by the authors. Licensee MDPI, Basel, Switzerland. This article is an open access article distributed under the terms and conditions of the Creative Commons Attribution (CC BY) license (http://creativecommons.org/licenses/by/4.0/).

Article

A Holistic Analysis for Landslide Susceptibility Mapping Applying Geographic Object-Based Random Forest: A Comparison between Protected and Non-Protected Forests

Zeinab Shirvani

Institute for Cartography, Department of Geosciences, Technische Universität Dresden, 01069 Dresden, Germany; zeinab.shirvani@tu-dresden.de or zeinab.shirvani@gmail.com; Tel.: +49-351-463-33568

Received: 1 December 2019; Accepted: 27 January 2020; Published: 29 January 2020

Abstract: Despite recent progress in landslide susceptibility mapping, a holistic method is still needed to integrate and customize influential factors with the focus on forest regions. This study was accomplished to test the performance of geographic object-based random forest in modeling the susceptibility of protected and non-protected forests to landslides in northeast Iran. Moreover, it investigated the influential conditioning and triggering factors that control the susceptibility of these two forest areas to landslides. After surveying the landslide events, segment objects were generated from the Landsat 8 multispectral images and digital elevation model (DEM) data. The features of conditioning factors were derived from the DEM and available thematic layers. Natural triggering factors were derived from the historical events of rainfall, floods, and earthquake. The object-based image analysis was used for deriving anthropogenic-induced forest loss and fragmentation. The layers of logging and mining were obtained from available historical data. Landslide samples were extracted from field observations, satellite images, and available database. A single database was generated including all conditioning and triggering object features, and landslide samples for modeling the susceptibility of two forest areas to landslides using the random forest algorithm. The optimal performance of random forest was obtained after building 500 trees with the area under the receiver operating characteristics (AUROC) values of 86.3 and 81.8% for the protected and non-protected forests, respectively. The top influential factors were the topographic and hydrologic features for mapping landslide susceptibility in the protected forest. However, the scores were loaded evenly among the topographic, hydrologic, natural, and anthropogenic triggers in the non-protected forest. The topographic features obtained about 60% of the importance values with the domination of the topographic ruggedness index and slope in the protected forest. Although the importance of topographic features was reduced to 36% in the non-protected forest, anthropogenic and natural triggering factors remarkably gained 33.4% of the importance values in this area. This study confirms that some anthropogenic activities such as forest fragmentation and logging significantly intensified the susceptibility of the non-protected forest to landslides.

Keywords: object-based; random forest; landslide susceptibility; conditioning factors; natural triggering factors; anthropogenic triggering factors; protected forest; non-protected forest

1. Introduction

Despite the importance of physical-resisting forces of forests to the propensity for landslide occurrence, human and non-human variables can accelerate the spatial probability of landslide occurrence through slope stability in a given area [1]. However, the holistic understanding of the importance of conditioning and triggering factors that control the susceptibility of forest areas to landslides has not been appropriately customized yet. Anthropogenic triggering factors may reduce the resisting forces of forests to landslides by deforestation [2–10], logging [4,11–16], and mining [17], or may increase the susceptibility of forest areas to landslides by the fragmentation induced by infrastructure development such as road-network expansion [4,5,16,18–23] with the consequences of mass movements and slope failures. Likewise, natural triggering factors such as earthquake [24–27], rainfall [28–32], and flooding [33,34] may increase the propensity for occurring landslides by reducing the resisting forces in forest areas. Meanwhile, a comparison between the importance of conditioning and triggering factors in protected and non-protected forests may reveal the effects of anthropogenic activities on the susceptibility of disturbed forests to landslides.

Various methods have been developed for assessing landslide susceptibility with respect to knowledge-driven approaches, physical and statistical models, and machine learning algorithms [35,36]. Knowledge-driven approaches are subjective, and determines the influence of a variable through an expert's opinion [37] that may affect the real expectation of landslide susceptibility [38]. Moreover, the physical models are appropriate for assessing the susceptibility of small areas to landslide in the presence of detailed geological, pedological, hydrological, and geomorphological information [39,40]. The statistical models are dependent on the input data characteristics, where any uncertainty in data may lead to a huge error in mapping the landslide susceptibility [40,41]. In contrast, machine learning applies algorithms for modeling through learning data, where their high ability in the estimation of a model has made them more popular for analyzing the landslide susceptibility at a regional scale [42] such as artificial neural networks (ANN) [43], decision trees (DT) [44], Bayesian network (BN) and naïve Bayes [45], support vector machines (SVM) [46], and random forest (RF) [42,47–53]. RF, as an ensemble machine learning algorithm, is known for its ability in handling both parametric and non-parametric variables, working with big data without any selection, reduction, or preprocessing, handling missing values automatically, avoiding the risk of over-fitting, self-testing using "out of bag" data, and yielding high satisfactory accuracy in modeling [54]. Furthermore, RF has achieved robust performance for the mapping of landslide susceptibility in comparison with the conventional statistical models such as weights-of-evidence [55], logistic regression [41,55,56], and generalized additive models (GAM) [55]; or even other machine learning techniques such as boosted regression trees [42,57], regression tree [58], ANN [56,59], and SVM [55,60]. For example, Vorpahl et al. [57] concluded that RF indicated a higher performance than statistical and other machine learning methods such as GAM, generalized linear models (GLM), the maximum entropy method (MEM), classification tree analysis, multivariate adaptive regression splines, and ANN for analyzing influential variables that control natural landslides in a montane tropical forest, South Ecuador. Likewise, Dou et al. [61] reported that RF performed higher overall efficiency than DT for mapping rainfall-induced landslide susceptibility at a regional scale in Japan.

Recent studies have criticized the current derived landslide susceptibility mapping in terms of applying similar geo-environmental factors over different regions and times [36,62,63], considering fixed effects of a variable [2,63] such as distance to roads [63] and land-use/land-cover derived from the current available images without assessing their dynamic changes [2,62]. However, the current land-cover may not reflect its actual status during the time a landslide occurs in a specific area [2,64], and human-induced triggering factors such as logging and road construction may reduce slope stability over time. For example, Wolter et al. [12] showed that landslide events were observed in forests that had been opened by logging activities or fragmented by road construction in the Chilliwack River Valley, British Columbia, and reported that other geo-environmental variables did not show significant effects on the slope instability.

The Hyrcanian ecoregion has been degraded by different human and natural triggering factors such as forest and rangeland conversion [65,66], forest fires [67,68], flooding [69], landslides [70–72], soil erosion [73], and climate hazards [74,75] in northeast (NE) Iran. Several studies have accomplished mapping landslide susceptibility in Hyrcanian forests, but have mostly focused on applying models and using common conditioning factors [51,76–79] with less attention given to the temporal dynamics of natural and anthropogenic triggering factors.

Less is known about the influence of different conditioning factors such as natural and particularly anthropogenic triggering factors for mapping landslide susceptibility in forest areas. For this purpose, a holistic approach needs to be developed to model the actual importance of conditioning and triggering factors, which control the susceptibility of protected and disturbed forests to landslides. Therefore, this research was designed to evaluate the performance of applying object-based random forest for mapping landslide susceptibility in a protected forest and a non-protected forest in NE Iran. Furthermore, we compared the importance of influential variables that control the susceptibility of these forests to landslides. Specifically, we aimed to find appropriate answers to the following questions: (i) Does object-based random forest show a satisfactory performance for modeling landslide susceptibility in protected and non-protected forests? (ii) Which conditioning and triggering factors are at the top in modeling the susceptibility of these two forest areas to landslides? and (iii) How do natural and anthropogenic triggering factors affect the susceptibility of protected and non-protected forests to landslides?

2. Materials and Methods

2.1. Description of Study Area

We selected a protected and a non-protected forest for analyzing landslide susceptibility in the eastern part of the Hyrcanian forests, southeast Caspian Sea, Iran. The largest Iranian National Park, Golestan, is assigned as a protected forest (approx. 500 out of 920 km^2) (Figure 1a). The protection of this park has taken place since 1957 and was registered by UNESCO as a biosphere reserve in 1976 as it contains fifty percent of the total of Iran's mammal species and above 1400 plant species registered by UNESCO [80,81]. A non-protected forest was selected in the neighborhood of this protected area (approx. 1500 km^2). This area has been affected by a variety of human activities such as deforestation [65], logging, mining, and road construction [82]. The annual rate of deforestation was reported at about 0.85% [65], the number of forest logging parcels increased to 400 (34,000 ha), the number of mines reached 12 plans (12,520 ha), and the length of roads increased from 120 to 1257 km between 1966 and 2016 (Figure 1b). The average elevation, slope, and rainfall of the two studied forests are about 1280 m, 30° and 600 mm, with the predominant forest type of *Quercus castaneafolia-Carpinus betulus*.

However, dominated lithology types of the protected and non-protected forests are Jl (limestone, oolitic-porous dolomitic limestone; Lar formation; Mesozoic era; Jurassic period) and Js (upper: shale, marl, sandstone, nodular Ls, Ammonite, Belemnite, and lower: shale, sandstone with thin-bedded limestone; Shemshak formation; Mesozoic era; Jurassic period), respectively.

2.2. Landslide Surveying

Landslide events were collected from different sources. The CORONA KH-4B image (Mission ID: 1110-1089Fore) of 1970 (~2 m) (https://corona.cast.uark.edu/atlas) and aerial photos (1:20,000) from 1966 were used to survey old landslides (~230 samples) mostly in the protected forest. The new landslides (~430 samples) were obtained from field observations, the available database [83], and high-resolution images of Google Earth for 2016 (Figure 1). In addition to these samples, about 210 and 1650 polygons of old and new landslides, which were mapped using Sentinel images by Shirvani et al. [72], were used along with the landslide samples for the two study areas (Figure 1). The size of the landslide samples ranged between 0.095 and 239.6 ha in the protected forest and between 0.018 and 60.82 ha in the non-protected forest.

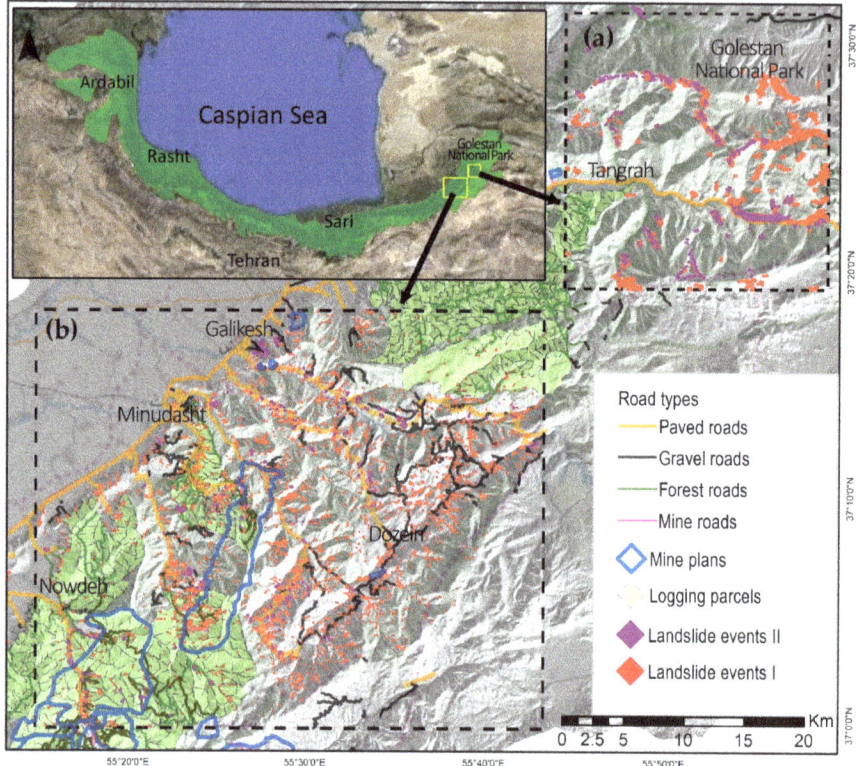

Figure 1. Location of the study areas in the Hyrcanian ecoregion in NE Iran: the Golestan National Park as a "protected forest" (**a**); and disturbed forests by mining, logging, and road building as a "non-protected forest" (**b**). Landslide events I were collected from different resources in the current research and the landslide events II were adopted from Shirvani et al. [72].

2.3. Image Segmentation for Generating Objects

A group of pixels that have similarity in their spectral and spatial properties is defined as an object [84]. The object-based paradigm has the ability to derive all possible features from the spectral, geometrical, contextual, and textural properties of either satellite images [84] or GIS (Geospatial Information Systems)-based data [85]. To obtain homogeneous objects, multi-resolution segmentation was implemented on the spectral bands of Landsat 8 of 2016 and SRTM (Shuttle Radar Topography Mission)-derived digital elevation model (30 m) [86]. After testing different scales by trial and error, the scale of 150 was selected with a higher weight for the near-infrared band (NIR) and the compactness and shape values of 0.8 and 0.2, respectively. The final segmentation was optimized using some digital elevation model (DEM) derivatives [72] such as slope, hillshade, terrain ruggedness index (TRI), and flow direction river (FDR). These image-segmented objects were assigned to calculate anthropogenic-induced deforestation and forest fragmentation from remote sensing data (Figure 2). Moreover, the summary statistics of conditioning and other triggering factors were calculated within each object segment.

2.4. Conditioning and Triggering Factors

We divided the driving forces of landslide susceptibility into conditioning and triggering factors based on the previous studies in the literature review [51,87] and the landslide characteristics of the study areas. Conditioning factors are cumulative events that show the potential of landslide occurrence, but do not necessarily trigger landslides [88], while triggering factors activate the landslides and increase the probability of occurrence by disturbing the balance between driving and resisting forces [89]. The selected conditioning variables included topographic, hydrologic, geology, soil, and vegetation layers. We generated topographic layers from the DEM; the hydrologic layers from the DEM and digital topographic maps (1:25,000); geological maps (1:100,000) from the Geological Survey and Mineral Explorations of Iran (GSI) and soil maps (1:100,000) from the Agriculture Department of Iran; and forest types were based on the thematic maps of forest management plans (Table 1). Furthermore, triggering factors were classified into natural (i.e., rainfall, earthquake, and flood) and anthropogenic (i.e., forest fragmentation, forest loss, logging, and mining). Rainfall intensity and earthquake magnitude were created by kriging [90] and inverse distance weighted (IDW) interpolation methods [91] from the historical events (Table 2). Flood frequency was calculated based on the frequency of flood occurrence along the main rivers within a specific catchment during the last five decades (Table 2).

We created the forest layer of 1970 from CORONA images and aerial photos using object-based nearest neighbor classification (Figure 2). A multispectral image was created from the original band of the images and two created channels using the sharpening and embossing filters. The segmentation was implemented through the multispectral resolution algorithm on these images. The possible classes were defined and some objects from each class were selected as training samples. Varieties of ancillary spectral and textural features were created to improve the accuracy of the classification. After optimizing the dimensions of the features, the standard nearest neighbor algorithm was applied for classifying the images to the forest and non-forest classes [82]. Moreover, the forest layer of 2016 was mapped from Landsat 8 using the object rule-based classification (Figure 2). After image segmentation as described earlier in Section 2.3, some objects from each class were selected. Then, different spectral, contextual, and textural features were derived from the main spectral bands of Landsat 8. The thresholds of non-forest classes from the forest class were determined by matching their values within the derived object features. The objects that had maximum difference values less than 0.52 were classified as water, residential areas, and grasslands; objects that had green vegetation index (GVI) [92] values that were negative and standard deviations derived from the gray-level co-occurrence matrix (GLCM) less than 21 were classified as dry-farming; the objects that had enhanced vegetation index2 [93] values higher than 0.98 and entropy derived from the GLCM less than 7 were classified as irrigated-farming; and the remaining unclassified objects were classified as forest [82]. The accuracy of the classifications was validated using the provided ground truth samples and confusion matrix [94], as shown in Table A1.

We calculated a number of forest loss and forest fragmentation metrics by comparing these two forest layers such as the rate of forest loss [65], edge density (ED), mean patch size (MPS), mean shape index (MSI), mean patch edge (MPE), mean perimeter-area ratio (MPAR), and number of patches (NumP) within each object [95] (Table 2). These metrics were included as triggering of landslide susceptibility along with the other variables. The total volumes of logging and weights of mining materials were calculated within each object from 1970 to 2016 as indicators of logging and mining intensity in the two studied forests.

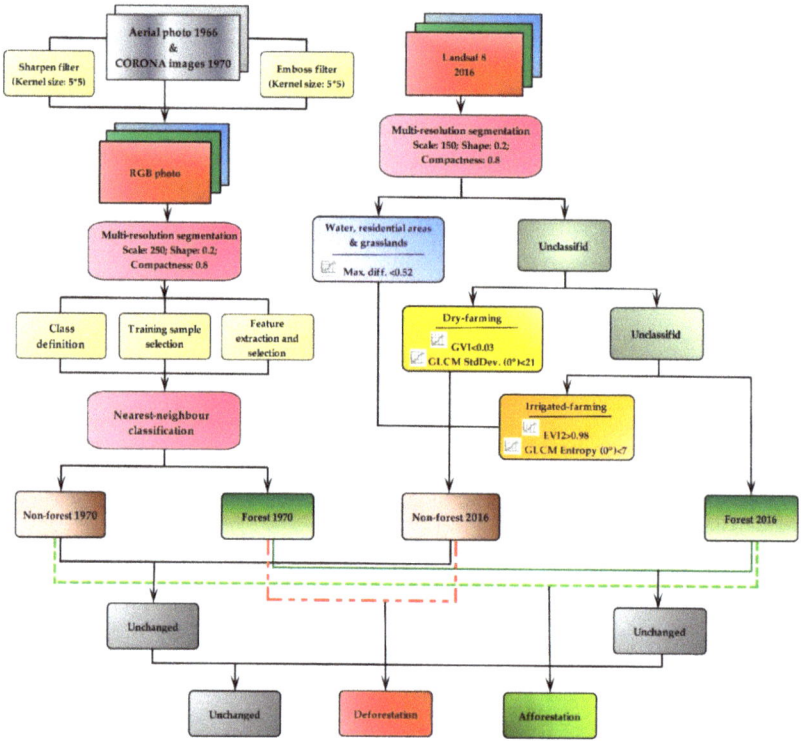

Figure 2. The process of discriminating forests from non-forests using object-based nearest neighbor and rule-based classifications by utilizing CORONA images/aerial photos of 1970 and Landsat 8 of 2016 over the protected and non-protected forests in NE, Iran (after Shirvani et al. [82]). Abbreviations: stdDev.: Standard deviation; Max. diff.: Maximum difference; GVI: Green Vegetation Index; EVI2: Enhanced vegetation index2; GLCM: Gray-level co-occurrence matrix.

2.5. Landslide Susceptibility

The object-based random forest approach was employed to assess the susceptibility of the protected and non-protected forests to landslides by contributing the conditioning and triggering factors as well as historical landslide samples.

Spatially, the landslide samples are joined to the objects of predictor variables. An object where over 50% of its area was affected by landslides was indicated as a landslide-affected object (LAO) and otherwise as a non-affected landslide object (NLAO). Roughly 20% of the LAO and NLAO objects were randomly selected for determining the importance of variables that control landslide susceptibility and modeling the spatial probability of landslide using a classification and regression trees (CART) procedure of RF [96,97].

RF is an ensemble-learning algorithm that builds several decision trees during the process of model formation. The training of each tree was carried out by bootstrap sampling from the generated dataset; about two-thirds of the samples were used for training a decision tree (in bag samples) and the remaining one-third was used to test the accuracy of the formed tree (out of bag (OOB) samples).

Multiple RFs were built to determine the optimal number of variables that needed to be applied for every splitting in each tree of the forest [98] in both study areas (Figure 3a,b). The OOB prediction was computed using the majority vote obtained from the OOB data for each object. The OOB error of an object was computed from the OOB prediction of that object. The results over all of the objects were used to calculate the error rate. The optimal performance of RF was determined with respect to the maximum area under the receiver operating characteristic (AUROC) [99,100] and the evaluating metrics of model performance such as sensitivity (Equation (1)), specificity (Equation (2)), precision (Equation (3)), and F-measure (Equation (4)) that were computed using the status of OOB errors including the objects that were labeled as LAO and also classified as LAO (TP); the objects that were labeled as NLAO and classified as NLAO (TN); the objects that were labeled as LAO but classified as NLAO (FN); and the objects that labeled as NLAO but classified as LAO (FP) [99,100].

$$\text{Sensitivity} = \frac{TP}{TP + FN} \tag{1}$$

$$\text{Specificity} = \frac{TN}{TN + FP} \tag{2}$$

$$\text{Precision} = \frac{TP}{TP + FP} \tag{3}$$

$$F1 = 2 \times \frac{Precision \times Sensitivity}{Precision + Sensitivity} \tag{4}$$

RF calculates the importance of each variable in the classification through Gini importance or Permutation importance [101]. The importance of variables was determined depending on the internal Gini method [102] in this study. The probability value of assigning an object to the class LAO—depending on a specific threshold—was indicated as the susceptibility of that object to the landslide. All objects were scored depending on the optimal-trained model for calculating their susceptibility to landslide from zero (very-low probability) to one (very-high probability) in both protected and non-protected forests.

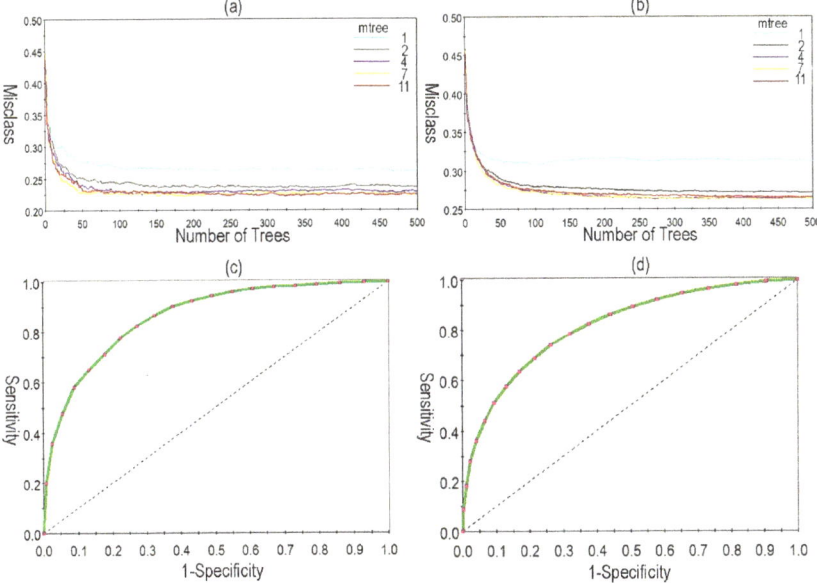

Figure 3. The optimal number of trees and the number of variables for splitting in each tree of the random forest based on the minimum misclassification error for mapping landslide susceptibility in the protected forest (**a**) and non-protected forest (**b**); the area under the ROC curve obtained from the out-of-bag error for testing the performance of random forest for mapping landslide susceptibility in the protected forest (**c**) and non-protected forest (**d**) in NE Iran.

Table 1. Conditioning factors for mapping landslide susceptibility used in NE Iran.

Category	Variables	Description	Sources
Conditioning factors of landslide			
Topographic	Elevation	The average of elevation (m) [103] in an object.	[42,47,50,52,57,58, 104–114]
	Slope (°)	The average of maximum changes in elevation value [115] within each object.	
	Aspect	The average of the slope direction [115] within each object.	
	Curvature	The average rate of changing in slope or aspect [116] within an object.	
	Plan curvature	The average values of the position of the curvature surface to the direction of slope perpendicularly within each object. The convex position indicates by positive values and concave position by negative values [117].	
	Profile curvature	The average of the amount of the curvature surface in the direction of maximum slope within each object. The convex surface indicates by negative values and concave surface by positive values [118].	
	Terrain convergence index (TCI)	TCI measures the intensity of the divergence or convergence within an object. Divergent surface indicates by positive values while convergent surface indicates by negative values [119]. $TCI = \left(\frac{1}{8}\sum_{i=1}^{8}\theta_i\right) - 90°$ θ: average degree between the direction of adjacent cells and the direction to the central cell.	
	Topographic position index (TPI)	TPI measures the difference between the elevation of the central point (z_0) against the average elevation ($\bar{z}$) in a specific radius (R) [120,121]. $TPI = z_0 - \bar{z}$ $\bar{z} = \frac{1}{n_R}\sum_{i \in R} z_i$. Positive values: higher position of the central points Negative values: lower position of the central points	
	Terrain ruggedness index (TRI)	TRI measures the heterogeneity in the landscape [122]. $TRI = \sqrt{|x|(max^2 - min^2)}$ max: maximum values of elevation within a 3 × 3 cell window min: minimum values of elevation within a 3 × 3 cell window.	
Hydrologic	Distance to river	Nearest distance to river based on Euclidean distance [123].	[42,50,52,61,76,104–106,108–114]
	River density	Magnitude of river (m) per hectare [124]	
	Topographic wetness index (TWI)	TWI measures topographic dimension of hydrological processes [116]. $TWI = ln\left(\frac{A_s}{\tan\beta}\right)$ A_s: Catchment area β: Slope gradient (degree) [116,120].	
	Stream power index (SPI)	SPI measures the erosive severity of a stream [116]. $SPI = A_s \times \tan\beta$ A_s: The area of a catchment β: Slope gradient (degree)	
	Sediment transport index (STI)	STI measures the erodability of a stream [125]. $STI = \left(\frac{A_s}{22.13}\right)^{0.6}\left(\frac{Sin\beta}{0.0896}\right)^{1.3}$ A_s: The area of a catchment β: Slope gradient (degree) [126].	
Geology	Lithology	Lithology units	[41,43,50–52,76,105, 106,113,127]
	Distance to faults	Nearest distance to the fault lines based on Euclidean distance [123].	
	Fault density	Magnitude of fault (m) per hectare [124].	
Soil	Soil texture	Soil textures	[50,76,106,107]
	Soil hydro group	Soil drainage	
Vegetation	Forest type	Dominant tree species [50] within an object.	[42,50,107,128,129]

Table 2. Triggering factors for mapping landslide susceptibility used in NE Iran.

Category		Variables	Description	Sources
Triggering factors of landslide				
Natural triggering factors		Rainfall	Long-term regional average annual raining data (mm/y) for 30 years interpolated by kriging [74,90].	[52,60,69,76, 106,109,127, 130]
		Earthquake	Long-term regional average of the magnitude of earthquakes (1970 to 2016) mapped by inverse distance weighting (IDW) interpolation method [91,131].	
		Flood frequency (FF)	FF measures the frequency of flood occurrence along the main rivers during last five decades within a specific catchment. $$FF = \frac{\sum_{i=1}^{n} L_i \times F_i}{A_S}$$ L_i: Length of a specific river (i) F_i: The frequency of flood for the long-term A_S: The area of the catchment	Current study
Anthropogenic triggering factors	Forest fragmentation	Patch density and size metrics [95]	Number of patches within an object (NumP) Mean patch size within an object (MPS) $$PS = \frac{TCA}{NumP}$$ NumP: Number of patches within an object TCA: Total area of patches in an object	Current study
		Edge metrics [95]	Edge density within an object (ED) $$ED = \frac{TE}{TCA}$$ TE: Total edge of patches within an object TCA: Total area of patches in an object Mean patch edge within an object (MPE) $$MPE = \frac{TE}{NumP}$$ NumP: Number of patches within an object TE: Total edge of patches within an object	
		Shape metrics [95]	Mean shape index (MSI) within an object $$MSI = \frac{TE}{\sqrt{TCA}}$$ TE: Total edge of patches within an object TCA: Total area of patches in an object Mean perimeter-area ratio (MPAR) within an object $$MPAR = \frac{ED}{NumP}$$ ED: Edge density within an object NumP: Number of patches within an object	
	Forest loss (FL)		$$FL = \frac{(F_a - F_b)}{(F_a * T)}$$ T: The time duration of each period F_a and F_b: The forest area at the beginning and end of each period [65].	Current study
	Logging		Total volume of logging (1966–2016) within an object	Current study
	Mining		Total weight of mining (1966–2016) within an object	Current study

3. Results

3.1. Summary of Model Validation

The evaluation measures indicated that object-based random forest showed good performance in assessing the importance of variables that control the susceptibility of protected and non-protected forests to landslides. The optimal number of variables for every splitting in each tree was obtained as 7 and 11 in the protected and non-protected forests, respectively, depending on the minimum misclassification error (Figure 3a,b). The optimal AUROC values of protected and non-protected forest obtained about 86.31 and 81.77% after the formation of 500 trees. The sensitivity values of 77.54 and 74.56% were obtained from mapping the landslide susceptibility in the protected and non-protected forests, as shown in Table 3.

Table 3. The results of the accuracy assessment of landslide susceptibility mapping based on the influential variables that control the occurrence of the landslide in the protected and non-protected forests.

	Metrics	Specificity (%)	Sensitivity (%)	Precision (%)	F1 Statistic (%)	Misclassification Rate (%)	AUROC [1] (%)
Landslide susceptibility	PF [2]	77.85	77.54	92.62	84.41	23.38	86.31
	NPF [3]	73.02	74.56	73.37	73.96	26.21	81.77

[1] AUROC: Area under the receiver operating characteristics; [2] PF: Protected forest; [3] NPF: Non-protected forest.

3.2. The Importance of Variables

The analysis of variable importance indicated that the top variables that controlled landslide susceptibility belonged to the topographic and hydrologic categories in both the protected and non-protected forests. The most influential variables were terrain ruggedness index (TRI) and river density with 19.54 and 6.07% of importance values in the protected and non-protected forests, respectively (Figure 4). The triggering variables had a significant influence on the landslide susceptibility in both regions; however, the score values of natural triggering factors (16.20%) were higher than the values of anthropogenic triggering factors (<1%) in the protected forest. On the other hands, anthropogenic factors (16.89%) such as forest fragmentation, logging, and mining activities recorded slightly higher score values than the natural triggering factors (16.50%) such as rainfall, flood, and earthquake in the non-protected forest (Figure 4). The geological variables recorded higher values in the non-protected forest in comparison to the protected forest. The type of forest variable showed a score value of 2.58% in the protected forest and a trivial importance in the non-protected forest. However, soil variables recorded total values of less than one percent for expressing landslide susceptibility in the two study areas.

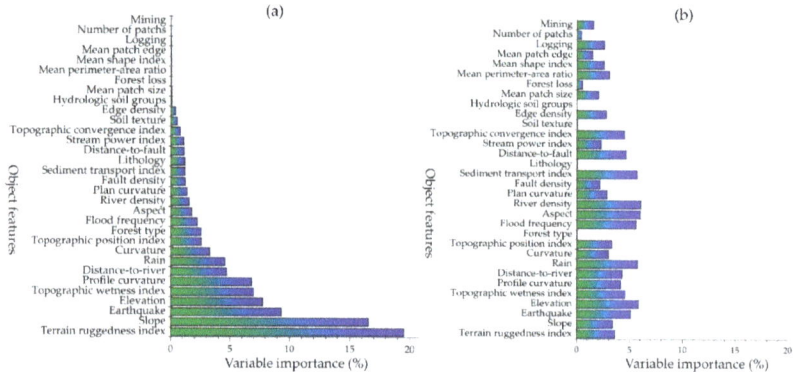

Figure 4. Comparison of the variable's importance in controlling the susceptibility of protected forest (**a**) and non-protected forest (**b**) to landslides in NE Iran. While topographic, hydrologic, and natural triggering factors were dominant variables in the protected forest, the anthropogenic triggering factors recorded higher importance values than the natural triggering factors with a total value close to the importance of topographic and hydrologic variables in the non-protected forest in NE Iran.

3.3. Landslide Susceptibility Mapping

The output maps indicate that the distribution of the landslide susceptibility of non-protected forest (0.51 ± 0.36) was higher than in the protected forest (0.34 ± 0.33) in the study area (Figure 5). The high susceptibility values of landslide were distributed in the east of the protected forest (Figure 5a), which resulted from the extremely rugged and steep surfaces as well as the magnitude of the occurred earthquake and hydrological variables such as topographic wetness index (TWI) and distance to river (Figure 4). Although different parts of non-protected forest were occupied by high values of landslide susceptibility, those forests affected by the interaction of hydrologic and topographic variables with anthropogenic and natural triggering factors received higher values of landslide susceptibility, particularly in the central and southern parts of the non-protected forest (Figure 5b).

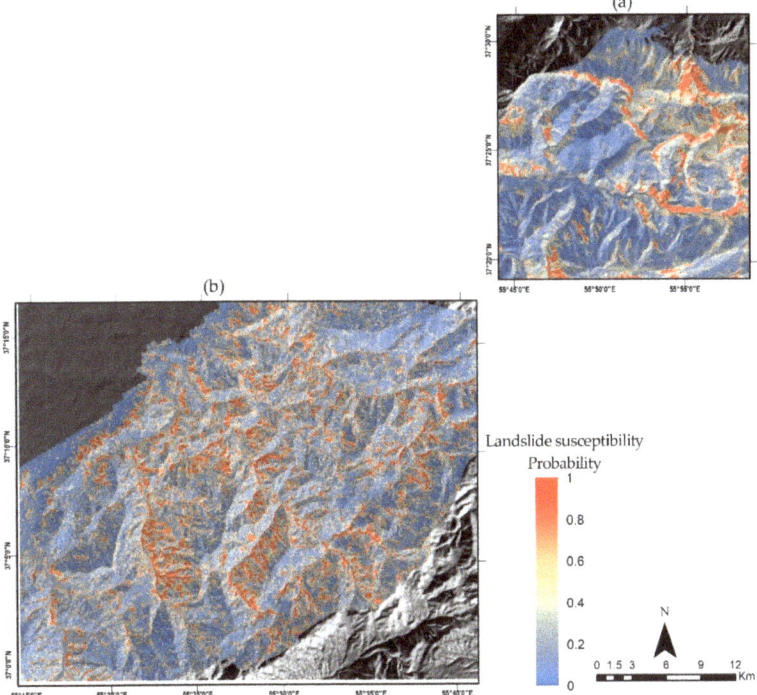

Figure 5. The landslide susceptibility maps of the protected forest (**a**) and the non-protected forest (**b**) in NE Iran. The maps show that the majority of high susceptibility values of landslide were distributed in the eastern parts of the protected forest (**a**) occupied by highly and extremely rugged or steep surfaces, while the high values of landslide susceptibility were distributed throughout the non-protected forest (**b**), particularly where anthropogenic and natural triggering factors interacted with the hydrologic and topographic variables.

4. Discussion

4.1. The Accuracy of Landslide Susceptibility Maps in the Protected and Non-Protected Forests

The results of the model assessment indicated the high accuracy of the obtained landslide susceptibility maps from the RF model with the contribution of influential conditioning and triggering variables for both the protected and non-protected forests. However, the landslide susceptibility map of the protected forest showed a higher AUROC value than the landslide susceptibility map of the non-protected forest (Table 3). The high performance of RF for landslide susceptibility mapping has also been verified in previous studies [41,42,57,58,60]. This study adds that the application of an object-based random forest resulted in a high accuracy of landslide susceptibility mapping, whereas the pixel-based random forest was the model of interest by the aforementioned researchers.

4.2. The Importance of Conditioning Factors for Mapping Landslide Susceptibility in Protected and Non-Protected Forests

Our analysis of comparing the influential variables revealed that the topographic factors obtained the highest scores for mapping landslide susceptibility in the protected forest; however, there was a relative balance between the scores of topographic, hydrologic, and triggering factors in the non-protected forest. The topographic features obtained about 60% of the total importance values in the protected forest; 36% of the values were assigned to the TRI (19.5%) and slope (16.5%) (Figure 4). The majority of landslide events fell in the old type in the protected forest, which are scattered in the steeped slopes and coarse rugged surfaces [72]. Furthermore, our analysis showed that the spatial probability of landslide significantly increased from 0.75 to 1 when the TRI increased from 14 to 27 (Figure 6a) and the slope increased from 25° to 51° (Figure 6b) in the protected forest. The high importance of the TRI [42,52] and slope [41,48,50,132] for mapping the landslide susceptibility has also been reported in several studies. Nevertheless, some research has addressed the low importance of slope for mapping landslide susceptibility [42,47,58].

Although topographic features gained about 36% of importance in the non-protected forest, their score was lower than the score of the topographic features in the protected forest. Both studied forests showed almost similar topographic characteristics; however, the aspect (Figure 7b) and elevation (Figure 7c) recorded higher scores among the topographic features in the non-protected forest (Figure 4). Likewise, several studies have confirmed the high importance of aspect and elevation for landslide susceptibility mapping [49–53].

The hydrological features obtained about 18% of scores with the top variables of TWI (7%) and distance to rivers (4.7%) in the protected forest. While in the non-protected forest, the importance of hydrological features increased to 28.5% with the top variables of river density (Figure 7a) and sediment transportation index (STI) (Figure 7e). We can infer from these results that it is likely that increasing human activities such as deforestation may cause changes in the hydrological system and increase the sediment [133,134] through the rivers with the consequences of increasing the susceptibility of landslide [11]. For example, Swanson and Dyrness [16] concluded that clear-cutting-induced landslides has substantially increased transported sediment materials in forest areas. The importance of the TWI [48,53,132] and distance to river [4,50,53] has also been reflected in earlier studies mapping landslide susceptibility.

The importance values of natural triggering factors were relatively equal between the two forests. The top variables of this category were earthquake (9.3%) and rainfall (4.6%) in the protected forest (Figure 4a), while all three variables roughly gained equal values in the non-protected forest (Figure 4b). Although the importance of natural triggering factors such as earthquake [52] and rainfall [49,52,53,78] has been reported for mapping landslide susceptibility, earthquakes trigger landslides by generating primary slips and intensifying liquefaction in the saturated soils [52]. The intensification of natural hazards due to human intervention can increase the landslide susceptibility, as the importance of flood in the mapping of landslide susceptibility increased from 2.3% in the protected forest to 5.6% in the non-protected forest.

Although anthropogenic triggering factors obtained less than one percent of importance in the protected forest, their importance was recorded at roughly 17% in the non-protected forest. The features of forest fragmentation (Figure 8) ranked the highest among the anthropogenic factors, which resulted from forest conversion and road-network expansion for logging, rural usages, and transporting mine materials in the non-protected forest since the 1970s [82]. For example, the length of the rural roads have increased from 113 to 752 km between 1970 and 2016, and about 245 and 155 km of logging and mine roads were built before 2016, respectively. All the fragmentation metrics showed higher values in the non-protected forest in comparison to the protected forest (Figures 4 and 8). Moreover, the importance of logging and mining was 2.6% and 1.5% in the non-protected forest, respectively. The number of parcels for timber harvesting increased from 0 to 404 between 1970 and 2016; the area of mining plans also expanded to 12,520 ha in the non-protected forest prior to 2016.

Most previous studies have frequently pointed to the anthropogenic triggering factors such as distance to roads [51,53,78,105,130], road density [76,105,135], land-use/land-cover types [42,47,49,58,78,106], and land-use changes [3,136,137] for the mapping of landslide susceptibility.

However, the current study explicitly localized and classified the significant anthropogenic triggering factors depending on the human footprint including forest fragmentation, forest conversion, timber harvesting, and mining within the forests. The influences of building forest roads [15,18,21–23], logging [4,11–15], deforestation [2–8], forest fragmentation [5], and mining [17] on the occurrence, frequency, and distribution of landslides have been demonstrated in the forest areas. For example, Guns and Vanacker [7] highlighted that anthropogenic activities such as forest conversion increased the occurrence of small landslides and sediment deposition in tropical forests. Borga et al. [18] concluded that forest roads changed the stream flows and increased the susceptibility of the forest to shallow landslides on steep slopes. Guthrie [13] reported that the frequency and density of landslides have significantly increased, following timber harvesting in the forested watersheds.

Although a number of studies have reported geological features as the main causes of increasing landslide susceptibility [49,51,76,132], our analysis revealed that the importance of these variables was lower than the topographic, hydrologic, and natural triggering factors in both the protected and non-protected forests as well as lower than the importance of anthropogenic triggering factors in the non-protected forest. Distance to faults with a value of 4.6% was the top variable of the geological features in the non-protected forest. In addition, some studies reported the low importance of lithology [41,50,58,78], but the high importance of distance to faults [58] for mapping landslide susceptibility.

Moreover, forest type did not show considerable importance for landslide susceptibility mapping in both forests. With respect to the importance of forest loss and forest fragmentation in the non-protected forest, we can argue that forest dynamics are superior to the forest type in landslide susceptibility mapping. Soil variables showed neutral influence on landslide susceptibility [50] in both forests.

This study indicated that the influential conditioning and triggering factors that control the susceptibility of the protected and non-protected forests to landslides are different. Likewise, some studies have verified the variety of landslide triggering factors for different regions [48,63]. The triggering factors of landslides have regional differences and the types of data in different study areas are not exactly the same.

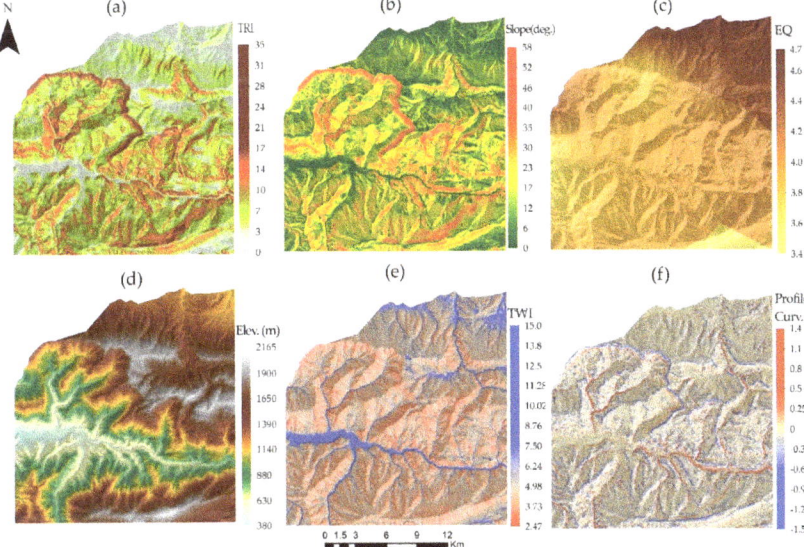

Figure 6. The layers of the top influential factors that control landslide susceptibility in the protected forest in NE Iran: terrain ruggedness index (TRI) (**a**); slope (**b**); earthquake (EQ) (**c**); elevation (Elev.) (**d**); topographic wetness index (TWI) (**e**); and profile curvature (Profile Curv.) (**f**).

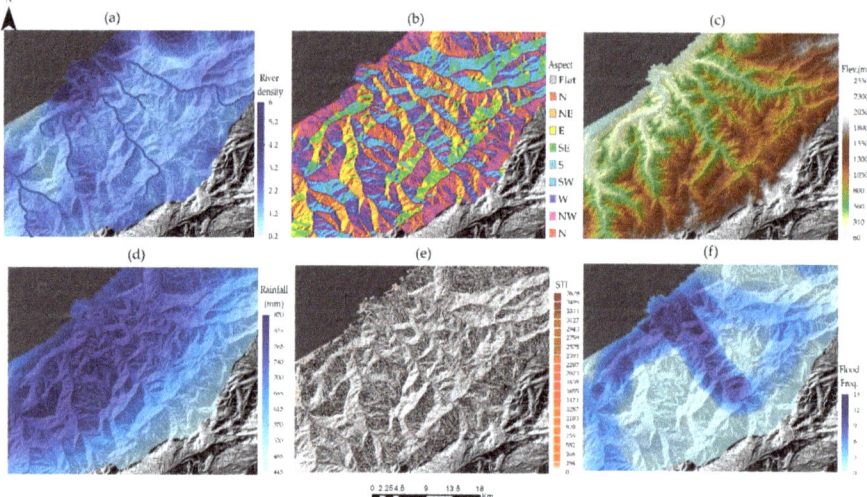

Figure 7. The layers of the top influential factors that control landslide susceptibility in the non-protected forest in NE Iran: river density (**a**); aspect (**b**); elevation (Elev.) (**c**); rainfall (**d**); sediment transport index (STI) (**e**); and flood (**f**).

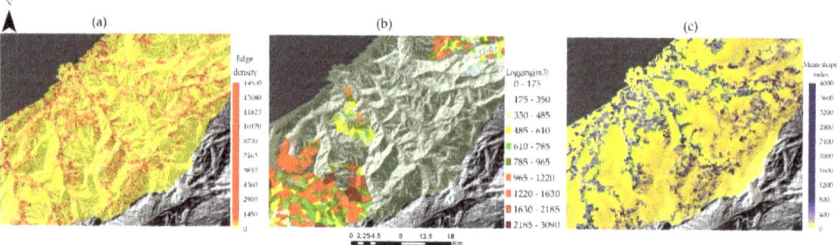

Figure 8. The top influential anthropogenic triggering factors for mapping landslide susceptibility in the non-protected forest, NE Iran: the edge density and mean shape index indicating the forest fragmentation induced by road-network expansion and forest conversion (**a,c**); and the aggregation of logging volumes (**b**) from 1970 to 2016.

The integration of random forest and an object-based approach yielded a good performance for mapping the landslide susceptibility in our forest regions. However, the comparison of the integration of other machine learning algorithms with the object-based approach needs to be considered to improve the best method to map the landslide susceptibility in the forest regions. Furthermore, this research used multiple conditioning and triggering factors to assess the susceptibility of forest areas to landslides. However, other factors may trigger landslide hazards such as ground water flow [138] in forest regions that need to be explored in the upcoming studies.

5. Conclusions

This study investigated the performance of a geographic object-based random forest for modeling the susceptibility of protected and non-protected forests to landslides. Various object features of conditioning (topographic, hydrologic, geologic, geology, soil, and vegetation) and triggering factors (rainfall, flood, earthquake, deforestation, forest fragmentation, logging, and mining) were applied as a database for the mapping of landslide susceptibility in the two forest areas using the random forest algorithm.

Although the random forest exhibited good performance for the mapping of landslide susceptibility in both the protected and non-protected forests, its sensitivity in the protected forest was higher than that in the non-protected forest. The influential variables controlling the susceptibility of these two forests to landslides were different. Approximately 88% of the susceptibility of protected forests were explained by the conditioning factors focusing on the topographic (60%) and hydrologic (18%) features. Moreover, triggering factors recorded 22% of importance, focusing on natural triggering factors (16%). The top five variables were TRI, slope, earthquake, elevation, and TCI for the mapping of landslide susceptibility in the protected forest. In contrast, the importance values were distributed among the object features of both the conditioning and triggering factors in the non-protected forests. While the importance of topographic factors has significantly decreased, the importance of triggering factors focusing on anthropogenic features has substantially increased from less than 1% in the protected forest to about 17%—focusing on forest fragmentation and logging—in the non-protected forest. Moreover, the effects of some features of hydrologic and natural triggering factors such as sediment transport index and flood frequency were amongst the top variables that control landslide susceptibility in the non-protected forest. The effects of these features could be caused or intensified by human activities such as deforestation, forest fragmentation, logging, and mining. These results provide managers and decision-makers with information in which to assess the consequences of developing destructive schemes such as road building, logging, and mining before any intervention in forest areas. The importance of geology and soil features was lower than the importance of other variables in the non-protected forest.

This study indicates that different forest areas can be affected by different conditioning and triggering factors that control their susceptibility to landslides. Consequently, there are no uniformly predefined influential variables for mapping landslide susceptibility in forest areas.

Author Contributions: Z.S. contributed to writing all parts of this article including the design of the work, image processing, and object-based classification using eCognition®, modeling of landslide susceptibility in R, and mapping in ArcGIS®. All authors have read and agreed to the published version of the manuscript.

Funding: This research received no external funding.

Acknowledgments: The author expresses her acknowledgment to the three anonymous reviewers whose comments/suggestions helped improve and clarify this manuscript. The author acknowledges support by the German Research Foundation and the Open Access Funding by the Publication Fund of the TU Dresden.

Conflicts of Interest: The author declare no conflicts of interest.

Appendix A

Table A1. The results of the accuracy assessment of forest and non-forest classification categories using aerial photos and Landsat images in NE, Iran.

Metric	Category	User's Accuracy		Producer's Accuracy	
Time		1966	2016	1966	2016
Method		NN [1]	RB [2]	NN	RB
Category	Forest	0.8102	0.96	0.9911	0.9320
	Non-forest	0.9841	0.93	0.7045	0.9588
Observed agreement		0.865	0.945	—	—
Kappa coefficient		0.7175	0.89	—	—

[1] NN: Nearest neighbor classification; [2] RB: rule-based classification using object-based image analysis.

References

1. Walker, L.R.; Shiels, A.B. *Landslide Ecology*; Cambridge University Press: New York, NY, USA, 2013.
2. Guns, M.; Vanacker, V. Forest cover change trajectories and their impact on landslide occurrence in the tropical Andes. *Environ. Earth Sci.* **2013**, *70*, 2941–2952. [CrossRef]
3. Reichenbach, P.; Busca, C.; Mondini, A.C.; Rossi, M. The influence of land use change on landslide susceptibility zonation: the Briga catchment test site (Messina, Italy). *Environ. Manag.* **2014**, *54*, 1372–1384. [CrossRef] [PubMed]
4. Chang, J.-C.; Slaymaker, O. Frequency and spatial distribution of landslides in a mountainous drainage basin: Western Foothills, Taiwan. *Catena* **2002**, *46*, 285–307. [CrossRef]
5. Yeh, C.-K.; Liaw, S.-C. Application of landscape metrics and a Markov chain model to assess land cover changes within a forested watershed, Taiwan. *Hydrol. Process.* **2015**, *29*, 5031–5043. [CrossRef]
6. Alcántara-Ayala, I.; Esteban-Chávez, O.; Parrot, J. Landsliding related to land-cover change: A diachronic analysis of hillslope instability distribution in the Sierra Norte, Puebla, Mexico. *Catena* **2006**, *65*, 152–165. [CrossRef]
7. Guns, M.; Vanacker, V. Shifts in landslide frequency–area distribution after forest conversion in the tropical Andes. *Anthropocene* **2014**, *6*, 75–85. [CrossRef]
8. Glade, T. Landslide occurrence as a response to land use change: A review of evidence from New Zealand. *Catena* **2003**, *51*, 297–314. [CrossRef]
9. Alcántara-Ayala, I.; Moreno, A.R. Landslide risk perception and communication for disaster risk management in mountain areas of developing countries: A Mexican foretaste. *J. Mt. Sci.* **2016**, *13*, 2079–2093. [CrossRef]
10. Saito, H.; Murakami, W.; Daimaru, H.; Oguchi, T. Effect of forest clear-cutting on landslide occurrences: Analysis of rainfall thresholds at Mt. Ichifusa, Japan. *Geomorphology* **2017**, *276*, 1–7. [CrossRef]
11. Imaizumi, F.; Sidle, R.C.; Kamei, R. Effects of forest harvesting on the occurrence of landslides and debris flows in steep terrain of central Japan. *Earth Surf. Process. Landforms* **2008**, *33*, 827–840. [CrossRef]
12. Wolter, A.; Ward, B.; Millard, T. Instability in eight sub-basins of the Chilliwack River Valley, British Columbia, Canada: A comparison of natural and logging-related landslides. *Geomorphology* **2010**, *120*, 123–132. [CrossRef]
13. Guthrie, R. The effects of logging on frequency and distribution of landslides in three watersheds on Vancouver Island, British Columbia. *Geomorphology* **2002**, *43*, 273–292. [CrossRef]
14. Goetz, J.N.; Guthrie, R.H.; Brenning, A. Forest harvesting is associated with increased landslide activity during an extreme rainstorm on Vancouver Island, Canada. *Nat. Hazards Earth Syst. Sci.* **2015**, *15*, 1311–1330. [CrossRef]
15. Wolfe, M.D.; Williams, J.W. Rates of Landsliding as Impacted by Timber Management Activities in Northwestern California. *Environ. Eng. Geosci.* **1986**, *23*, 53–60. [CrossRef]
16. Swanson, F.J.; Dyrness, C.T. Impact of clear-cutting and road construction on soil erosion by landslides in the western Cascade Range, Oregon. *Geology* **1975**, *3*, 393.
17. Arca, D.; Kutoğlu, H.Ş.; Becek, K. Landslide susceptibility mapping in an area of underground mining using the multicriteria decision analysis method. *Environ. Monit. Assess.* **2018**, *190*, 725. [CrossRef]
18. Borga, M.; Tonelli, F.; Fontana, G.D.; Cazorzi, F. Evaluating the influence of forest roads on shallow landsliding. *Ecol. Model.* **2005**, *187*, 85–98. [CrossRef]
19. Haigh, M.J.; Rawat, J.; Rawat, M.; Bartarya, S.; Rai, S. Interactions between forest and landslide activity along new highways in the Kumaun Himalaya. *For. Ecol. Manag.* **1995**, *78*, 173–189. [CrossRef]
20. Fransen, P.J.B.; Phillips, C.J.; Fahey, B.D. Forest road erosion in New Zealand: Overview. *Earth Surf. Process. Landforms* **2001**, *26*, 165–174.
21. Anderson, M.G. Road-cut slope topography and stability relationships in St Lucia, West Indies. *Appl. Geogr.* **1983**, *3*, 105–114. [CrossRef]
22. Sidle, R.C.; Pearce, A.J.; O'Loughlin, C.L. *Hillslope Stability and Land Use*; American Geophysical Union (AGU): Washington, DC, USA, 1985.
23. Douglas, I. Natural and man-made erosion in the humid tropics of Australia, Malaysia and Singapore. *Inter. Assoc. Sci. Hydrol.* **1967**, *75*, 17–30.
24. Keefer, D.K. Landslides caused by earthquakes. *Geol. Soc. Am. Bull.* **1984**, *95*, 406–421. [CrossRef]

25. Dai, F.; Xu, C.; Yao, X.; Xu, L.; Tu, X.; Gong, Q. Spatial distribution of landslides triggered by the 2008 Ms 8.0 Wenchuan earthquake, China. *J. Asian Earth Sci.* **2011**, *40*, 883–895. [CrossRef]
26. Parker, R.N.; Densmore, A.L.; Rosser, N.J.; De Michele, M.; Li, Y.; Huang, R.; Whadcoat, S.; Petley, D.N. Mass wasting triggered by the 2008 Wenchuan earthquake is greater than orogenic growth. *Nat. Geosci.* **2011**, *4*, 449–452. [CrossRef]
27. Zhao, B.; Wang, Y.-S.; Luo, Y.-H.; Li, J.; Zhang, X.; Shen, T. Landslides and dam damage resulting from the Jiuzhaigou earthquake (8 August 2017), Sichuan, China. *R. Soc. Open Sci.* **2018**, *5*, 171418. [CrossRef]
28. Iverson, R.M. Landslide triggering by rain infiltration. *Water Resour. Res.* **2000**, *36*, 1897–1910. [CrossRef]
29. Saito, H.; Nakayama, D.; Matsuyama, H. Relationship between the initiation of a shallow landslide and rainfall intensity—duration thresholds in Japan. *Geomorphology* **2010**, *118*, 167–175. [CrossRef]
30. Hong, Y.; Hiura, H.; Shino, K.; Sassa, K.; Suemine, A.; Fukuoka, H.; Wang, G. The influence of intense rainfall on the activity of large-scale crystalline schist landslides in Shikoku Island, Japan. *Landslides* **2005**, *2*, 97–105. [CrossRef]
31. Ibsen, M.-L.; Casagli, N. Rainfall patterns and related landslide incidence in the Porretta-Vergato region, Italy. *Landslides* **2004**, *1*, 143–150. [CrossRef]
32. Crosta, G.B.; Frattini, P. Rainfall-induced landslides and debris flows. *Hydrol. Process.* **2008**, *22*, 473–477. [CrossRef]
33. Larsen, M.C.; Wieczorek, G.F.; Latrubesse, E. Geomorphic effects of large debris flows and flash floods, northern Venezuela, 1999. *Zeitschrift für Geomorphologie Neue Folge, Supplementband* **2006**, *145*, 147–175.
34. Larsen, I.J.; Montgomery, D.R. Landslide erosion coupled to tectonics and river incision. *Nat. Geosci.* **2012**, *5*, 468–473. [CrossRef]
35. Guzzetti, F.; Carrara, A.; Cardinali, M.; Reichenbach, P. Landslide hazard evaluation: A review of current techniques and their application in a multi-scale study, Central Italy. *Geomorphology* **1999**, *31*, 181–216. [CrossRef]
36. Pourghasemi, H.R.; Yansari, Z.T.; Panagos, P.; Pradhan, B. Analysis and evaluation of landslide susceptibility: a review on articles published during 2005–2016 (periods of 2005–2012 and 2013–2016). *Arab. J. Geosci.* **2018**, *11*, 193. [CrossRef]
37. Dahal, R.K.; Hasegawa, S.; Nonomura, A.; Yamanaka, M.; Masuda, T.; Nishino, K. GIS-based weights-of-evidence modelling of rainfall-induced landslides in small catchments for landslide susceptibility mapping. *Environmen. Geol.* **2008**, *54*, 311–324. [CrossRef]
38. Hjort, J.; Luoto, M. Statistical Methods for Geomorphic Distribution Modeling. In *Treatise on Geomorphology: Quantitative Modeling of Geomorphology*; Shroder, J., Ed.; Academic Press: Amsterdam, The Netherlands, 2013; pp. 59–73.
39. Canli, E.; Thiebes, B.; Petschko, H.; Glade, T. Comparing physically-based and statistical landslide susceptibility model outputs-a case study from Lower Austria. In Proceedings of the EGU General Assembly Conference, Vienna, Austria, 12–17 April 2015.
40. Van Westen, C.J.; Castellanos, E.; Kuriakose, S.L. Spatial data for landslide susceptibility, hazard, and vulnerability assessment: An overview. *Eng. Geol.* **2008**, *102*, 112–131. [CrossRef]
41. Trigila, A.; Iadanza, C.; Esposito, C.; Scarascia-Mugnozza, G. Comparison of Logistic Regression and Random Forests techniques for shallow landslide susceptibility assessment in Giampilieri (NE Sicily, Italy). *Geomorphology* **2015**, *249*, 119–136. [CrossRef]
42. Park, S.; Kim, J. Landslide Susceptibility Mapping Based on Random Forest and Boosted Regression Tree Models, and a Comparison of Their Performance. *Appl. Sci.* **2019**, *9*, 942. [CrossRef]
43. Pradhan, B.; Buchroithner, M.F. Comparison and Validation of Landslide Susceptibility Maps Using an Artificial Neural Network Model for Three Test Areas in Malaysia. *Environ. Eng. Geosci.* **2010**, *16*, 107–126. [CrossRef]
44. Park, I.; Lee, S. Spatial prediction of landslide susceptibility using a decision tree approach: a case study of the Pyeongchang area, Korea. *Int. J. Remote. Sens.* **2014**, *35*, 6089–6112. [CrossRef]
45. Pham, B.T.; Pradhan, B.; Bui, D.T.; Prakash, I.; Dholakia, M. A comparative study of different machine learning methods for landslide susceptibility assessment: A case study of Uttarakhand area (India). *Environ. Model. Softw.* **2016**, *84*, 240–250. [CrossRef]

46. Peng, L.; Niu, R.; Huang, B.; Wu, X.; Zhao, Y.; Ye, R. Landslide susceptibility mapping based on rough set theory and support vector machines: A case of the Three Gorges area, China. *Geomorphology* **2014**, *204*, 287–301. [CrossRef]
47. Catani, F.; Lagomarsino, D.; Segoni, S.; Tofani, V. Landslide susceptibility estimation by random forests technique: sensitivity and scaling issues. *Nat. Hazards Earth Syst. Sci.* **2013**, *13*, 2815–2831. [CrossRef]
48. Lagomarsino, D.; Tofani, V.; Segoni, S.; Catani, F.; Casagli, N. A Tool for Classification and Regression Using Random Forest Methodology: Applications to Landslide Susceptibility Mapping and Soil Thickness Modeling. *Environ. Model. Assess.* **2017**, *22*, 201–214. [CrossRef]
49. Chen, W.; Sun, Z.; Han, J. Landslide Susceptibility Modeling Using Integrated Ensemble Weights of Evidence with Logistic Regression and Random Forest Models. *Appl. Sci.* **2019**, *9*, 171. [CrossRef]
50. Kim, J.-C.; Lee, S.; Jung, H.-S.; Lee, S. Landslide susceptibility mapping using random forest and boosted tree models in Pyeong-Chang, Korea. *Geocarto Int.* **2018**, *33*, 1000–1015. [CrossRef]
51. Pourghasemi, H.R.; Kerle, N. Random forests and evidential belief function-based landslide susceptibility assessment in Western Mazandaran Province, Iran. *Environ. Earth Sci.* **2016**, *75*, 23. [CrossRef]
52. Zhang, K.; Wu, X.; Niu, R.; Yang, K.; Zhao, L. The assessment of landslide susceptibility mapping using random forest and decision tree methods in the Three Gorges Reservoir area, China. *Environ. Earth Sci.* **2017**, *76*. [CrossRef]
53. Taalab, K.; Cheng, T.; Zhang, Y. Mapping landslide susceptibility and types using Random Forest. *Big Earth Data* **2018**, *2*, 159–178. [CrossRef]
54. Salford Systems Ltd. Salford Predictive Modeller: Introduction to Random Forests. Available online: https://www.minitab.com/uploadedFiles/Content/Products/SPM/IntroRF.pdf (accessed on 9 September 2019).
55. Goetz, J.; Brenning, A.; Petschko, H.; Leopold, P. Evaluating machine learning and statistical prediction techniques for landslide susceptibility modeling. *Comput. Geosci.* **2015**, *81*, 1–11. [CrossRef]
56. Sevgen, E.; Kocaman, S.; Nefeslioglu, H.A.; Gokceoglu, C. A Novel Performance Assessment Approach using Photogrammetric Techniques for Landslide Susceptibility Mapping with Logistic Regression, ANN and Random Forest. *Sensors* **2019**, *19*, 3940. [CrossRef] [PubMed]
57. Vorpahl, P.; Elsenbeer, H.; Märker, M.; Schröder, B.; Maerker, M. How can statistical models help to determine driving factors of landslides? *Ecol. Model.* **2012**, *239*, 27–39. [CrossRef]
58. Chen, W.; Xie, X.; Wang, J.; Pradhan, B.; Hong, H.; Bui, D.T.; Duan, Z.; Ma, J. A comparative study of logistic model tree, random forest, and classification and regression tree models for spatial prediction of landslide susceptibility. *Catena* **2017**, *151*, 147–160. [CrossRef]
59. Dou, J.; Yamagishi, H.; Pourghasemi, H.R.; Yunus, A.P.; Song, X.; Xu, Y.; Zhu, Z. An integrated artificial neural network model for the landslide susceptibility assessment of Osado Island, Japan. *Nat. Hazards* **2015**, *78*, 1749–1776. [CrossRef]
60. Ada, M.; San, B.T. Comparison of machine-learning techniques for landslide susceptibility mapping using two-level random sampling (2LRS) in Alakir catchment area, Antalya, Turkey. *Nat Hazards* **2018**, *90*, 237–263. [CrossRef]
61. Dou, J.; Yunus, A.P.; Bui, D.T.; Merghadi, A.; Sahana, M.; Zhu, Z.; Chen, C.-W.; Khosravi, K.; Yang, Y.; Pham, B.T. Assessment of advanced random forest and decision tree algorithms for modeling rainfall-induced landslide susceptibility in the Izu-Oshima Volcanic Island, Japan. *Sci. Total. Environ.* **2019**, *662*, 332–346. [CrossRef]
62. Reichenbach, P.; Rossi, M.; Malamud, B.D.; Mihir, M.; Guzzetti, F. A review of statistically-based landslide susceptibility models. *Earth-Sci. Rev.* **2018**, *180*, 60–91. [CrossRef]
63. Xiao, L.; Zhang, Y.; Peng, G. Landslide Susceptibility Assessment Using Integrated Deep Learning Algorithm along the China-Nepal Highway. *Sensors* **2018**, *18*, 4436. [CrossRef]
64. Schmaltz, E.M.; Steger, S.; Glade, T. The influence of forest cover on landslide occurrence explored with spatio-temporal information. *Geomorphology* **2017**, *290*, 250–264. [CrossRef]
65. Shirvani, Z.; Abdi, O.; Buchroithner, M.F.; Pradhan, B. Analysing Spatial and Statistical Dependencies of Deforestation Affected by Residential Growth: Gorganrood Basin, Northeast Iran. *Land Degrad. Dev.* **2017**, *28*, 2176–2190. [CrossRef]
66. Abdi, O.; Shirvani, Z.; Buchroithner, M.F. Visualization and quantification of significant anthropogenic drivers influencing rangeland degradation trends using Landsat imagery and GIS spatial dependence models: A case study in Northeast Iran. *J. Geogra. Sci.* **2018**, *28*, 1933–1952. [CrossRef]

67. Abdi, O.; Kamkar, B.; Shirvani, Z.; Teixeira da Silva, J.A.; Buchroithner, M.F. Spatial-statistical analysis of factors determining forest fires: a case study from Golestan, Northeast Iran. *Geomat. Nat. Hazards Risk* **2018**, *9*, 267–280. [CrossRef]
68. Abdi, O. Climate-Triggered Insect Defoliators and Forest Fires Using Multitemporal Landsat and TerraClimate Data in NE Iran: An Application of GEOBIA TreeNet and Panel Data Analysis. *Sensors* **2019**, *19*, 3965. [CrossRef] [PubMed]
69. Mirzaei, G.; Soltani, A.; Soltani, M.; Darabi, M. An integrated data-mining and multi-criteria decision-making approach for hazard-based object ranking with a focus on landslides and floods. *Environ. Earth Sci.* **2018**, *77*, 581. [CrossRef]
70. Jarjani, A.; Akbari, H.; Hosseini, S.A.; Abdi, O. Investigation of Landslide Ranger Zoning using Analytical Hierarchy Process in GIS Environment (Case Study: Azadshahr Kohmian Forestry Design). *J. Watershed Manag. Res.* **2018**, *10*, 197–207.
71. Kornejady, A.; Ownegh, M.; Rahmati, O.; Bahremand, A. Landslide susceptibility assessment using three bivariate models considering the new topo-hydrological factor: HAND. *Geocarto Int.* **2018**, *33*, 1155–1185. [CrossRef]
72. Shirvani, Z.; Abdi, O.; Buchroithner, M. A Synergetic Analysis of Sentinel-1 and -2 for Mapping Historical Landslides Using Object-Oriented Random Forest in the Hyrcanian Forests. *Remote Sens.* **2019**, *11*, 2300. [CrossRef]
73. Mousavinejad, S.H.; Habashi, H.; Kiani, F.; Shataee, S.; Abdi, O. Evaluation of soil erosion using imagery SPOT5 satellite in Chehel chi catchment of Golestan Province. *Wood Forest Sci. Technol.* **2017**, *24*, 73–86.
74. Abdi, O.; Shirvani, Z.; Buchroithner, M.F. Spatiotemporal drought evaluation of Hyrcanian deciduous forests and semi-steppe rangelands using moderate resolution imaging spectroradiometer time series in Northeast Iran. *Land Degrad. Dev.* **2018**, *29*, 2525–2541. [CrossRef]
75. Abdi, O.; Shirvani, Z.; Buchroithner, M.F. Forest drought-induced diversity of Hyrcanian individual-tree mortality affected by meteorological and hydrological droughts by analyzing moderate resolution imaging spectroradiometer products and spatial autoregressive models over northeast Iran. *Agric. For. Meteorol.* **2019**, *275*, 265–276. [CrossRef]
76. Mohammady, M.; Pourghasemi, H.R.; Pradhan, B. Landslide susceptibility mapping at Golestan Province, Iran: A comparison between frequency ratio, Dempster–Shafer, and weights-of-evidence models. *J. Asian Earth Sci.* **2012**, *61*, 221–236. [CrossRef]
77. Pourghasemi, H.R.; Rossi, M. Landslide susceptibility modeling in a landslide prone area in Mazandarn Province, north of Iran: a comparison between GLM, GAM, MARS, and M-AHP methods. *Theor. Appl. Climatol.* **2017**, *130*, 609–633. [CrossRef]
78. Arabameri, A.; Pradhan, B.; Rezaei, K.; Lee, S.; Sohrabi, M. An ensemble model for landslide susceptibility mapping in a forested area. *Geocarto Int.* **2019**, *77*, 1–26. [CrossRef]
79. Nohani, E.; Moharrami, M.; Sharafi, S.; Khosravi, K.; Pradhan, B.; Pham, B.T.; Lee, S.; M. Melesse, A. Landslide Susceptibility Mapping Using Different GIS-Based Bivariate Models. *Water* **2019**, *11*, 1402. [CrossRef]
80. Akhani, H.; Ziegler, H. Photosynthetic pathways and habitats of grasses in Golestan National Park (NE Iran), with an emphasis on the C4-grass dominated rock communities. *Phytocoenologia* **2002**, *32*, 455–501. [CrossRef]
81. Talebi, K.S.; Sajedi, T.; Pourhashemi, M. *Forests of Iran*; Springer Science and Business Media LLC: Berlin/Heidelberg, Germany, 2014.
82. Shirvani, Z.; Abdi, O.; Buchroithner, M.F. A New Analysis Approach for Long-Term Variations of Forest Loss, Fragmentation and Degradation Resulting from Road-Network Expansion Using Landsat Time-Series and OBIA. *Land Degrad. Dev.* **2019**. [CrossRef]
83. Iranian Landslide Working Party (ILWP). *Iranian landslides list*; Forest, Rangeland and Watershed Association: Tehran, Iran, 2007.
84. Navulur, K. *Multispectral Image Analysis Using the Object-Oriented Paradigm*; CRC Press/Taylor & Francis: Boca Raton, FL, USA, 2007; ISBN 1420043064.
85. Hay, G.J.; Castilla, G. *Geographic Object-Based Image Analysis (GEOBIA): A new name for a new discipline*; Springer Science and Business Media LLC: Berlin/Heidelberg, Germany, 2008; pp. 75–89.
86. Earth Resources Observation and Science (EROS) Center. Shuttle Radar Topography Mission (SRTM) 1 Arc-Second Global. Available online: https://www.usgs.gov/centers/eros (accessed on 18 July 2019).

87. Abella, E.A.C.; Van Westen, C.J. Generation of a landslide risk index map for Cuba using spatial multi-criteria evaluation. *Landslides* **2007**, *4*, 311–325. [CrossRef]
88. Dikau, R. *Landslide Recognition: Identification, Movement and Causes*; Wiley: Chichester, UK, 1996; ISBN 0471964778.
89. Waltham, A.C. *Foundations of Engineering Geology*; Blackie Academic & Professional Publishing: Glasgow, UK, 1994.
90. Oyana, T.J.; Margai, F. *Spatial Analysis: Statistics, Visualization, and Computational Methods*; CRC Press/Taylor & Francis: Boca Raton, FL, USA, 2015; ISBN 0429069367.
91. ESRI. How IDW works—Help | ArcGIS for Desktop. Available online: http://desktop.arcgis.com/en/arcmap/10.3/tools/3d-analyst-toolbox/how-idw-works.htm (accessed on 14 August 2019).
92. Aguilar, F.J.; Nemmaoui, A.; Aguilar, M.A.; Chourak, M.; Zarhloule, Y.; Lorca, A.G. A Quantitative Assessment of Forest Cover Change in the Moulouya River Watershed (Morocco) by the Integration of a Subpixel-Based and Object-Based Analysis of Landsat Data. *Forest* **2016**, *7*, 23. [CrossRef]
93. Jiang, Z.; Huete, A.R.; Didan, K.; Miura, T. Development of a two-band enhanced vegetation index without a blue band. *Remote. Sens. Environ.* **2008**, *112*, 3833–3845. [CrossRef]
94. Gómez, D.; Biging, G.; Montero, J. Accuracy statistics for judging soft classification. *Int. J. Remote Sens.* **2008**, *29*, 693–709. [CrossRef]
95. McGarigal, K.; Marks, B.J. *FRAGSTATS: Spatial Pattern Analysis Program for Quantifying Landscape Structure*; USDA Forest Service: WA, USA, 1995; Volume 351, p. 122. [CrossRef]
96. Breiman, L. Random forests. *Mach. Learn.* **2001**, *45*, 5–32. [CrossRef]
97. Liaw, A.; Wiener, M. Classification and regression by randomforest. *R News* **2002**, *2*, 18–22.
98. Cutler, A.; Cutler, D.R.; Stevens, J.R. Random Forests. In *Ensemble Machine Learning*; Zhang, C., Ma, Y., Eds.; Springer US: Boston, MA, USA, 2012; pp. 157–175. ISBN 978-1-4419-9325-0.
99. Witten, I.H.; Frank, E.; Hall, M.A. *Data Mining: Practical Machine Learning Tools and Techniques*, 3rd ed.; Morgan Kaufmann: Burlington, MA, USA, 2011; ISBN 978-0-12-374856-0.
100. Fawcett, T. An introduction to ROC analysis. *Pattern Recognit. Lett.* **2006**, *27*, 861–874. [CrossRef]
101. Louppe, G.; Wehenkel, L.; Sutera, A.; Geurts, P. Understanding variable importances in forests of randomized trees. In Proceedings of the Advances in Neural Information Processing Systems 26. 27th annual conference on neural information processing systems (NIPS 2013), Lake Tahoe, NV, USA, 5–10 December 2013; Burges, C.J.C., Bottou, L., Welling, M., Ghahramani, Z., Weinberger, K.Q., Eds.; 2013; pp. 431–439.
102. Strobl, C.; Boulesteix, A.-L.; Zeileis, A.; Hothorn, T. Bias in random forest variable importance measures: Illustrations, sources and a solution. *BMC Bioinform.* **2007**, *8*, 25. [CrossRef] [PubMed]
103. Hutchinson, M. A new procedure for gridding elevation and stream line data with automatic removal of spurious pits. *J. Hydrol.* **1989**, *106*, 211–232. [CrossRef]
104. Lee, C.F.; Li, J.; Xu, Z.W.; Dai, F.C. Assessment of landslide susceptibility on the natural terrain of Lantau Island, Hong Kong. *Environ. Earth Sci.* **2001**, *40*, 381–391. [CrossRef]
105. Yesilnacar, E.; Topal, T. Landslide susceptibility mapping: A comparison of logistic regression and neural networks methods in a medium scale study, Hendek region (Turkey). *Eng. Geol.* **2005**, *79*, 251–266. [CrossRef]
106. Pradhan, B.; Oh, H.-J.; Buchroithner, M. Weights-of-evidence model applied to landslide susceptibility mapping in a tropical hilly area. *Geomat. Nat. Hazards Risk* **2010**, *1*, 199–223. [CrossRef]
107. Yeon, Y.-K.; Han, J.-G.; Ryu, K.H. Landslide susceptibility mapping in Injae, Korea, using a decision tree. *Eng. Geol.* **2010**, *116*, 274–283. [CrossRef]
108. Zhu, A.-X.; Wang, R.; Qiao, J.; Qin, C.-Z.; Chen, Y.; Liu, J.; Du, F.; Lin, Y.; Zhu, T. An expert knowledge-based approach to landslide susceptibility mapping using GIS and fuzzy logic. *Geomorphology* **2014**, *214*, 128–138. [CrossRef]
109. Pham, B.T.; Bui, D.T.; Dholakia, M.B.; Prakash, I.; Pham, H.V.; Mehmood, K.; Le, H.Q. A novel ensemble classifier of rotation forest and Naïve Bayer for landslide susceptibility assessment at the Luc Yen district, Yen Bai Province (Viet Nam) using GIS. Geomatics. *Nat. Hazards Risk* **2017**, *8*, 649–671. [CrossRef]
110. Chen, W.; Shirzadi, A.; Shahabi, H.; Bin Ahmad, B.; Zhang, S.; Hong, H.; Zhang, N. A novel hybrid artificial intelligence approach based on the rotation forest ensemble and naïve Bayes tree classifiers for a landslide susceptibility assessment in Langao County, China. *Geomat. Nat. Hazards Risk* **2017**, *8*, 1955–1977. [CrossRef]

111. Chen, W.; Pourghasemi, H.R.; Naghibi, S.A. A comparative study of landslide susceptibility maps produced using support vector machine with different kernel functions and entropy data mining models in China. *Bull. Eng. Geol. Environ.* **2018**, *77*, 647–664. [CrossRef]
112. Zhou, C.; Yin, K.; Cao, Y.; Ahmed, B.; Li, Y.; Catani, F.; Pourghasemi, H.R. Landslide susceptibility modeling applying machine learning methods: A case study from Longju in the Three Gorges Reservoir area, China. *Comput. Geosci.* **2018**, *112*, 23–37. [CrossRef]
113. Nguyen, V.V.; Pham, B.T.; Vu, B.T.; Prakash, I.; Jha, S.; Shahabi, H.; Shirzadi, A.; Ba, D.N.; Kumar, R.; Chatterjee, J.M.; et al. Hybrid Machine Learning Approaches for Landslide Susceptibility Modeling. *Forest* **2019**, *10*, 157. [CrossRef]
114. He, H.; Hu, D.; Sun, Q.; Zhu, L.; Liu, Y. A Landslide Susceptibility Assessment Method Based on GIS Technology and an AHP-Weighted Information Content Method: A Case Study of Southern Anhui, China. *ISPRS Int. J. Geo-Inf.* **2019**, *8*, 266. [CrossRef]
115. Burrough, P.A.; McDonnell, R.A. *Principles of Geographical Information Systems*; Oxford University Press: Oxford, UK, 1998.
116. Moore, I.D.; Grayson, R.B.; Ladson, A.R. Digital terrain modelling: A review of hydrological, geomorphological, and biological applications. *Hydrol. Process.* **1991**, *5*, 3–30. [CrossRef]
117. Zevenbergen, L.W.; Thorne, C.R. Quantitative analysis of land surface topography. *Earth Surf. Process. Landforms* **1987**, *12*, 47–56. [CrossRef]
118. Shary, P.A. Land surface in gravity points classification by a complete system of curvatures. *Math. Geol.* **1995**, *27*, 373–390. [CrossRef]
119. Claps, P.; Fiorentino, M.; Oliveto, G. Informational entropy of fractal river networks. *J. Hydrol.* **1994**, *187*, 145–156. [CrossRef]
120. Wilson, J.P.; Gallant, J.C. *Terrain Analysis: Principles and Applications*; Wiley: New York, NY, USA, 2000; ISBN 0471321885.
121. Weiss, A. Topographic position and landforms analysis. In Proceedings of the ESRI User Conference 2001, San Diego, CA, USA, 9–13 July 2001; Volume 200.
122. Riley, S.J.; DeGloria, S.D.; Elliot, R. Index that quantifies topographic heterogeneity. *Int. J. Sci.* **1999**, *5*, 23–27.
123. ESRI. Understanding Euclidean distance analysis—Help | ArcGIS for Desktop. Available online: http://desktop.arcgis.com/en/arcmap/10.3/tools/spatial-analyst-toolbox/understanding-euclidean-distance-analysis.htm (accessed on 20 August 2019).
124. ESRI. Kernel Density—Help | ArcGIS for Desktop. Available online: http://desktop.arcgis.com/en/arcmap/10.3/tools/spatial-analyst-toolbox/kernel-density.htm (accessed on 21 August 2019).
125. Moore, I.D.; Burch, G.J. Sediment Transport Capacity of Sheet and Rill Flow: Application of Unit Stream Power Theory. *Water Resour. Res.* **1986**, *22*, 1350–1360. [CrossRef]
126. Moore, I.D.; Wilson, J.P. Length-slope factors for the Revised Universal Soil Loss Equation: Simplified method of estimation. *J. Soil Water Conserv.* **1992**, *47*, 423–428.
127. Pham, B.T.; Bui, D.T.; Prakash, I.; Dholakia, M. Hybrid integration of Multilayer Perceptron Neural Networks and machine learning ensembles for landslide susceptibility assessment at Himalayan area (India) using GIS. *Catena* **2017**, *149*, 52–63. [CrossRef]
128. Kadavi, P.R.; Lee, C.-W.; Lee, S. Application of Ensemble-Based Machine Learning Models to Landslide Susceptibility Mapping. *Remote Sens.* **2018**, *10*, 1252. [CrossRef]
129. Pourghasemi, H.R.; Gayen, A.; Park, S.; Lee, C.-W.; Lee, S. Assessment of Landslide-Prone Areas and Their Zonation Using Logistic Regression, LogitBoost, and NaïveBayes Machine-Learning Algorithms. *Sustainability* **2018**, *10*, 3697. [CrossRef]
130. Wang, Q.; Li, W.; Chen, W.; Bai, H. GIS-based assessment of landslide susceptibility using certainty factor and index of entropy models for the Qianyang County of Baoji city, China. *J. Earth Syst. Sci.* **2015**, *124*, 1399–1415. [CrossRef]
131. Saputra, A.; Gomez, C.; Hadmoko, D.S.; Sartohadi, J. Coseismic landslide susceptibility assessment using geographic information system. *Geoenviron. Disasters* **2016**, *3*, 77. [CrossRef]
132. Hong, H.; Liu, J.; Bui, D.T.; Pradhan, B.; Acharya, T.D.; Pham, B.T.; Zhu, A.-X.; Chen, W.; Bin Ahmad, B. Landslide susceptibility mapping using J48 Decision Tree with AdaBoost, Bagging and Rotation Forest ensembles in the Guangchang area (China). *Catena* **2018**, *163*, 399–413. [CrossRef]

133. Gomi, T.; Sidle, R.C. Bed load transport in managed steep-gradient headwater streams of southeastern Alaska. *Water Resour. Res.* **2003**, *39*, 77. [CrossRef]
134. Constantine, J.A.; Pasternack, G.B.; Johnson, M.L. Logging effects on sediment flux observed in a pollen-based record of overbank deposition in a northern California catchment. *Earth Surf. Process. Landforms* **2005**, *30*, 813–821. [CrossRef]
135. Luo, X.; Lin, F.; Zhu, S.; Yu, M.; Zhang, Z.; Meng, L.; Peng, J. Mine landslide susceptibility assessment using IVM, ANN and SVM models considering the contribution of affecting factors. *PLoS ONE* **2019**, *14*, e0215134. [CrossRef] [PubMed]
136. Jamal, M.; Mandal, S. Monitoring forest dynamics and landslide susceptibility in Mechi–Balason interfluves of Darjiling Himalaya, West Bengal using forest canopy density model (FCDM) and Landslide Susceptibility Index model (LSIM). *Model. Earth Syst. Environ.* **2016**, *2*, 1–17. [CrossRef]
137. Meneses, B.M.; Pereira, S.; Reis, E. Effects of different land use and land cover data on the landslide susceptibility zonation of road networks. *Nat. Hazards Earth Syst. Sci.* **2019**, *19*, 471–487. [CrossRef]
138. Gattinoni, P.; Scesi, L. Lanslide hydrogeological susceptibility of Maierato (Vibo Valentia, Southern Italy). *Nat. Hazards* **2013**, *66*, 629–648. [CrossRef]

© 2020 by the author. Licensee MDPI, Basel, Switzerland. This article is an open access article distributed under the terms and conditions of the Creative Commons Attribution (CC BY) license (http://creativecommons.org/licenses/by/4.0/).

Article

Competition and Burn Severity Determine Post-Fire Sapling Recovery in a Nationally Protected Boreal Forest of China: An Analysis from Very High-Resolution Satellite Imagery

Lei Fang [1], Ellen V. Crocker [2], Jian Yang [2] Yan Yan [3], Yuanzheng Yang [3] and Zhihua Liu [1,*]

1. CAS Key Laboratory of Forest Ecology and Management, Institute of Applied Ecology, Chinese Academy of Sciences, Shenyang 110016, China; fanglei@iae.ac.cn
2. Department of Forestry and Natural Resources, TP Cooper Building, University of Kentucky, Lexington, KY 40546, USA; e.crocker@uky.edu (E.V.C.); jian.yang@uky.edu (J.Y.)
3. Key Laboratory of Environment Change and Resources Use in Beibu Gulf, Guangxi Teacher Education University, Nanning 530001, China; yanyan@gxtc.edu.cn (Y.Y.); yangyz@gxtc.edu.cn (Y.Y.)
* Correspondence: liuzh@iae.ac.cn; Tel.: +86-024-8397-0327

Received: 12 February 2019; Accepted: 1 March 2019; Published: 13 March 2019

Abstract: Anticipating how boreal forest landscapes will change in response to changing fire regime requires disentangling the effects of various spatial controls on the recovery process of tree saplings. Spatially explicit monitoring of post-fire vegetation recovery through moderate resolution Landsat imagery is a popular technique but is filled with ambiguous information due to mixed pixel effects. On the other hand, very-high resolution (VHR) satellite imagery accurately measures crown size of tree saplings but has gained little attention and its utility for estimating leaf area index (LAI, m^2/m^2) and tree sapling abundance (TSA, seedlings/ha) in post-fire landscape remains untested. We compared the explanatory power of 30 m Landsat satellite imagery with 0.5-m WorldView-2 VHR imagery for LAI and TSA based on field sampling data, and subsequently mapped the distribution of LAI and TSA based on the most predictive relationships. A random forest (RF) model was applied to assess the relative importance and causal mechanisms of spatial controls on tree sapling recovery. The results showed that pixel percentage of canopy trees (PPCT) derived from VHR imagery outperform all Landsat-derived spectral indices for explaining variance of LAI (R^2_{VHR} = 0.676 vs. $R^2_{Landsat}$ = 0.427) and TSA (R^2_{VHR} = 0.508 vs. $R^2_{Landsat}$ = 0.499). The RF model explained an average of 55.5% (SD = 3.0%, MSE = 0.382, N = 50) of the variation of estimated LAI. Understory vegetation coverage (competition) and post-fire surviving mature trees (seed sources) were the most important spatial controls for LAI recovery, followed by burn severity (legacy effect), topographic factors (environmental filter) and nearest distance to unburned area (edge effect). These analyses allow us to conclude that in our study area, mitigating wildfire severity and size may increase forest resilience to wildfire damage. Given the easily-damaged seed banks and relatively short seed dispersal distance of coniferous trees, reasonable human help to natural recovery of coniferous forests is necessary for severe burns with a large patch size, particularly in certain areas. Our research shows the VHR WorldView-2 imagery better resolves key characteristics of forest landscapes like LAI and TSA than Landsat imagery, providing a valuable tool for land managers and researchers alike.

Keywords: forest recovery; resilience; leaf area index; tree sapling abundance; burn severity; boreal forests; WorldView-2

1. Introduction

Wildfire has been well recognized as a crucial process governing the dynamics of forest structure, composition and function in boreal forests [1,2]. Severe burns can cause abrupt changes to forest ecosystem by killing living plants, consuming organic matter, and altering biophysical environments [3–5]. In general, forest ecosystems are resilient and able to recover key functional and compositional attributes after disturbances [6–8]. However, the resilience of boreal forests is decreasing due to a misalignment between changing fire regimes and vulnerable ecosystem states [9–11]. Biome transitions in boreal forests, including shifts from forests to treeless steppes and an increase in deciduous trees, are thought to be exacerbated by changing fire regimes [12–15]. In addition, both fire frequency and severity in boreal ecosystems is predicted to increase in the future [16–19]. Research on the spatial controls of forest restoration post-fire is essential for understanding how boreal ecosystems respond to changing fire regimes and advancing ecologically sound policies in boreal fire management and restoration.

Tree sapling regeneration is strongly correlated to the future successional trajectory of burned forests and thus can be used to anticipate future compositional and structural dynamics [20,21]. Structural attributes of regeneration observed in the field, such as abundance, biomass, and species composition, can be used to determine the strength of post-fire forest recovery [21–24]. Field-based inventory approaches can provide first-hand information, but this is a labor-intensive and time-consuming process with limited utility for long-term monitoring over a broad spatial scale. On the other hand, remote sensing is a novel and cost-effective technique for monitoring forest ecosystems [25], as well as for evaluating fire-related characteristics such as burn severity [26,27], burned area [28], and monitoring post-fire vegetation recovery [29–31].

Satellite imagery for monitoring post-fire vegetation dynamics assumes that the remotely sensed surface reflectance can capture spectral signal variations that correspond to vegetation recovery. Vegetation indices were designed to reflect vegetation photosynthesis and were widely accepted as surrogates of canopy attributes, biomass, or tree coverage across global forest ecosystems [30,32]. However, spectral variation as exhibited in satellite imagery is not a biophysical parameter that can precisely characterize structure or function of forest ecosystems [33]. Forest recovery is typically described ambiguously in terms of changes in vegetation index or greenness in remote sensing literatures, however little attention has been paid to investigate whether such remotely sensed spectral information can reflect key forest structures in newly established boreal forest stands [34,35]. On the other hand, structural attributes rather than spectral variations are the critical indicators of forest ecosystem functions (i.e., carbon sequestration and water budgets) [36,37], and are thus more important for assessing forest recovery for management and scientific perspectives.

Tree sapling abundance (TSA), the number of tree saplings per unit area, is a crucial attribute of the post-fire forest stands as it determines the future successional trajectory of forests [38,39]. Burned areas with high TSA have a higher possibility of developing as forests through succession and recovery processes, while burned areas with few tree saplings may take a longer time to approach canopy closure. The spatial distribution of TSA is thus a critical reference for directing artificial management measures that aim to promote forest restoration post-fire. Leaf area index (LAI) is another structural attribute with critical importance to forest productivity and biomass accumulation through influencing forest canopy photosynthesis [40,41]. LAI can be measured rapidly in the field and is one of the few forest structures that is well-linked to optical remote sensing at various spatial scales [42,43].

Spatially monitoring these two structural attributes is essential for understanding the functional dynamics of post-fire forest ecology but presents a challenge for traditional remote sensing approaches. In post-fire landscapes, tree sapling distribution usually exhibits a high degree of heterogeneity as a result of interactions among seed availability (legacy effects), filter effects from environmental factors, inter- and intraspecific competition and edge effects [23,44]. Previous studies aimed at disentangling these effects were dependent on field investigations but very high-resolution (VHR) satellite imagery, which can provide fine details of post-fire landscapes (<1 m), is a promising approach that has not previously been utilized for tree saplings. The high resolution of VHR imagery is significant in allowing for accurate assessment of tree sapling crown size, but very few studies have investigated the utility of VHR imagery for monitoring forest structural attributes in newly reestablished stands.

Based on this, we sought to address the following questions: (1) How do the performances of VHR and Landsat imagery differed for delineating the spatial distribution of TSA and LAI in the early post-fire landscape; (2) which spatial controls (interspecific competition, legacy effects, environmental filters or edge effects) are most important in the recovery of LAI and TSA post-fire; and (3) how do these spatial control factors influence (inhibiting or facilitating) spatial distribution of LAI and TSA post-fire?

2. Materials and Methods

2.1. Study Area

The study was conducted in Huzhong National Natural Reserve of Great Xing'an Mountains (Figure 1), where primary forests are well protected since all logging has been prohibited since 1958. In this area, post-fire forests follow a natural recovery process and successional trajectory without artificial seeding or plantation. A strict ban on logging of natural forests has been enforced across the Great Xing'an Mountains since April 2014, thus historical fires in the nature reserve will provide valuable insight for understanding how resilient this forest is in responding to wildfire and for refining post-fire management practices. The particular fire (Lat/Lon: 122°50′E, 51°53′N) we studied was ignited by lightning on 17 June 2000 and lasted for seven days [45]. It burned an area approximately 7735 ha, most of which was moderate-high severity with high tree mortality rates [26,39].

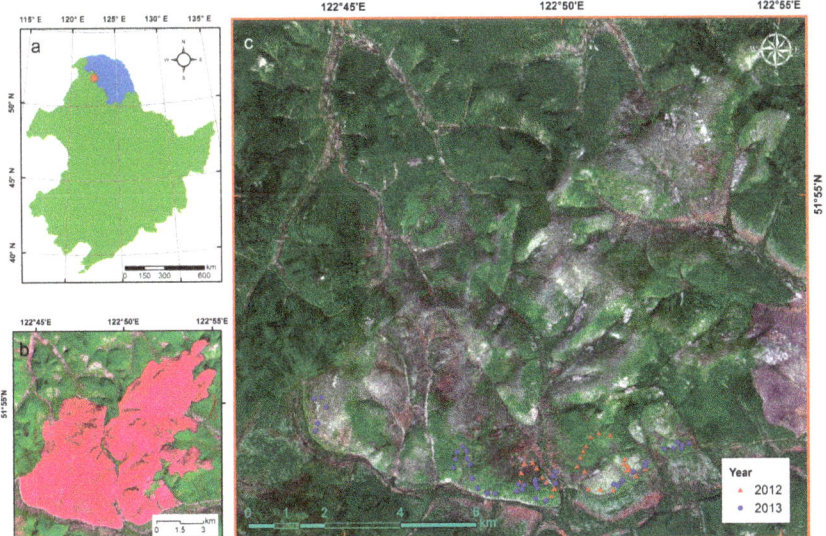

Figure 1. Spatial location of the study area (**a**, red dot) in Great Xing'an Mountains (**a**, in blue) of northeastern China (**a**, in green). The false color Landsat image (**b**, R: TM7, G: TM4, B: TM3) acquired on 24 May 2002 exhibits an overview of two-year post-fire landscape, while the WorldView-2 image (**c**) shows the post-fire landscape after 14 years recovery. The red and blue dots are field plots explored in 2012 and 2013, respectively.

The pre-fire forests were dominated by Dahurian larch (*Larix gmelinii*), a deciduous conifer, and a few deciduous broadleaf species of birch (*Betula platyphylla*) and aspen (*Populus davidiana* and *Populus suaveolens*) [46]. Some evergreen conifers including Scotch pine (*Pinus sylvestris var. mongolica*) and Korean spruce (*Picea koraiensis*) were also sparsely distributed. The understory layer was the primary fuel bed and source of ladder fuels [45], which consisted of evergreen shrubs (e.g., *Pinus pumila*, *Ledum linnaeus* and *Vaccinium vitisidaea*), deciduous shrubs (e.g., *Betulafruticosa* and *Rhododendron dauricum linnaeus*) and some herbaceous plants (e.g., *Chamaenerion angustifolium*, *Carex appendiculata* and *Rubus linnaeus*) that varied with edaphic and topographic conditions [47]. This area has a relatively short growing seasons (150 days) and frost-free period (~80–100 days), with a mean annual temperature of approximately −4.7 °C and mean annual precipitation of about 460 mm [26]. The topography is mountainous with elevation ranges from 760 m to 1300 m.

2.2. Field Data Collection

For field data collection, we focused our attention on two structural attributes of post-fire vegetation, TSA and LAI, which can be easily evaluated in the field. In summers of 2012 and 2013, our field crews collected 70 plots of sapling inventory data for both TSA and LAI (Figure 1), with plot size set to 30 m × 30 m in order to match the spatial resolution of Landsat imagery and to minimize potential scale issues. This will improve the accuracy of linkage between in situ measurement and Landsat spectral properties. All sites were at least 150 m from roads to eliminate edge effect and were selected according to severity and topographic positions. Within each plot, we set three 5 m × 5 m quadrants along the diagonal (~42 m) to survey sapling stems with a height greater than 1.5 m (Figure 2). We did not distinguish between tree species (although most were white birch and larch) as they were not differentiable in VHR imagery due to very similar spectral signature and highly mixed in situ. All saplings were counted based on the number of stems (>1.5 m) because asexual resprouting of white birch is common. TSA was calculated as the mean number of saplings and was normalized to saplings per hectare.

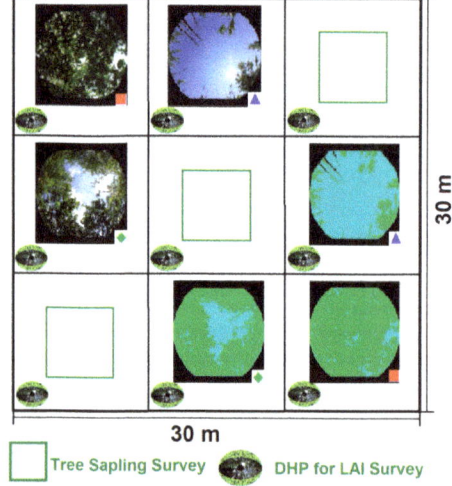

Figure 2. Layout of sampling plot. Three 5 m × 5 m quadrates are used for tree sapling abundance survey. Nine fish-eye photos were used for estimation of leaf area index. Three upward digital hemispherical photography (DHP, top-left) pictures show examples of in situ canopy measurement, while tree processed pictures (low-right) show corresponding classification results (see marks) in CAN-EYE software. The black areas in mid-right picture (blue triangle) represent masked dead standing stems in corresponding DHP picture (top-middle).

LAI was measured using a digital hemispherical photography (DHP) system using a Canon EOS 60D Digital Single Lens Reflex camera and a Sigma 8-mm F3.5 EX DG Fisheye lens. We followed the user manual of CAN-EYE software (Version 6.3.3, National Institute of Agronomical Research, Toulouse, France) to calibrate our DHP system and derive the parameters for establishing the projection functions [48]. Using a 1920 × 1280 pixels parameter for digital photograph, our calibration process generated accurate projection functions that are comparable with results as shown in user manual (see Appendix A). The DHP system was installed to a tripod with a fixed 0.70 m height. Within each sampling plot, we took nine upward digital photographs distributed as shown in Figure 2. Following the recommendation of the user manual, the limit of the circle of interest was set as 60° in CAN-EYE software to avoid mixed pixels effects in the blended photograph edges. Other input parameters were set as either default values or derived from photographs automatically. Given this procedure, dead standing stems with large diameter had the potential to influence the LAI estimation, so we manually masked them during the image process in CAN-EYE software to reduce the overestimation. The canopy of surviving trees was retained, however, because their greenness was involved in Landsat spectral signals and not possible to isolate.

2.3. Remote Sensing for Estimating Vegetation Coverage

2.3.1. Remote Sensing Data Process

The cloud-free WorldView-2 (DigitalGlobe. Inc., Herndon, VA, USA, Figure 1) image was acquired on 1 June 2014 at 2 m spatial resolution including four multispectral bands (Blue 450–510 nm; Green 510–580 nm, Red 705–745 nm, and near infrared (NIR) 860–1040 nm) and 0.5 m spatial resolution for panchromatic band (450–800 nm). We used the Pan Sharpening toolbox in ENVI 5.3 software (Harris Geospatial Solution. Inc., Washington, DC, USA) to perform image fusion, and improved the spatial resolution of multispectral bands to 0.5 m. The WorldView-2 imagery had been geo-spatially projected using the Universal Transverse Mercator (UTM) coordinate system based on WGS-84 ellipsoid with a nominal positioning accuracy of 3.5 m. It is accurately overlapped with Landsat imageries, which have 30-m spatial resolution.

To produce time-synchronous comparisons between Landsat and field survey, the spectral indices were extracted according to the year when the TSA and LAI data were collected in the field. In addition, we used the same field data for the Landsat-derived indices of 2014 in order to form an unbiased comparison with WorldView-2 data. We obtained 8 Landsat-5 TM, Landsat-7 ETM+, and Landsat-8 OLI surface reflectance images (path-row: 122/24 and 121/24, time: 2010, 2012–2014) from the EROS Science Processing Architecture (ESPA) system of the USGS (Table 1). Atmospheric and topographic correction had been carried out for these before distribution. Due to the lack of cloud-free days over the study area during the time periods assessed, we used "cloud masking" to remove cloud and shadow pixels [49], and then carried out image mosaic to fill data-gap for each year. Although the intra-annual Landsat imageries have very close observation dates, we still applied the iteratively re-weighted multivariate alteration detection (IR-MAD) approach to minimize spectral inconsistency among Landsat sensors [50,51]. The Landsat-5 imagery of 2011 was used as the reference imagery for band-by-band radiometric normalization. The mean correlation coefficients (for 6 bands) of four target imageries were each higher than 0.95, indicating good performance of IR-MAD for radiometric normalization.

Table 1. Detailed information of Landsat scenes. Invalid pixels are cloud, shadow and null value pixels.

Satellite	Sensor	Acquisition Date	Path-Row	Percentage of Invalid Pixels	Usage
Landsat-5	TM	5 September 2011	122-24	3.6%	Reference Image
Landsat-7	ETM+	30 August 2012	122-24	21.7%	Spectral Indices
Landsat-7	ETM+	8 September 2012	121-24	33.1%	Spectral Indices
Landsat-8	OLI	25 August 2013	122-24	36.8%	Spectral Indices
Landsat-8	OLI	3 September 2013	121-24	13.3%	Spectral Indices
Landsat-8	OLI	21 August 2014	121-24	33.2%	Spectral Indices
Landsat-7	ETM+	29 August 2014	121-24	35.0%	Spectral Indices
Landsat-7	ETM+	5 September 2014	122-24	31.0%	Spectral Indices

2.3.2. Landsat-Derived Spectral Indices

Vegetation indices developed from remote-sensed Red and NIR bands, such as the Normalized Differenced Vegetation Index (NDVI), are often used as indicators of vegetation recovery in many forest ecosystems [31]. But saturation issues with vegetation indices are widely reported [34,52], especially in areas covered with dense vegetation, which may limit the strength of vegetation indices for detecting structural changes related to forest recovery [30]. Spectral indices derived from the shortwave infrared (SWIR) bands, which are sensitive to water content in vegetation foliage, are proving to be good indicators of post-fire recovery in many forest ecosystems [34,53,54]. In addition, components of Tasseled Cap (TC) transformation (i.e., brightness, greenness and wetness) and their modified spectral indices are also commonly used for monitoring forest disturbance and forest recovery [55]. Different indices may provide different perspectives of forest recovery. Therefore, it is critical to elucidate which Landsat-derived spectral indices have the feasibility to depict post-fire tree sapling attributes in our ecosystem. Such efforts can be helpful to provide an in-depth understanding of what remote sensing indices mean in the field. To comprehensively investigate the performance of Landsat-derived spectral indices on estimating TSA and LAI, we calculated nine commonly-used spectral indices (Table 2) to analyze their explanatory strength.

Table 2. Formulas of nine Landsat-derived spectral indices. Spectral Index Abbreviation: NDVI—Normalized Difference of Vegetation Index; EVI—Enhanced Vegetation Index [56]; SAVI—Soil Adjusted Vegetation Index [52]; MSAVI—Modified Soil Adjusted Vegetation Index [52]; NBR—Normalized Burn Ratio [54]; NDMI—Normalized Difference Moisture Index [57]; TCA—Tasseled Cap Angle [55]; TCW—Tasseled Cap Wetness. Note: spectral bands are different between Landsat5/7 and Landsat 8.

Spectral Index	Spectral Index
$NDVI = \frac{\rho_{NIR}-\rho_{Red}}{\rho_{NIR}+\rho_{Red}}$	$NBR = \frac{\rho_{NIR}-\rho_{SWIR2}}{\rho_{NIR}+\rho_{SWIR2}}$
$EVI = \frac{2.5\times(\rho_{NIR}-\rho_{Red})}{\rho_{NIR}+6\times\rho_{Red}-7.5\times\rho_{Blue}+1}$	$NDMI = \frac{\rho_{NIR}-\rho_{SWIR1}}{\rho_{NIR}+\rho_{SWIR1}}$
$EVI2 = \frac{2.5\times(\rho_{NIR}-\rho_{Red})}{\rho_{NIR}+2.4\times\rho_{Red}+1}$	$TCA = \arctan(\frac{TC_{Greenness}}{TC_{Brightness}})$
$SAVI = \frac{\rho_{NIR}-\rho_{Red}}{1.5\times(\rho_{NIR}+\rho_{Red}+0.5)}$	$TCW = 0.1446\times \rho_{Blue} + 0.1761\times\rho_{Green} + 0.3322\times\rho_{Red} + 0.3396\times\rho_{NIR} - 0.621\times\rho_{SWIR1} - 0.4186\times\rho_{SWIR2} - 3.3828$
$MSAVI = 0.5\times(2\times\rho_{NIR}+1-\sqrt{(2\times\rho_{NIR}+1)^2 - 8\times(\rho_{NIR}-\rho_{Red})})$	

2.3.3. Image Textures of WorldView-2 Imagery

The WorldView-2 has limited spectral bands that cannot sufficiently support to calculate much spectral indices, but it offers image textures to improve the land cover mapping [58,59] and estimation of forest structures [60,61], yet its performance in estimating attributes of sapling tree was not verified. Using the panchromatic band of WorldView-2 imagery, we calculated 13 image texture variables, which consist of five occurrence measures and eight co-occurrence measures. The occurrence measures directly use the number of occurrences of each gray level within a given moving window for calculating summaries of statistics (i.e., Range, Mean, Variance, Entropy and Skewness). The co-occurrence measures are second-order statistical calculations involving the spatial relationships among neighboring pixels. The grey level co-occurrence matrix, which is a function of both the angular relationship and distance between a central pixel and its neighboring pixels, was used to calculate eight statistical measures including mean, variance, homogeneity, contrast, dissimilarity, entropy, angular second moment (ASM), and correlation.

Calculation of image texture is influenced by the window size as it determines the number of surrounding pixels. Wood, et al. [62] suggested that a small window size for image texture analysis may best match the spatial scale at which the forest structure varies. We tested three kinds of window size, including 5 × 5, 15 × 15 and 25 × 25, but did not find notable differences for subsequent analysis. We chose a 5 × 5 window size for all texture variable calculations because it matches the spatial size of a single tree sapling and since it was the least costly in computing time. We did not calculate image texture for Landsat imagery because our field plots had identical spatial size with Landsat pixel, which means the measured TSA and LAI can only reflect within-pixel (30 m) variations. All image texture variables were computed using ENVI 5.3 software (Harris Geospatial Solution. Inc., Herndon, VA, USA).

2.3.4. Land Cover Mapping of WorldView-2 Imagery

We used the per-pixel classification method to obtain the land cover information from the 0.5 m WorldView-2 imagery. According to our field knowledge and visual interpretation of the imagery, we used a classification scheme consisting of 9 popular objects in the field during the preliminary analysis. We used the support vector machine (SVM) approach to carry out the land cover classification. The SVM is a supervised classifier that has been widely applied for land cover and land use mapping through the local to global spatial scales [63,64]. Training pixels were selected with very high confidence based on our knowledge. We chose over 3000 training pixels for each land cover type in order to provide sufficient model training and avoid overfitting problem [65]. We calculated the Jeffries–Matusita distance (JMD) to evaluate the separability of training pixels, defined as:

$$JMD_{xy} = 2\left(1 - e^{-B}\right)$$

where B is the Bhattacharyya distance, which is usually used for indicating dissimilarity between two distributions:

$$B = \frac{1}{8}(x-y)^t \left(\frac{\sum x + \sum y}{2}\right)^{-1}(x-y) + \frac{1}{2}ln\left(\frac{\left|\frac{\sum x + \sum y}{2}\right|}{\sqrt{|\sum x| \times |\sum y|}}\right)$$

In this equation, x and y are two vectors of spectral and textural signatures and $\sum x$ and $\sum y$ are covariance matrix of x and y respectively [66]. The JMD will be close to 2 when spectral and textural signatures of two classes are completely different (high separability), and close to 0 when spectral and textural signatures are identical (low separability).

Classification accuracy was evaluated based on additional pixels (~1000 pixels for each class) that were independently selected from the training pixels. Using the confusion matrix approach [67], we calculated the Cohen's Kappa coefficient to evaluate overall classification accuracy, and used the omission error and commission error to evaluate accuracy of individual class. The ability to distinguish tree saplings from the surviving trees was desired to refine our research. However, we found low separability between these two classes (Table 3). We also found low separability between grasslands and shrublands and therefore combined those classes for the further analysis (but reported accuracies evaluated based on the nine-class scheme).

2.4. Statistical Analysis

2.4.1. Compare Performance of Landsat and WorldView-2 on Predicting TSA and LAI

To match the spatial size of field data, we conducted spatial aggregation to obtain statistical measures for information derived from WorldView-2 data. For the land cover map, we used a moving window (60 × 60 pixels), whose spatial size is consistent with spatial resolution of Landsat imagery, to compute the percentage of pixels that were classified as the canopy trees (i.e., tree saplings and mature trees), denoted as PPCT. For image texture, we conducted spatial aggregation to obtain the mean value for each texture variable. Given high collinearity among image texture variables, pairwise Pearson correlation coefficients (r) calculated from the "Hmisc" package in R 3.4.1 (R Development Core Team 2017, Boston, MA, USA) were used as the criteria based on randomly sampled 300 pixels.

We used a forward stepwise method combined with variance inflation factors (VIF) functions in the "car" package to drop variables with high collinearity by applying a stringent threshold of 3, following recommendation of Zuur, et al. [68]. Landsat spectral indices often show collinearity as their computation may use the same spectral bands. Because of this we only retained two co-occurrence texture variables (i.e., mean and ASM) and one occurrence texture variable (i.e., Entropy) as they represented low correlations ($|r| < 0.50$). We retained all spectral indices for the further analysis to allow us to investigate the usage of a spectral index in monitoring post-fire vegetation dynamics.

The presence of spatial autocorrelation in dependent variables may violate the assumption that all observations are independent, which will inflate the significance and affect the coefficient estimates. Using functions in the "ape" package [69], we calculated the Moran's I index to examine spatial autocorrelations in our measured TSA and LAI respectively. Moran's I is similar to a correlation coefficient ranging between −1 and 1. Higher positive values indicate greater similarity, while lower negative values indicate stronger dissimilarity. Autocorrelation was found significant at $\alpha < 0.05$ level, but very weak for both TSA (Moran's I = 0.18) and LAI (Moran's I = 0.20). It suggested our field observations on TSA and LAI have low spatial dependence, and the subsequent analysis will not be strongly influenced by spatial autocorrelation.

The coefficient of determination (R^2) was used to evaluate the explanatory power of remotely sensed variables for variations of field-measured TSA and LAI. Because the LAI is partially determined by tree density, it is not surprising that our LAI data represented very high correlation ($r = 0.783$, $p < 0.01$) with TSA data. We retained both variables for the remote sensing analysis as they reflect different perspectives of forest structure and may be sensitive to different spectral bands. Before parameterizing the regression model, we applied the Shapiro-Wilk test and histogram approach to detect whether the normality assumption is violated [70]. If the null hypothesis (i.e., samples came from a normally distributed population) was rejected at $\alpha < 0.05$ level, we applied logarithm transformation to mitigate skew. Homogeneity of variance was evaluated using the Fligner–Killeen test [71]. The test results indicated that the null hypothesis (i.e., all populations variances are equal) was not violated in any regression model. Outliers (Cook's distance greater than four times of the mean) were detected and removed. The Landsat indices and WorldView-2 variables with the highest R^2 were used to predict TSA and LAI through the post-fire landscape.

2.4.2. Evaluation of Relative Importance Using Random Forest Model

We used the random forest (RF) model in the "randomForest" package to evaluate the relative importance of spatial controls on determining TSA and LAI. The RF model is a machine learning algorithm with advantages for dealing with nonlinear relationships, multi-collinearity, and complex interactions without imposing assumptions on data distribution [72]. Based on a bootstrap subsampling (bagging) scheme, the RF model can generate multiple regression trees with low variance that can be combined for an accurate prediction. The best split of node is chosen from the random subsets of predictor variables to ensure that all predictors are tested. For each individual tree, the remaining (i.e., out-of-bag, OOB) data that not drawn into training subset is used for unbiased model validation [73]. We used the variance explained (R^2) to evaluate the goodness of fit of RF model for training data. It is calculated based on the internal OOB error rate in terms of mean of squared residuals (MSE):

$$MSE = n^{-1} \sum_{1}^{n} (y_i - \hat{y}_i^{OOB})^2$$

where $\hat{y}_i^{OOB}$ is the average of the OOB predictions for observation i. The R^2 is calculated as:

$$R^2 = \frac{MSE}{\hat{\sigma}_y^2}$$

The most important variable will cause highest degradation on model fit when it is omitted.

Predictors were spatial variables representing gradients of burn severity, topography and understory vegetation abundance (Table 3). Burn severity was evaluated based on a quadratic correlation model (R^2 = 0.856), which was developed from the difference between the normalized burned ratio (dNBR) and field-sampled severity measures [26]. We applied a 30 m resolution digital elevation model (DEM) to generate topographic variables reflecting elevation, topographic relief, topographic wetness, and the total solar radiation of growing season (April to October). Understory vegetation abundance was derived from WorldView-2 land cover map. Similar to PPCT, we calculated the coverage of understory vegetation (i.e., shrubland and grassland) using a 30 m × 30 m moving window aggregation approach. We also found that the shadow pixels largely reflected mature trees in our WorldView-2 imagery. We used the coverage of tree shadow as a surrogate to represent the number of surviving mature trees post-fire. This is similar to Berner, et al. [74], who applied the tree shadow coverage as a surrogate of tree biomass. To understand the edge effects of unburned areas on post-fire recovery, we also calculated the nearest distance to the edge of unburned area as a predictor. The remote sensed TSA and LAI were used as the response variables for two RF models respectively. We generated random samplings based on a 150 m sampling space to balance the spatial autocorrelation and the maximum sampling number (~500 points). To reduce the risk of stochastic errors and create stable model outputs, we carried out 50 RF modeling trials independently and used the average value as the final result.

Table 3. Description of predictors used in random forest models.

Predictor	Category	Description
dNBR	Legacy effect	Indicator of burn severity, calculated based on bi-temporal difference of NBR. Higher dNBR values represent higher tree mortality and more combustion of surface organic matters.
Shadow	Legacy effect	Surrogate amount of surviving trees post-fire. This is derived from the 0.5 m land cover map of WorldView-2. Higher shadow coverage indicates more surviving trees (and likely higher seed availability).
Elevation	Topography filter	Altitude of a given site.
TPI	Topography filter	Topographic position index. A positive TPI value indicates a higher altitude than neighborhood pixels, while a negative TPI value indicates a lower altitude than surrounding areas. A TPI value of 0 indicates a flat area or an area near mid-slope.
TWI	Topography filter	Topographic wetness index. TWI values typically range from 3 to 30. Higher TWI values indicate high soil moisture potential.
Solar radiation	Topography filter	Incoming solar radiation from a raster surface during the growing season. Higher values indicate higher exposure to solar radiation. Southern-facing slopes usually have higher solar radiation.
Understory coverage	Competition	Percentage of pixels classified as understory (grasslands and shrublands) in land cover map of WorldView-2 for each 30 m × 30 m site. Higher understory coverage indicates more space occupied by understory plants. Understory plants do not include tree sapling here.
Nearest Distance	Edge Effect	The nearest distance to unburned areas. It reflects the potential of a given site to receive seed source from unburned areas with proximity to unburned forests indicating a higher likelihood of receiving seeds.

3. Results

3.1. Accuracy Assessment of WorldView-2 Classification

Relatively high pairwise JMD values (Table 4) indicated very high spectral dissimilarity among the most of land cover classes. The overall accuracy of classification was 81.3% and the Cohen's Kappa coefficient was 0.790. Mature trees had strong spectral similarity with tree saplings, resulting in a high rate of commission errors (41.6%) for mature tree classification and high omission error (66.5%) for tree sapling classification. Similarly, spectral separability between shrublands and grasslands was also relatively low, preventing us from distinguishing the two classes in subsequent analyses. Mature trees and tree saplings together occupied about 23.6% of the total burned area, while shrublands and grasslands occupied approximately 51.9% of the total burned area (Figure 3). Bare rock, where vegetation has difficulty establishing, accounted about 10.2% of the total burned area. The remaining 14.3% consisted of bare soil, water bodies, shadow areas and moss.

Table 4. Pairwise Jeffries–Matusita distance (JMD, values within brackets) and confusion matrix for evaluating intra-classes separability and classification accuracies respectively. The bold represents the number of verified samples.

Class	Mature Tree	Tree Sapling	Bare Soil	Bare Rock	Grass-Land	Shrub-Land	Water Body	Shadow Area	Moss
Mature Tree	**958**	671			0	0		16	
Tree Sapling	(1.83)	**342**							
Bare Soil	(2.00)	(2.00)	**919**		1	59	1		
Bare Rock	(2.00)	(2.00)	103 (1.99)	**1029**					
Grassland	(1.97)	6 (2.00)	3 (2.00)	12 (1.99)	**403**	10			
Shrubland	42 (1.98)	1 (2.00)	6 (1.95)	(2.00)	661 (1.46)	**960**			43
Water Body	(2.00)	(2.00)	(2.00)	(2.00)	(2.00)	(2.00)	**1082**	109	
Shadow Area	(1.99)	(2.00)	(2.00)	(2.00)	(1.98)	(2.00)	(1.86)	**908**	
Moss	(2.00)	(2.00)	(1.97)	(2.00)	(2.00)	4 (2.00)	(2.00)	(2.00)	**1000**

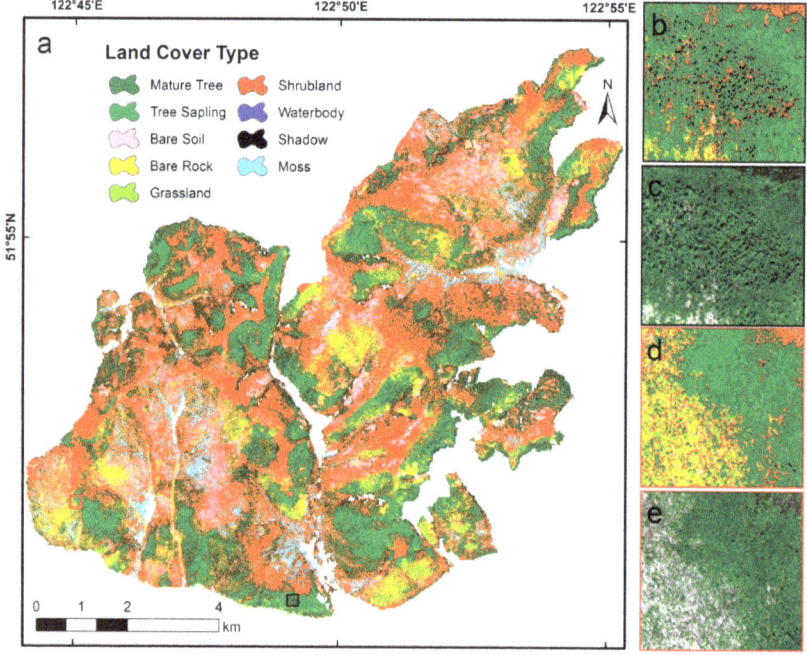

Figure 3. Land cover mapping of the burned area (**a**) based on WorldView-2 imagery. Small windows (**b**–**e**) show zoomed views of land cover maps (**b**,**d**) compared to the RGB imageries (**c**,**e**) for two sites (black box for **b**,**c**, red box from **d**,**e**) respectively.

3.2. Correlations between Remote Sensed Variables and LAI and TSA

All Landsat-derived indices exhibited significant correlations with LAI and TSA (Table 5). In two Landsat cases, the EVI2 (R^2 = 0.427, RMSE = 0.348, Figure 4a) and NBR (R^2 = 0.489, RMSE = 0.331, Figure 4d) were found the most explanative Landsat-derived indices for LAI, while the NBR (R^2 = 0.499, RMSE = 0.953, Figure 4b) and NDMI (R2 = 0.478, RMSE = 0.983, Figure 4e) were the most explanative variables for TSA. Spectral indices derived from SWIR bands (e.g., NBR and NDMI) generally performed better than vegetation indices (e.g., NDVI, EVI, and SAVI). The PPCT variable of

WorldView-2 (VHR) explained the highest proportions of variance for LAI ($R^2 = 0.676$, RMSE = 0.257, Figure 4c) and TSA ($R^2 = 0.508$, RMSE = 0.977, Figure 4f) among all variables, but image texture represented relatively weak correlations (Table 5).

Table 5. Coefficients of linear regression models for evaluating relationships between remotely sensed variables and leaf area index (LAI) and tree sapling abundance (TSA). Landsat-derived spectral index abbreviations see footnote of Table 2. PPCT—Pixel Percentage of Canopy Tree; Co_Means—Means of Co-occurrence Texture; ASM—Angular Second Moment. RMSE—Root Mean Square Error.

Satellite Imagery	Predictor	LAI $^\epsilon$			TSA $^\epsilon$		
		R^2 †	RMSE	Outliers	R^2	RMSE	Outliers
Landsat (2010–2013)	EVI $^\epsilon$	0.283 **	0.390	3	0.409 **	1.048	3
	EVI2	0.427 **	0.348	4	0.299 **	1.111	3
	MSAVI $^\epsilon$	0.282 **	0.396	2	0.251 **	1.188	1
	NBR	0.418 **	0.354	2	0.499 **	0.953	5
	NDMI	0.337 **	0.376	2	0.369 **	1.077	3
	NDVI $^\epsilon$	0.376 **	0.367	3	0.316 **	1.13	2
	SAVI $^\epsilon$	0.339 **	0.380	2	0.311 **	1.134	2
	TCA	0.405 **	0.354	4	0.351 **	1.076	4
	TCW	0.331 **	0.379	3	0.385 **	1.052	4
Landsat (2014)	EVI	0.171 **	0.412	3	0.173 **	1.196	3
	EVI2	0.377 **	0.359	2	0.322 **	1.09	2
	MSAVI $^\epsilon$	0.394 **	0.357	3	0.333 **	1.081	3
	NBR	0.489 **	0.331	5	0.450 **	1.022	3
	NDMI	0.385 **	0.360	3	0.478 **	0.983	4
	NDVI $^\epsilon$	0.374 **	0.360	2	0.324 **	1.088	2
	SAVI $^\epsilon$	0.368 **	0.362	2	0.304 **	1.105	2
	TCA	0.455 **	0.339	3	0.395 **	1.037	3
	TCW	0.333 **	0.379	3	0.384 **	1.053	4
WorldView-2 (2014)	PPCT $^\epsilon$	0.676 **	0.257	4	0.508 **	0.977	2
	Entropy $^\epsilon$	0.008	0.437	3	0.038	1.282	3
	Co_Means	0.008	0.437	4	0.002	1.275	3
	ASM $^\epsilon$	0.089 *	0.418	4	0.005	1.316	2

$^\epsilon$: Model fits using logarithmic transformation; †: Significance code: ** $p < 0.01$; * $p < 0.05$.

The spatial distributions of LAI (Figure 5a) and TSA (Figure 5b) 14 years post-fire were mapped based on the correlation models with highest R^2 in Table 5. We found about 51.1% (~4098 ha) of the burned area exhibited high LAI recovery (greater than 1), which corresponded to approximately 10,790~74,380 sapling/ha TSA. About 14.6% (~1169 ha) of the burned area exhibited moderate LAI recovery (between 0.5 and 1) of 4770~10,680 sapling/ha, and approximately 2745 ha of the burned area exhibited recovered poor recovery (LAI < 0.5, and TSA < 4630 sapling/ha).

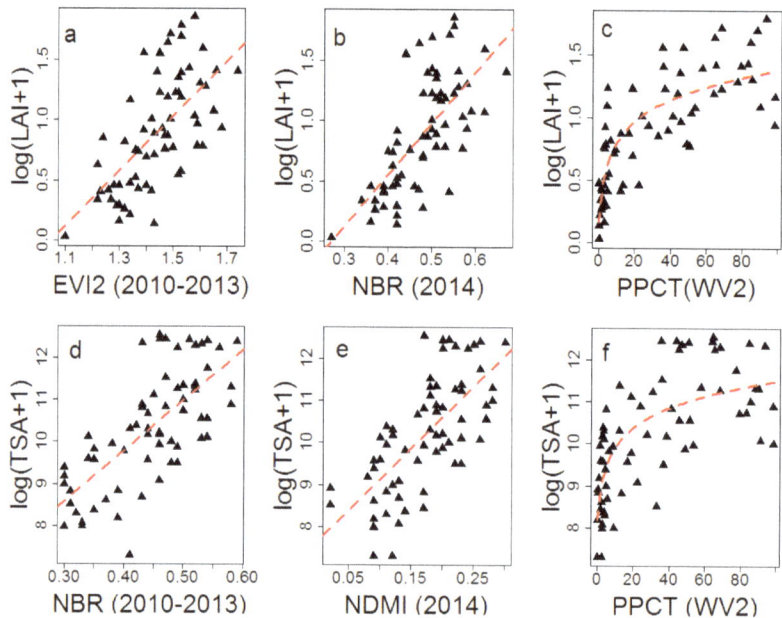

Figure 4. Scatter plots depict the best relationships (Table 4) between leaf area index (LAI) and two Landsat indices (**a**,**b**), and WroldView-2 case (**c**); and relationships between Tree sapling abundance (TSA) and two Landsat indices (**d**,**e**), and WroldView-2 case (**f**).

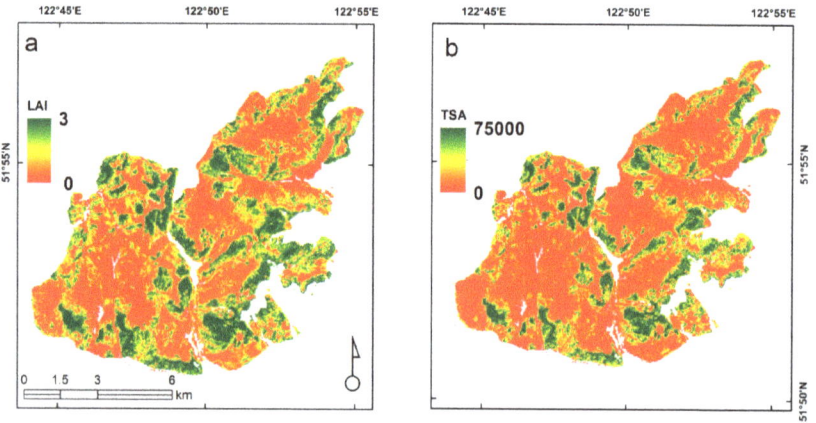

Figure 5. Two maps show spatial distribution of Leaf Area Index (LAI, **a**) and Tree Sapling Abundance (TSA, **b**) based on Pixel Percentage of Canopy Tree (PPCT) derived from WorldView-2 image.

3.3. Relative Importance of Predictors to LAI Recovery

Because WorldView-2 derived PPCT was the most predictive indicator for both LAI and TSA, and LAI and TSA are highly correlated with each other, the RF model produces the same results for both LAI and TSA. Thus we only represented the RF model results for LAI. The 50 RF models explained a maximum of 62.4% (Mean R^2 = 55.5%, SD = 3.0%, MSE = 0.382, N = 50) of the variation in estimated LAI. The coverage of understory vegetation and shadow pixels were the top-two most important predictor variables, decreasing MSE about 43.3% (SD = 4.6%) and 42.89% (SD = 3.8%) respectively when incorporated in RF models (Figure 6). The dNBR contributed about a 23.4% (SD = 2.9%) decrease in MSE, while contributions of topography variables ranged from 9.6% to 16.7%, followed by Solar Radiation (SD = 3.7%), Topographic Position Index (TPI) (SD = 2.3%), TWI (SD = 3.9%) and Elevation (SD = 2.1%). The edge effect was found to be the least important factor, contributing about a 7.2% (SD = 3.2%) decrease in MSE.

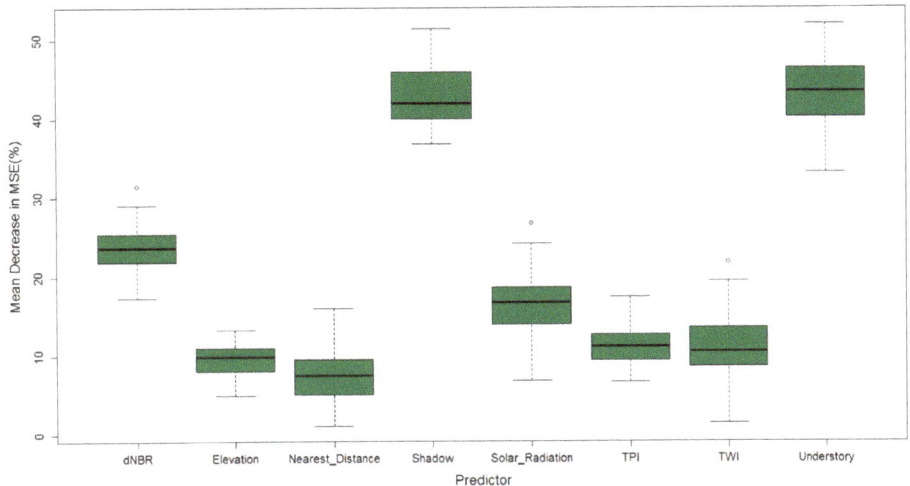

Figure 6. Relative importance from 50 random forest models, measured as the normalized difference between the mean square errors (MSE) when permuting the out-of-bag portion of the data and the MSE when permuting given variable. Magnitude of decrease of in MSE indicates relative importance of predictor variable (see Table 3 for variable abbreviations).

4. Discussion

4.1. Remotely Sensed Estimation of LAI and TSA

Our field investigation suggests that the LAI is highly correlated with TSA in early post-fire landscapes. Given that surveying TSA in the field is time-consuming and labor intensive, utilizing a DHP system can be a more efficient way to monitor post-fire tree sapling recovery. We found that all Landsat-derived spectral indices exhibited statistically significant correlations with both LAI and TSA, but the variance explained was usually less than 50%. EVI2 and NBR were the most predictive spectral indices for LAI and TSA respectively, but we found that NBR performed consistently well for simulating LAI and TSA in two Landsat analysis cases. This suggests that NBR may be the best choice for monitoring post-fire vegetation dynamics in our study area. A similar finding was also reported by Chen, et al. [75], who found that NBR is the most sensitive indicator of post-fire vegetation recovery in the Black Hills National Forest of the USA, suggesting that the value of NBR extends beyond our boreal forest study system. Previous studies have also shown that NBR is better than NDVI for quantifying burn severity [26,76], further supported by Kennedy, Yang and Cohen [54] and White, Wulder, Hermosilla, Coops and Hobart [53], who concluded that the NBR is a very useful indicator for detecting fire disturbance and monitoring post-fire vegetation trajectory.

Although Landsat-derived spectral indices exhibited significant correlations with the LAI and TSA, we found that the WorldView-2 (VHR) imagery performed even better. PPCT had a stronger explanatory power for LAI than for TSA because our DHP measurement of LAI consisted of canopy foliage of both surviving trees and tree saplings. The in-field upward canopy photograph results closely match PPCT derived from WorldView-2 imagery, further confirming the value of this technique. However, PPCT cannot wholly reflect TSA as there were surviving trees that exhibited disproportionately large canopy projection areas in the imagery when compared to tree saplings. The low spectral separability of surviving trees and tree saplings limited the applicability of WorldView-2 imagery for monitoring TSA.

Image textures are commonly used as auxiliary inputs to provide spatial features that spectral information cannot characterize. Many previous studies have reported that including image textures as predictors can improve the estimation of forest structural parameters such as tree diameter [77], stand density [61] and crown diameter [78]. However, image textures did not exhibit strong explanatory power to either LAI or TSA in our study. One reason is that crown shadow is an important factor in representing spatial variation as reflected in the imagery [61]. In contrast to the coarse surface of mature forests, the post-fire landscape where dominated by tree sapling that did not represent very strong shadow effects (contrasts) in the imagery (Figure 3c). In addition, the value of textural features has no explicit relationship to spectral features and the same value of textural attributes may represent very different land surface properties. For example, post-fire landscapes usually exhibit high homogeneity, however similar homogeneity values may be corresponding to different land cover compositions.

4.2. The Response of Forest Recovery to Spatial Controls

We found that including surrogates of intra-specific competition (i.e., understory coverage) and density of surviving trees (i.e., shadow coverage) greatly improved model fit. According to partial dependence curves, we found a negative effect of understory (grass and shrub species) coverage on tree sapling regeneration (Figure 7a). Understory species usually act as pioneers in post-disturbance landscapes and rapidly occupy available spaces and resources. In addition, some understory species are resistant to wildfire and recover rapidly. By comparison, tree saplings may require longer timespans to establish; they will compete with understory species for available resources and may be limited by seed availability.

Tree sapling regeneration (estimated by LAI) is very sensitive to shadow coverage, suggesting that even a few surviving trees can strongly promote post-fire recovery (Figure 7b). The nearest distance to unburned area, which reflects the importance of seed inputs from out of the burned area, significantly impacted LAI recovery with areas near the edge of boundaries having a higher opportunity to receive seeds from unburned areas. Our results showed that edge effects extend for a distance of approximately 700 m for this fire (Figure 7h), which can partially explain why the core area exhibited poor tree recovery. Trees that survive fire are thus critical for supporting reestablishment of tree community post-burn [79,80]. Nonetheless, if the number of surviving trees was over a certain threshold, canopy closure was negatively affected because our study area was dominated by mature larch forests, which have wide tree spacing and relatively open canopies. Surviving trees may prevent the establishment of saplings, likely through resource competition.

Burn severity is often described as a legacy effect as high severity fires have a long-term impact on forest ecosystems through removing surface organic layers and creating canopy openings through killing nearly all large trees. Burn severity is well-reported to have adverse effects on forest restoration in boreal forests [81] and subalpine forest ecosystems [82]. Our results suggest that high severity burns decrease tree sapling regeneration post-fire, consistent with previous findings in similar ecosystems. In our study wildfire exerted limited effects on tree recovery when dNBR was lower than 0.730 (Figure 7c), which is related to low severity (dNBR < 0.418) and moderate severity (dNBR < 0.942) levels according to published thresholds from Fang and Yang [26]. Surviving trees contributed a large part of canopy foliage at low and moderate burn severity areas and combining with tree sapling foliage to result in high post-fire LAI under low and moderate burn severities. We found that LAI recovery decreases dramatically with dNBR increasing, implying that moderate to high severity fires have profound effects on diminishing forest resilience. Severe fire behavior may cause crown fires in our study area [45], which will destroy areal seed banks in serotinous cones of coniferous trees [8]. At the same time, severe surface fires will consume fine fuels such as litters and soil organic materials, create exposure of soil and base rock, and destroy surface seed beds. Given the key roles of coniferous tree species on maintaining the unique functions of boreal forest communities [9,14,23], recovery of coniferous tree species has raised increasing concerns [10,21]. Coniferous tree species have relatively shorter seed dispersal distance (e.g., larch < 150 m, and Scotch pine ~ 500 m) than broadleaf trees (e.g., birch and aspen > 1000 m) due to heavier seed mass [83,84]. Many studies found that Great Xing'an boreal forests will experience severe fire activities that were characterized as more frequent and large burns with higher severity along with climatic changes and fuel accumulation [18,85,86]. It is not difficult to speculate that coniferous forests may experience more recovery limitations, which has been validated in boreal forests of Central Siberian [84] and North America [8,11]. Besides, very high severity fires may have also killed broadleaf tree species and understory species whose roots would otherwise provide a source of asexual reproduction even if above-ground portions are killed, creating opportunities for primary succession.

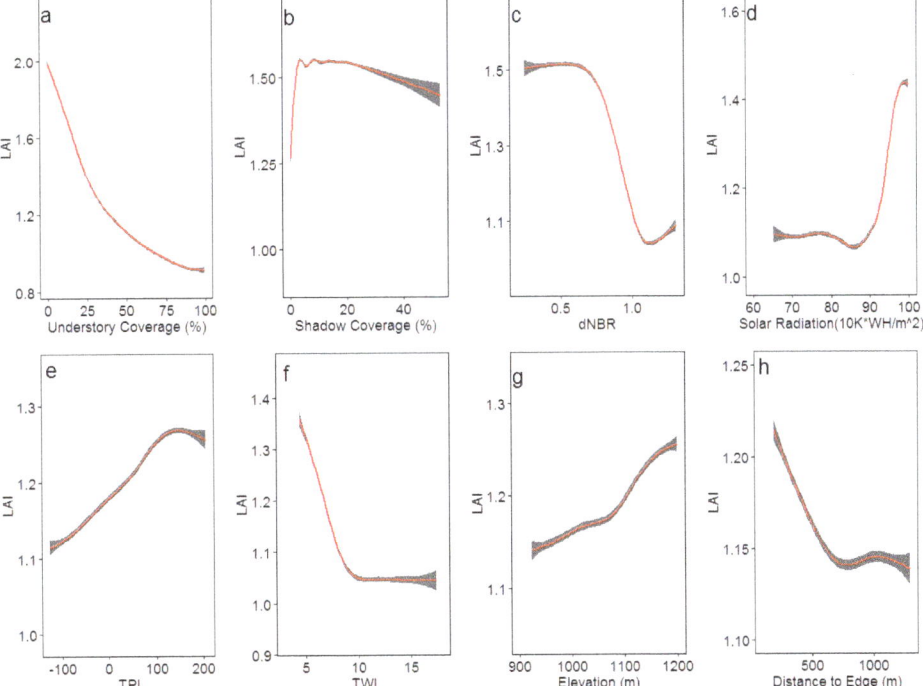

Figure 7. Partial dependence plots of a-h show influences of eight selected spatial controls (variable names see x-axis) on forest recovery in terms of leaf area index (LAI). Variable abbreviations were described in Table 3.

Topography is usually considered an environmental filter because it theoretically determines the drainage, thermal, solar radiation, surface roughness and other site factors which are usually exert considerable influence on tree species distribution and growth [87,88]. Total solar radiation during the growing season is typically the most important topographical factor and we found that LAI increased significantly when solar radiation exceeded 90,000 WH/m^2 (Figure 7d). At our study site, southern slopes and flat areas favored broad leaf tree species, such as white birch and aspen, that can regenerate through roots sprouting [21,47]. The TPI, TWI and elevation shared similar relative importance. Higher TPI indicates higher slope position (Figure 7e), and thereby sites that tend to have good drainage and open space to accept seed rain, especially low weight seeds of white birch and aspen dispersed via wind.

Our results suggest that TWI is inversely related to LAI (Figure 7f) while elevation has a positive influence LAI (Figure 7g). TWI is known to have a strong direct relationship with soil moisture in boreal forests ecosystems [89] and our findings are consistent with Cai, Yang, Liu, Hu and Weisberg [21], who reported similar relationships between soil moisture, elevation and broad leaf tree sapling abundance through field investigation. However, while Cai, Yang, Liu, Hu and Weisberg [21] found moist soil valley bottoms favorable for larch saplings, our study finds them to be more limited in terms of tree recovery than more elevated locations. One explanation is that at our site permafrost is usually present in those moist valleys [90], and the seasonal thaw may prevent seed germination. In addition, moss (*Sphagnum cuspidatulum*) and grass (*Carex appendiculata*) is often thick in moist valley bottoms and can prevent larch seed from encountering the mineral soil necessary for their establishment.

5. Conclusions

Our study is one of very few to investigate the utility of VHR satellite imagery for monitoring post-fire forest structural attributes (specifically tree sapling regeneration) in terms of LAI and TSA. We found that WorldView-2 VHR imagery outperforms Landsat imagery but still has some limitations. In particular, the high price of VHR imagery as well as the cost of associated computing resources may limit its practical usage. However, this may change in the future as costs are reducing for both UAV Photogrammetry and the launch of VHR satellites, potentially increasing accessibility by those managing forest resources [91,92], including the monitoring of post-fire forest recovery [93]. Our study shows that spectral variations can provide more useful information than image textures for the retrieval of parameters of tree saplings. Tree saplings provide very similar spectral signatures to surviving mature trees in our ecosystem, but this limitation may be less significant in other ecosystems exhibiting vegetation phenology differences or other unique characteristics that can be reflected in VHR imagery. Recognizing surviving mature trees in post-fire landscapes is of critical significance for quantifying post-fire landscape resilience. Our work suggests that shadow analysis can provide useful information for identifying surviving trees and this method merits further study as quick post-fire assessment of surviving trees through VHR imagery may provide useful information for forest managers, such as identifying locations where artificial regeneration may be needed.

Given that seed banks of coniferous tree species are easily damaged, and their seed dispersal distances are relatively short, our results suggested that combined natural forest restoration and proper human help (i.e., aerial seeding) for coniferous forests is necessary in our study area, especially following wildfires that burned severely across large geographic areas. In these areas, northern slopes and valley bottoms where solar radiation is low may require extra attentions from forest managers. While our work also suggests that understory plants can inhibit tree sapling recovery, a more thorough assessment of the effects of understory plants (e.g., soil and water retention) is required for ecological sound design and planning to assist land managers in making decisions about where across the landscape to prioritize reforestation efforts. Decreasing severity and size of wildfires through fuel management is known to improve the resilience of forest ecosystems [94–96]. Although edge effects did not exhibit a strong effect in our study due to continuous distribution of this high severity fire [26], islands of unburned patch may play important roles in accelerating natural forest recovery post-fire.

Author Contributions: L.F. and J.Y. conceived the research design; L.F. and Y.Y. (Yuanzheng Yang) conducted the field data collection; L.F., E.C., J.Y. and Z.L. conducted analysis and wrote the manuscript. L.F. and Y.Y. (Yan Yan) conducted the analysis of satellite data.

Funding: This study is financially funded by the National Key R&D Program of China (2017YFA0604403), the National Natural Science Foundation of China (Project No.31500387, 31470517 and 31800395), and CAS Pioneer Hundred Talents Program.

Acknowledgments: The authors appreciate Yongzhi Liu, Hongxin Zhang, Hailong Zhang and Changhe Hu, who are working for the administrative agencies of Huzhong Nutural Reserve and Huzhong Forestry Bureau, for their assistance during field data collection. We appreciate two anonymous reviewers for comments that improved this manuscript.

Conflicts of Interest: The authors declare no conflict of interest. The funding sponsors had no role in the design of the study; the collection, analyses, or interpretation of data; the writing of the manuscript; or the decision to publish the results.

Appendix A

The calibration results of our digital hemispherical photography (DHP) system (Panel 1), and results cited from reference document provided by CAN-EYE software (Panel 2). This is a proof to verify that our DHP system can generate reliable model outputs in CAN-EYE software. Note that two DHP systems use different cameras, fisheye lens and picture sizes, thus they have different parameters for characterizing optical center position and projection function.

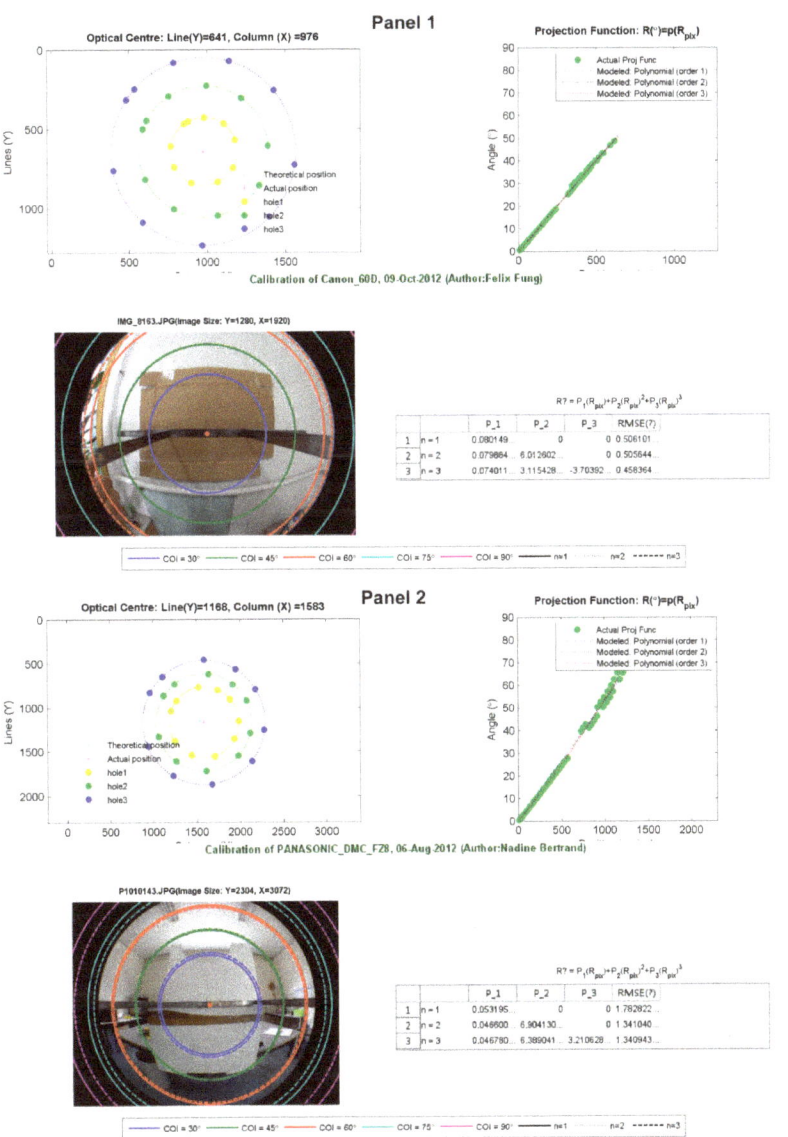

References

1. Bond-Lamberty, B.; Peckham, S.D.; Ahl, D.E.; Gower, S.T. Fire as the dominant driver of central Canadian boreal forest carbon balance. *Nature* **2007**, *450*, 89–92. [CrossRef] [PubMed]
2. Rogers, B.M.; Soja, A.J.; Goulden, M.L.; Randerson, J.T. Influence of tree species on continental differences in boreal fires and climate feedbacks. *Nat. Geosci.* **2015**, *8*, 228–234. [CrossRef]
3. Keane, R.E.; Agee, J.K.; Fule, P.; Keeley, J.E.; Key, C.; Kitchen, S.G.; Miller, R.; Schulte, L.A. Ecological effects of large fires on US landscapes: Benefit or catastrophe? *Int. J. Wildland Fire* **2008**, *17*, 696–712. [CrossRef]
4. Kasischke, E.S.; Johnstone, J.F. Variation in postfire organic layer thickness in a black spruce forest complex in interior Alaska and its effects on soil temperature and moisture. *Can. J. For. Res.* **2005**, *35*, 2164–2177. [CrossRef]
5. Bowd, E.J.; Banks, S.C.; Strong, C.L.; Lindenmayer, D.B. Long-term impacts of wildfire and logging on forest soils. *Nat. Geosci.* **2019**. [CrossRef]
6. Anderson-Teixeira, K.J.; Miller, A.D.; Mohan, J.E.; Hudiburg, T.W.; Duval, B.D.; Delucia, E.H. Altered dynamics of forest recovery under a changing climate. *Glob. Chang. Biol.* **2013**, *19*, 2001–2021. [CrossRef] [PubMed]
7. Seidl, R.; Rammer, W.; Spies, T.A. Disturbance legacies increase the resilience of forest ecosystem structure, composition, and functioning. *Ecol. Appl.* **2014**, *24*, 2063–2077. [CrossRef] [PubMed]
8. Johnstone, J.F.; Allen, C.D.; Franklin, J.F.; Frelich, L.E.; Harvey, B.J.; Higuera, P.E.; Mack, M.C.; Meentemeyer, R.K.; Metz, M.R.; Perry, G.L. Changing disturbance regimes, ecological memory, and forest resilience. *Front. Ecol. Environ.* **2016**, *14*, 369–378. [CrossRef]
9. Chapin, F.; McGuire, A.; Ruess, R.; Hollingsworth, T.; Mack, M.; Johnstone, J.; Kasischke, E.; Euskirchen, E.; Jones, J.; Jorgenson, M. Resilience of Alaska's boreal forest to climatic change. *Can. J. For. Res.* **2010**, *40*, 1360–1370. [CrossRef]
10. Johnstone, J.F.; Hollingsworth, T.N.; Chapin, F.S.; Mack, M.C. Changes in fire regime break the legacy lock on successional trajectories in Alaskan boreal forest. *Glob. Chang. Biol.* **2010**, *16*, 1281–1295. [CrossRef]
11. Hart, S.J.; Henkelman, J.; McLoughlin, P.D.; Nielsen, S.E.; Truchon-Savard, A.; Johnstone, J.F. Examining forest resilience to changing fire frequency in a fire-prone region of boreal forest. *Glob. Chang. Biol.* **2018**. [CrossRef]
12. Scheffer, M.; Hirota, M.; Holmgren, M.; Van Nes, E.H.; Chapin, F.S., 3rd. Thresholds for boreal biome transitions. *Proc. Natl. Acad. Sci. USA* **2012**, *109*, 21384–21389. [CrossRef]
13. Wolken, J.M.; Hollingsworth, T.N.; Rupp, T.S.; Chapin, F.S.; Trainor, S.F.; Barrett, T.M.; Sullivan, P.F.; McGuire, A.D.; Euskirchen, E.S.; Hennon, P.E.; et al. Evidence and implications of recent and projected climate change in Alaska's forest ecosystems. *Ecosphere* **2011**, *2*, 1–35. [CrossRef]
14. Beck, P.S.; Goetz, S.J.; Mack, M.C.; Alexander, H.D.; Jin, Y.; Randerson, J.T.; Loranty, M. The impacts and implications of an intensifying fire regime on Alaskan boreal forest composition and albedo. *Glob. Chang. Biol.* **2011**, *17*, 2853–2866. [CrossRef]
15. Johnstone, J.F.; Chapin, F.S. Fire Interval Effects on Successional Trajectory in Boreal Forests of Northwest Canada. *Ecosystems* **2006**, *9*, 268–277. [CrossRef]
16. Wang, X.; Thompson, D.K.; Marshall, G.A.; Tymstra, C.; Carr, R.; Flannigan, M.D. Increasing frequency of extreme fire weather in Canada with climate change. *Clim. Chang.* **2015**, *130*, 573–586. [CrossRef]
17. de Groot, W.J.; Flannigan, M.D.; Cantin, A.S. Climate change impacts on future boreal fire regimes. *For. Ecol. Manag.* **2013**, *294*, 35–44. [CrossRef]
18. Liu, Z.H.; Yang, J.; Chang, Y.; Weisberg, P.J.; He, H.S. Spatial patterns and drivers of fire occurrence and its future trend under climate change in a boreal forest of Northeast China. *Glob. Chang. Biol.* **2012**, *18*, 2041–2056. [CrossRef]
19. Turetsky, M.R.; Kane, E.S.; Harden, J.W.; Ottmar, R.D.; Manies, K.L.; Hoy, E.; Kasischke, E.S. Recent acceleration of biomass burning and carbon losses in Alaskan forests and peatlands. *Nat. Geosci.* **2010**, *4*, 27–31. [CrossRef]
20. Johnstone, J.F.; Chapin Iii, F.S.; Foote, J.; Kemmett, S.; Price, K.; Viereck, L. Decadal observations of tree regeneration following fire in boreal forests. *Can. J. For. Res.* **2004**, *34*, 267–273. [CrossRef]
21. Cai, W.; Yang, J.; Liu, Z.; Hu, Y.; Weisberg, P.J. Post-fire tree recruitment of a boreal larch forest in Northeast China. *For. Ecol. Manag.* **2013**, *307*, 20–29. [CrossRef]

22. Liu, Z.; Yang, J. Quantifying ecological drivers of ecosystem productivity of the early-successional boreal Larix gmelinii forest. *Ecosphere* **2014**, *5*, art84. [CrossRef]
23. Brown, C.D.; Liu, J.; Yan, G.; Johnstone, J.F. Disentangling legacy effects from environmental filters of postfire assembly of boreal tree assemblages. *Ecology* **2015**, *96*, 3023–3032. [CrossRef]
24. Mack, M.C.; Treseder, K.K.; Manies, K.L.; Harden, J.W.; Schuur, E.A.G.; Vogel, J.G.; Randerson, J.T.; Chapin, F.S. Recovery of Aboveground Plant Biomass and Productivity After Fire in Mesic and Dry Black Spruce Forests of Interior Alaska. *Ecosystems* **2008**, *11*, 209–225. [CrossRef]
25. Masek, J.G.; Hayes, D.J.; Joseph Hughes, M.; Healey, S.P.; Turner, D.P. The role of remote sensing in process-scaling studies of managed forest ecosystems. *For. Ecol. Manag.* **2015**, *355*, 109–123. [CrossRef]
26. Fang, L.; Yang, J. Atmospheric effects on the performance and threshold extrapolation of multi-temporal Landsat derived dNBR for burn severity assessment. *Int. J. Appl. Earth Obs. Geoinf.* **2014**, *33*, 10–20. [CrossRef]
27. French, N.H.F.; Kasischke, E.S.; Hall, R.J.; Murphy, K.A.; Verbyla, D.L.; Hoy, E.E.; Allen, J.L. Using Landsat data to assess fire and burn severity in the North American boreal forest region: An overview and summary of results. *Int. J. Wildland Fire* **2008**, *17*, 443–462. [CrossRef]
28. Parks, S.A. Mapping day-of-burning with coarse-resolution satellite fire-detection data. *Int. J. Wildland Fire* **2014**, *23*, 215–223. [CrossRef]
29. Yang, J.; Pan, S.; Dangal, S.; Zhang, B.; Wang, S.; Tian, H. Continental-scale quantification of post-fire vegetation greenness recovery in temperate and boreal North America. *Remote Sens. Environ.* **2017**, *199*, 277–290. [CrossRef]
30. Veraverbeke, S.; Gitas, I.; Katagis, T.; Polychronaki, A.; Somers, B.; Goossens, R. Assessing post-fire vegetation recovery using red–near infrared vegetation indices: Accounting for background and vegetation variability. *ISPRS J. Photogramm. Remote Sens.* **2012**, *68*, 28–39. [CrossRef]
31. Goetz, S.J.; Fiske, G.J.; Bunn, A.G. Using satellite time-series data sets to analyze fire disturbance and forest recovery across Canada. *Remote Sens. Environ.* **2006**, *101*, 352–365. [CrossRef]
32. Lu, D.; Chen, Q.; Wang, G.; Liu, L.; Li, G.; Moran, E. A survey of remote sensing-based aboveground biomass estimation methods in forest ecosystems. *Int. J. Digit. Earth* **2014**, *9*, 63–105. [CrossRef]
33. Glenn, E.; Huete, A.; Nagler, P.; Nelson, S. Relationship between remotely-sensed vegetation indices, canopy attributes and plant physiological processes: What vegetation indices can and cannot tell us about the landscape. *Sensors* **2008**, *8*, 2136–2160. [CrossRef]
34. Chu, T.; Guo, X. Remote Sensing Techniques in Monitoring Post-Fire Effects and Patterns of Forest Recovery in Boreal Forest Regions: A Review. *Remote Sens.* **2014**, *6*, 470–520. [CrossRef]
35. Gitas, I.; Mitri, G.; Veraverbeke, S.; Polychronaki, A. Advances in remote sensing of post-fire vegetation recovery monitoring-a review. In *Remote Sensing of Biomass-Principles and Applications*; InTech: Vienna, Austria, 2012.
36. Nolan, R.H.; Lane, P.N.J.; Benyon, R.G.; Bradstock, R.A.; Mitchell, P.J. Changes in evapotranspiration following wildfire in resprouting eucalypt forests. *Ecohydrology* **2014**, *7*, 1363–1377. [CrossRef]
37. Bond-Lamberty, B.; Wang, C.; Gower, S.T. Net primary production and net ecosystem production of a boreal black spruce wildfire chronosequence. *Glob. Chang. Biol.* **2004**, *10*, 473–487. [CrossRef]
38. Alexander, H.D.; Mack, M.C.; Goetz, S.; Loranty, M.M.; Beck, P.S.A.; Earl, K.; Zimov, S.; Davydov, S.; Thompson, C.C. Carbon Accumulation Patterns During Post-Fire Succession in Cajander Larch (Larix cajanderi) Forests of Siberia. *Ecosystems* **2012**, *15*, 1065–1082. [CrossRef]
39. Cai, W.H.; Yang, J. High-severity fire reduces early successional boreal larch forest aboveground productivity by shifting stand density in north-eastern China. *Int. J. Wildland Fire* **2016**, *25*. [CrossRef]
40. Chen, J.M.; Rich, P.M.; Gower, S.T.; Norman, J.M.; Plummer, S. Leaf area index of boreal forests: Theory, techniques, and measurements. *J. Geophys. Res. Atmos.* **1997**, *102*, 29429–29443. [CrossRef]
41. Bond-Lamberty, B.; Wang, C.; Gower, S.T.; Norman, J. Leaf area dynamics of a boreal black spruce fire chronosequence. *Tree Physiol.* **2002**, *22*, 993–1001. [CrossRef]
42. Chen, J.M.; Cihlar, J. Retrieving leaf area index of boreal conifer forests using Landsat TM images. *Remote Sens. Environ.* **1996**, *55*, 153–162. [CrossRef]
43. Jonckheere, I.; Fleck, S.; Nackaerts, K.; Muys, B.; Coppin, P.; Weiss, M.; Baret, F. Review of methods for in situ leaf area index determination: Part I. Theories, sensors and hemispherical photography. *Agric. For. Meteorol.* **2004**, *121*, 19–35. [CrossRef]

44. Hollingsworth, T.N.; Johnstone, J.F.; Bernhardt, E.L.; Chapin III, F.S. Fire severity filters regeneration traits to shape community assembly in Alaska's boreal forest. *PLoS ONE* **2013**, *8*, e56033. [CrossRef]
45. Fang, L.; Yang, J.; White, M.; Liu, Z. Predicting Potential Fire Severity Using Vegetation, Topography and Surface Moisture Availability in a Eurasian Boreal Forest Landscape. *Forests* **2018**, *9*, 130. [CrossRef]
46. Li, X.; He, H.S.; Wu, Z.; Liang, Y.; Schneiderman, J.E. Comparing effects of climate warming, fire, and timber harvesting on a boreal forest landscape in Northeastern China. *PLoS ONE* **2013**, *8*, e59747. [CrossRef]
47. Wang, C.; Gower, S.T.; Wang, Y.; Zhao, H.; Yan, P.; Lamberty, B.P. The influence of fire on carbon distribution and net primary production of boreal Larix gmelinii forests in north-eastern China. *Glob. Chang. Biol.* **2001**, *7*, 719–730. [CrossRef]
48. Weiss, M.; Baret, F. *CAN_EYE V6.4.91 USER MANUAL*. Available online: https://www6.paca.inra.fr/can-eye/content/download/3052/30819/version/4/file/CAN_EYE_User_MaManu.pdf (accessed on 5 March 2019).
49. Zhu, Z.; Woodcock, C.E. Object-based cloud and cloud shadow detection in Landsat imagery. *Remote Sens. Environ.* **2012**, *118*, 83–94. [CrossRef]
50. Canty, M.J.; Nielsen, A.A. Automatic radiometric normalization of multitemporal satellite imagery with the iteratively re-weighted MAD transformation. *Remote Sens. Environ.* **2008**, *112*, 1025–1036. [CrossRef]
51. Roy, D.P.; Kovalskyy, V.; Zhang, H.K.; Vermote, E.F.; Yan, L.; Kumar, S.S.; Egorov, A. Characterization of Landsat-7 to Landsat-8 reflective wavelength and normalized difference vegetation index continuity. *Remote Sens. Environ.* **2016**, *185*, 57–70. [CrossRef]
52. Qi, J.; Chehbouni, A.; Huete, A.; Kerr, Y.; Sorooshian, S. A modified soil adjusted vegetation index. *Remote Sens. Environ.* **1994**, *48*, 119–126. [CrossRef]
53. White, J.C.; Wulder, M.A.; Hermosilla, T.; Coops, N.C.; Hobart, G.W. A nationwide annual characterization of 25 years of forest disturbance and recovery for Canada using Landsat time series. *Remote Sens. Environ.* **2017**, *194*, 303–321. [CrossRef]
54. Kennedy, R.E.; Yang, Z.; Cohen, W.B. Detecting trends in forest disturbance and recovery using yearly Landsat time series: 1. LandTrendr—Temporal segmentation algorithms. *Remote Sens. Environ.* **2010**, *114*, 2897–2910. [CrossRef]
55. Gómez, C.; White, J.C.; Wulder, M.A. Characterizing the state and processes of change in a dynamic forest environment using hierarchical spatio-temporal segmentation. *Remote Sens. Environ.* **2011**, *115*, 1665–1679. [CrossRef]
56. Jiang, Z.; Huete, A.; Didan, K.; Miura, T. Development of a two-band enhanced vegetation index without a blue band. *Remote Sens. Environ.* **2008**, *112*, 3833–3845. [CrossRef]
57. Jin, S.; Sader, S.A. Comparison of time series tasseled cap wetness and the normalized difference moisture index in detecting forest disturbances. *Remote Sens. Environ.* **2005**, *94*, 364–372. [CrossRef]
58. Puissant, A.; Hirsch, J.; Weber, C. The utility of texture analysis to improve per-pixel classification for high to very high spatial resolution imagery. *Int. J. Remote Sens.* **2005**, *26*, 733–745. [CrossRef]
59. Lu, D.; Weng, Q. A survey of image classification methods and techniques for improving classification performance. *Int. J. Remote Sens.* **2007**, *28*, 823–870. [CrossRef]
60. Beguet, B.; Guyon, D.; Boukir, S.; Chehata, N. Automated retrieval of forest structure variables based on multi-scale texture analysis of VHR satellite imagery. *ISPRS J. Photogramm. Remote Sens.* **2014**, *96*, 164–178. [CrossRef]
61. Kayitakire, F.; Hamel, C.; Defourny, P. Retrieving forest structure variables based on image texture analysis and IKONOS-2 imagery. *Remote Sens. Environ.* **2006**, *102*, 390–401. [CrossRef]
62. Wood, E.M.; Pidgeon, A.M.; Radeloff, V.C.; Keuler, N.S. Image texture as a remotely sensed measure of vegetation structure. *Remote Sens. Environ.* **2012**, *121*, 516–526. [CrossRef]
63. Mountrakis, G.; Im, J.; Ogole, C. Support vector machines in remote sensing: A review. *ISPRS J. Photogramm. Remote Sens.* **2011**, *66*, 247–259. [CrossRef]
64. Gong, P.; Wang, J.; Yu, L.; Zhao, Y.; Zhao, Y.; Liang, L.; Niu, Z.; Huang, X.; Fu, H.; Liu, S. Finer resolution observation and monitoring of global land cover: First mapping results with Landsat TM and ETM+ data. *Int. J. Remote Sens.* **2013**, *34*, 2607–2654. [CrossRef]
65. Beleites, C.; Neugebauer, U.; Bocklitz, T.; Krafft, C.; Popp, J. Sample size planning for classification models. *Anal. Chim. Acta* **2013**, *760*, 25–33. [CrossRef]
66. Bruzzone, L.; Roli, F.; Serpico, S. An extension to multiclass cases of the Jeffries–Matusita distance. *IEEE Trans. Geosci. Remote Sens.* **1995**, *33*, 1318–1321. [CrossRef]

67. Congalton, R.G. A review of assessing the accuracy of classifications of remotely sensed data. *Remote Sens. Environ.* **1991**, *37*, 35–46. [CrossRef]
68. Zuur, A.F.; Ieno, E.N.; Elphick, C.S. A protocol for data exploration to avoid common statistical problems. *Methods Ecol. Evol.* **2010**, *1*, 3–14. [CrossRef]
69. Paradis, E.; Claude, J.; Strimmer, K. APE: Analyses of phylogenetics and evolution in R language. *Bioinformatics* **2004**, *20*, 289–290. [CrossRef]
70. Royston, P. Approximating the Shapiro-Wilk W-Test for non-normality. *Stat. Comput.* **1992**, *2*, 117–119. [CrossRef]
71. Conover, W.J.; Johnson, M.E.; Johnson, M.M. A comparative study of tests for homogeneity of variances, with applications to the outer continental shelf bidding data. *Technometrics* **1981**, *23*, 351–361. [CrossRef]
72. Prasad, A.M.; Iverson, L.R.; Liaw, A. Newer classification and regression tree techniques: Bagging and random forests for ecological prediction. *Ecosystems* **2006**, *9*, 181–199. [CrossRef]
73. Cutler, D.R.; Edwards, T.C., Jr.; Beard, K.H.; Cutler, A.; Hess, K.T.; Gibson, J.; Lawler, J.J. Random forests for classification in ecology. *Ecology* **2007**, *88*, 2783–2792. [CrossRef]
74. Berner, L.T.; Beck, P.S.A.; Loranty, M.M.; Alexander, H.D.; Mack, M.C.; Goetz, S.J. Cajander larch (Larix cajanderi) biomass distribution, fire regime and post-fire recovery in northeastern Siberia. *Biogeosciences* **2012**, *9*, 3943–3959. [CrossRef]
75. Chen, X.; Vogelmann, J.E.; Rollins, M.; Ohlen, D.; Key, C.H.; Yang, L.; Huang, C.; Shi, H. Detecting post-fire burn severity and vegetation recovery using multitemporal remote sensing spectral indices and field-collected composite burn index data in a ponderosa pine forest. *Int. J. Remote Sens.* **2011**, *32*, 7905–7927. [CrossRef]
76. Epting, J.; Verbyla, D.; Sorbel, B. Evaluation of remotely sensed indices for assessing burn severity in interior Alaska using Landsat TM and ETM+. *Remote Sens. Environ.* **2005**, *96*, 328–339. [CrossRef]
77. Ozdemir, I.; Karnieli, A. Predicting forest structural parameters using the image texture derived from WorldView-2 multispectral imagery in a dryland forest, Israel. *Int. J. Appl. Earth Obs. Geoinf.* **2011**, *13*, 701–710. [CrossRef]
78. Song, C.; Dickinson, M.B.; Su, L.; Zhang, S.; Yaussey, D. Estimating average tree crown size using spatial information from Ikonos and QuickBird images: Across-sensor and across-site comparisons. *Remote Sens. Environ.* **2010**, *114*, 1099–1107. [CrossRef]
79. Meigs, G.; Krawchuk, M. Composition and Structure of Forest Fire Refugia: What Are the Ecosystem Legacies across Burned Landscapes? *Forests* **2018**, *9*, 243. [CrossRef]
80. Jõgiste, K.; Korjus, H.; Stanturf, J.A.; Frelich, L.E.; Baders, E.; Donis, J.; Jansons, A.; Kangur, A.; Köster, K.; Laarmann, D. Hemiboreal forest: Natural disturbances and the importance of ecosystem legacies to management. *Ecosphere* **2017**, *8*. [CrossRef]
81. Johnstone, J.F.; Chapin, F.S. Effects of Soil Burn Severity on Post-Fire Tree Recruitment in Boreal Forest. *Ecosystems* **2006**, *9*, 14–31. [CrossRef]
82. Chambers, M.E.; Fornwalt, P.J.; Malone, S.L.; Battaglia, M.A. Patterns of conifer regeneration following high severity wildfire in ponderosa pine–dominated forests of the Colorado Front Range. *For. Ecol. Manag.* **2016**, *378*, 57–67. [CrossRef]
83. Zhao, F.; Qi, L.; Fang, L.; Yang, J. Influencing factors of seed long-distance dispersal on a fragmented forest landscape on Changbai Mountains, China. *Chin. Geogr. Sci.* **2015**, *26*, 68–77. [CrossRef]
84. Tautenhahn, S.; Lichstein, J.W.; Jung, M.; Kattge, J.; Bohlman, S.A.; Heilmeier, H.; Prokushkin, A.; Kahl, A.; Wirth, C. Dispersal limitation drives successional pathways in Central Siberian forests under current and intensified fire regimes. *Glob. Chang. Biol.* **2016**, *22*, 2178–2197. [CrossRef]
85. Fang, L.; Yang, J.; Zu, J.; Li, G.; Zhang, J. Quantifying influences and relative importance of fire weather, topography, and vegetation on fire size and fire severity in a Chinese boreal forest landscape. *For. Ecol. Manag.* **2015**, *356*, 2–12. [CrossRef]
86. Wu, Z.; He, H.S.; Yang, J.; Liu, Z.; Liang, Y. Relative effects of climatic and local factors on fire occurrence in boreal forest landscapes of northeastern China. *Sci. Total Environ.* **2014**, *493*, 472–480. [CrossRef]
87. Franklin, J.; McCullough, P.; Gray, C. Terrain variables used for predictive mapping of vegetation communities in Southern California. In *Terrain Analysis: Principles and Applications*; Wiley: New York, NY, USA, 2000; pp. 331–353.

88. Siegert, C.; Levia, D.; Hudson, S.; Dowtin, A.; Zhang, F.; Mitchell, M. Small-scale topographic variability influences tree species distribution and canopy throughfall partitioning in a temperate deciduous forest. *For. Ecol. Manag.* **2016**, *359*, 109–117. [CrossRef]
89. Sörensen, R.; Zinko, U.; Seibert, J. On the calculation of the topographic wetness index: Evaluation of different methods based on field observations. *Hydrol. Earth Syst. Sci. Discuss.* **2006**, *10*, 101–112. [CrossRef]
90. Bai, X.; Yang, J.; Tao, B.; Ren, W. Spatio-Temporal Variations of Soil Active Layer Thickness in Chinese Boreal Forests from 2000 to 2015. *Remote Sens.* **2018**, *10*, 1225. [CrossRef]
91. Bagaram, M.B.; Giuliarelli, D.; Chirici, G.; Giannetti, F.; Barbati, A. UAV remote sensing for biodiversity monitoring: Are forest canopy gaps good covariates? *Remote Sens.* **2018**, *10*, 1397.
92. Torresan, C.; Berton, A.; Carotenuto, F.; Di Gennaro, S.F.; Gioli, B.; Matese, A.; Miglietta, F.; Vagnoli, C.; Zaldei, A.; Wallace, L. Forestry applications of UAVs in Europe: A review. *Int. J. Remote Sens.* **2017**, *38*, 2427–2447. [CrossRef]
93. Aicardi, I.; Garbarino, M.; Lingua, A.; Lingua, E.; Marzano, R.; Piras, M. Monitoring Post-Fire Forest Recovery Using Multitemporal Digital Surface Models Generated from Different Platforms. *Earsel Eproc.* **2016**, *15*, 1–8.
94. Stephens, S.L.; Moghaddas, J.J.; Edminster, C.; Fiedler, C.E.; Haase, S.; Harrington, M.; Keeley, J.E.; Knapp, E.E.; McIver, J.D.; Metlen, K.; et al. Fire treatment effects on vegetation structure, fuels, and potential fire severity in western U.S. forests. *Ecol. Appl. Publ. Ecol. Soc. Am.* **2009**, *19*, 305–320. [CrossRef]
95. Stevens-Rumann, C.; Shive, K.; Fulé, P.; Sieg, C.H. Pre-wildfire fuel reduction treatments result in more resilient forest structure a decade after wildfire. *Int. J. Wildland Fire* **2013**, *22*, 1108. [CrossRef]
96. Lydersen, J.M.; Collins, B.M.; Brooks, M.L.; Matchett, J.R.; Shive, K.L.; Povak, N.A.; Kane, V.R.; Smith, D.F. Evidence of fuels management and fire weather influencing fire severity in an extreme fire event. *Ecol. Appl. Publ. Ecol. Soc. Am.* **2017**, *27*, 2013–2030. [CrossRef]

© 2019 by the authors. Licensee MDPI, Basel, Switzerland. This article is an open access article distributed under the terms and conditions of the Creative Commons Attribution (CC BY) license (http://creativecommons.org/licenses/by/4.0/).

Article

Spatio-Temporal Variations of Carbon Use Efficiency in Natural Terrestrial Ecosystems and the Relationship with Climatic Factors in the Songnen Plain, China

Bo Li, Fang Huang *, Lijie Qin, Hang Qi and Ning Sun

Key Laboratory of Geographical Processes and Ecological Security in Changbai Mountains,
Ministry of Education, School of Geographical Sciences, Northeast Normal University, Renmin Street No.5268, Changchun 130024, China; lib250@nenu.edu.cn (B.L.); qinlj953@nenu.edu.cn (L.Q.); qih102@nenu.edu.cn (H.Q.); sunn322@nenu.edu.cn (N.S.)
* Correspondence: huangf835@nenu.edu.cn; Tel.: +86-431-8509-9550

Received: 30 September 2019; Accepted: 25 October 2019; Published: 27 October 2019

Abstract: The Songnen Plain (SNP) is an important grain production base, and is designated as an ecological red-line as a protected area in China. Natural ecosystems such as the ecological protection barrier play an important role in maintaining the productivity and sustainability of farmland. Carbon use efficiency (CUE), defined as the ratio of net primary productivity (NPP) to gross primary productivity (GPP), represents the ecosystem capacity of transferring carbon from the atmosphere to terrestrial biomass. The understanding of the CUE of natural ecosystems in protected farmland areas is vital to predicting the impact of global change and human disturbances on carbon budgets and evaluating ecosystem functions. To date, the changes in CUE at different time scales and their relationships with climatic factors have yet to be fully understood. CUE and the response to land surface phenology are also deserving attention. In this study, variations in ecosystem CUE in the SNP during 2001–2015 were investigated using Moderate-Resolution Imaging Spectroradiometer (MODIS) GPP and NPP data products estimated using the Carnegie-Ames-Stanford approach (CASA) model. The relationships between CUE and phenological and climate factors were explored. The results showed that ecosystem CUE fluctuated over time in the SNP. The lowest and highest CUE values mainly occurred in May and October, respectively. At seasonal scale, average CUE followed a descending order of Autumn > Summer > Spring. The CUE of mixed forest was greater than that of other ecosystems at both monthly and seasonal scales. Land surface phenology plays an important role in the regulation of CUE. The earlier start (SOS), the later end (EOS) and longer length (LOS) of the growing season would contribute increasing of CUE. Precipitation and temperature affected CUE positively in most areas of the SNP. These findings help explain the CUE of natural ecosystems in the protected farmland areas and improve our understanding of ecosystem carbon allocation dynamics in temperate semi-humid to semi-arid transitional region under climate and phenological fluctuations.

Keywords: carbon use efficiency; Phenology; climate factors; MODIS GPP/NPP; Songnen Plain

1. Introduction

In recent decades, driven by intensive human activity and climate change, the function of terrestrial ecosystem has been disturbed and continuously degraded on regional and global scales. The increasing levels of atmospheric CO_2 concentrations and climate change have highlighted the need for a better understanding of terrestrial carbon cycling and its responses to climate change. Gross primary production (GPP) represents the capacity of the plants in an ecosystem to capture energy and carbon [1]. Net Primary Productivity (NPP) is defined as the amount of atmospheric carbon that is captured by plants and transformed into biomass [2]. The GPP is the sum of NPP and autotrophic respiration (Ra), and Ra plus heterotrophic respiration (Rh) comprises ecosystem respiration. GPP, NPP and Ra are the most important and highly related constituents of carbon cycling. The carbon fixed by photosynthesis is allocated to a variety of usages in plants, including growth and maintenance respirations and biomass accumulation [3]. About 50–70% of the carbon fixation is returned to the environment through Ra [4,5]. Carbon allocation among plant processes (e.g., respiration, biomass production) and organs (e.g., leaves, stem) is a key process in the carbon cycle because it determines the residence time and location of carbon in the ecosystem [6,7]. For example, the residence times of the carbon used for maintenance respiration and the carbon allocated to the structural biomass of organs are drastically different, ranging from a few hours to a few years [6]. Therefore, the allocation process of carbon is highly relevant to understanding ecosystem carbon stock and carbon cycles [6].

Carbon use efficiency (CUE) is defined as the ratio of NPP to GPP, which indicates the ecosystem capacity in transferring CO_2 into biomass and carbon sequestration [8]. CUE is an important functional parameter of ecosystems and can be used for comparing carbon cycle differences in various ecosystems [9]. The index is intuitive and easy to compare between different vegetation types, and to apply to different time scales [7]. A higher CUE indicates a higher growth transfer per unit of carbon sequestration. In practice, GPP usually represents the total amount of carbon captured through photosynthesis, and NPP is the net carbon stored in plant after the reduction of GPP through by plant respiration [1]. CUE is also a measure of how GPP is partitioned into NPP and Ra [7]. Less Ra may result in larger carbon reserve accumulation. Hence, CUE is related to photosynthetic process, and it is also regarded as an important indicator for characterizing ecosystem functions. How efficiently an ecosystem is able to convert GPP into plant and soil storage greatly determines the carbon sequestration of terrestrial ecosystems, so CUE changes strongly affect ecosystem carbon budgets [10]. Quantitative analysis of spatial-temporal changes of CUE and its influencing factors will help better understand the effects of climate change on carbon processes of ecosystems [11].

Satellite remote sensing provides critical information for investigating large-scale and long-term variability of ecosystem CUE. Piao et al. [12] demonstrated that CUE of different vegetation differed greatly from the south temperate to the tropic ecoregions based on a global forest C-flux database, and found that the spatial patterns of forest annual Ra at the global scale were largely controlled by temperature. Zhang et al. [1] reported that CUE exhibited a pattern depending on the climatic characteristics-based upon Moderate-Resolution Imaging Spectroradiometer (MODIS)-derived NPP and GPP data. He et al. [13] investigated spatial variations in CUE from different models and analyzed the responses of CUE to precipitation and temperature. Tang et al. [3] established a global database of site-year CUE based on field observations for five ecosystem types and diagnosed the spatial variability of CUE with climate and other environmental factors (e.g., soil variables). Two prominent gradients of CUE in ecosystem types and latitude were found worldwide. CUE varied with ecosystem types, being the highest in wetland and lowest in grassland. CUE decreased with latitude, showing the lowest values in tropics, and the highest CUE were found in higher-latitude regions. The above studies were based on annual scales and advanced the knowledge of understanding the global pattern of CUE. However, monthly scale analysis of CUE has rarely been studied.

From individual plants to an entire ecosystem, phenology directly or indirectly regulates carbon fluxes (e.g., photosynthesis and respiration) between the land surface and the atmosphere [14] through altering physiological and structural characteristics, including photosynthetic rate, canopy conductance and albedo [14–16]. Vegetation phenological changes are closely related to spatial-temporal dynamics of carbon cycle [17]. The change in the length of the growing season may have an important impact on vegetation growth, which will cause changes in the GPP and NPP [18]. CUE and its relationships with land surface phenology (LSP) deserve attention.

The Songnen Plain (SNP), located in temperate semi-humid to semi-arid transition ecological fragile zone in Northeast China, is highly sensitive to global change. As a key agricultural area and important grain commodity base, the SNP is among the designated ecological red-lines as protected farmland area in China. The natural terrestrial ecosystem acts as an ecological protection barrier for the croplands in the SNP. The productivity and sustainability of terrestrial ecosystems are vital to maintaining regional and national food and ecological security. Due to the combined effects of vulnerable physical conditions and excessive human activities, the SNP suffered from high risk of land degradation during the past century. Concerns for the aggravation of desertification have led to many measures and management actions for ecological and environmental protection. The trend of desertification and exacerbation has gradually slowed down [19]. Previous studies have mostly focused on the land cover/use change and effects of agricultural activities on environment [20]. In addition, most studies only focused on the condition of protected farmland, while ignoring the productivity and sustainability of natural terrestrial ecosystems around it. There is a lack of reports about the spatial-temporal patterns of ecosystem-level CUE and their response to phenology and climate change in the SNP region. This study attempts to fill in the gaps in the knowledge regarding biotic and abiotic impacts on CUE of the SNP region.

The objectives of this study are to: (1) estimate CUE of different ecosystems and investigate their monthly and seasonal changes based on MODIS GPP and NPP data from 2001 to 2015; (2) explain how phenology and climatic factors contribute to variations in ecosystem CUE, in order to improve our understanding of the carbon budget in temperate semi-arid and semi-humid transitional zone ecosystems and their driving mechanisms.

2. Materials and Methods

2.1. Study Area

The SNP is located in the central part of Northeast China, in a range of 121°38′ to 128°33′E, 42°49′ to 49°12′N, with a total area of 22.35×10^4 km^2 (Figure 1). It is an alluvial plain situated in the central Songliao Basin between the Xiaoxing'an and Changbai Mountains, through which the Songhua River and the Nenjiang River flow [10]. The SNP belongs to a temperate continental semi-humid and semi-arid monsoon climate zone, characterized by four seasons with a hot, rainy summer and a cold, dry winter and with significant windy days. The annual precipitation is between 350 and 800 mm [11]. Soils are fertile with chernozem, meadow soil, and black soil widely distributed. From west to east, the natural ecosystem type is typical grassland, meadow steppe and forest steppe, respectively. It is an important ecological protection barrier in the Northeast.

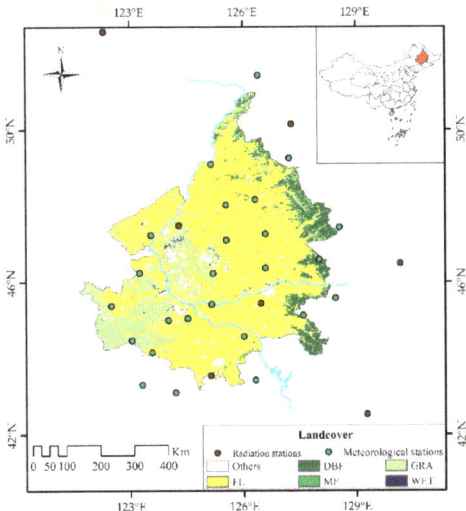

Figure 1. Study Area. FL: farmland; DBF: deciduous broad-leaved forest; MF: mixed forest; GRA: grassland and WET: wetland.

2.2. Data and Processing

In this study, CUE at the monthly and seasonal scale derived from MODIS data products and the ancillary data were used to explore the spatial-temporal variations of CUE of natural ecosystems and their responses to climate and LSP changes. The main steps are as follows: (1) estimating NPP at monthly scale by the CASA (Carnegie-Ames-Stanford approach) model; (2) calculating monthly CUE of different ecosystems and performing the trend analysis; (3) extracting LSP metrics and analyzing the effects of phenology and climatic factors on the variations of ecosystem CUE by the correlation and partial correlation analysis methods. Figure 2 illustrates the technical approach of this study.

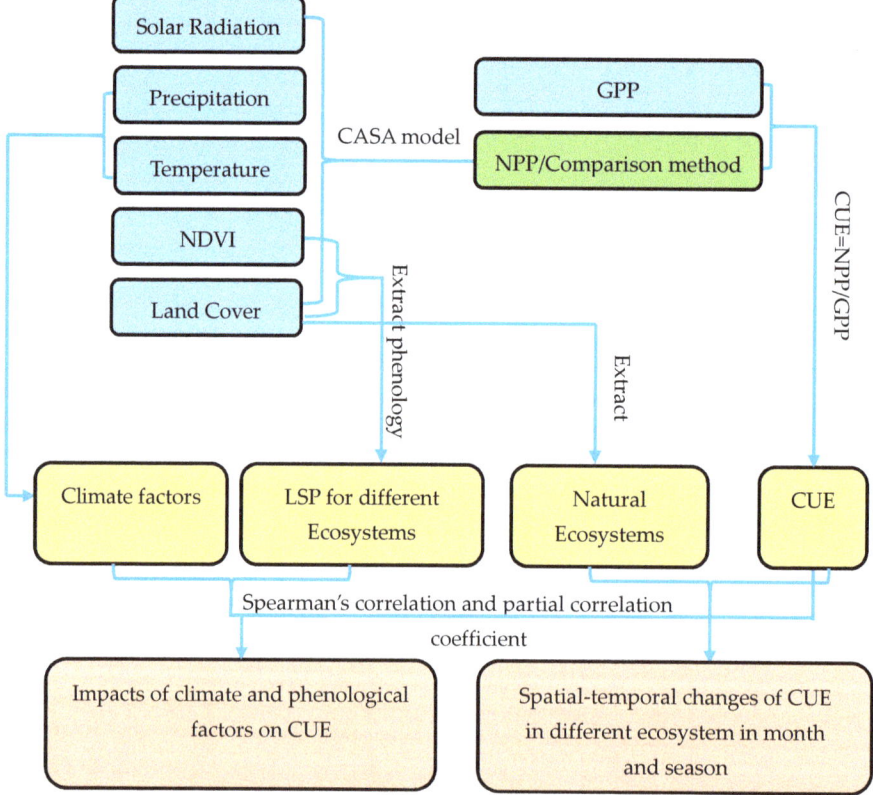

Figure 2. The flow chart of the research approach.

MODIS and Meteorological Data

The MOD17A2H version 6 GPP product is a cumulative 8-day composite of values based on the radiation use efficiency concept, which can be potentially used as an input to data models to compute energy, carbon, water cycle processes and biogeochemistry of vegetation [12]. Monthly Normalized Difference Vegetation Index (NDVI) spatial distribution data set was obtained from the MOD13A2. For the calculation of NPP, the Maximum Value Composite (MVC) was used to synthesize the monthly NDVI data. The phenological parameter extraction was based on 16-day data. Each time period of the MODIS data included 6 images.

The MODIS land cover type dataset (MCD12Q1) was downloaded from the Land Processes Distributed Active Archive Center (LP DAAC) (https://lpdaac.usgs.gov/) [13]. In this study, land cover data from 2001 to 2015 were adopted. According to the International Geosphere-Biosphere Program

(IGBP) class scheme and the regional characteristics of the SNP, natural ecosystem included the following four types, deciduous broad-leaved forest (DBF), mixed forest (MF), grassland (GRA) and wetland (WET).

The total monthly radiation was obtained from 7 radiation stations (Figure 1). The remaining meteorological data of 29 meteorological stations within and near the SNP from 2001 to 2015 were acquired from the National Meteorological Information Center (http://data.cma.cn/). Most studies suggest that precipitation and temperature be the main meteorological factors affecting CUE [3,13]. In this study, monthly precipitation (mm) and temperature (°C) were computed from May to November in order to explore the relationship between CUE and climate factors in growing seasons. MODIS, meteorological and other data were all resampled at a resolution of 1 km. Table 1 summarizes and describes the characteristics of the data and sources.

Table 1. Description of the data used in this study.

Dataset	Source	Temporal Resolution	Time Range	Spatial Resolution
GPP	MOD17	8-day composite	2001–2015	1 km
Land cover	MCD12Q1	annual	2001–2015	1 km
NDVI	MOD13	16-day composite	2001–2015	0.25 km
Temperature	http://data.cma.cn/	monthly	2001–2015	/
Precipitation	http://data.cma.cn/	monthly	2001–2015	/
Solar radiation	http://data.cma.cn/	monthly	2001–2015	/

2.3. Estimating NPP with the CASA Model

We estimated the monthly vegetation NPP of the SNP based on the CASA model, which considers the physiological and ecological characteristics of vegetation and the environmental conditions related to growth [14]. Vegetation NPP estimation was derived by using vegetation cover type, NDVI, monthly average temperature, total precipitation and solar radiation [15]. The basic calculation formula of the CASA model is as follows [21]:

$$NPP(x,t) = SOL(x,t) \times FPAR(x,t) \times 0.5 \times T_{\varepsilon 1}(x,t) \times T_{\varepsilon 2}(x,t) \times W_\varepsilon(x,t) \times \varepsilon_{max} \quad (1)$$

where SOL(x,t) is the total solar radiation at pixel x for month t. FPAR(x,t) is the fraction of photosynthetically active radiation absorbed by vegetation. 0.5 indicates the proportion of solar active radiation (0.4–0.7 μm) that can be utilized by vegetation to the total solar radiation. $T_{\varepsilon 1}(x,t)$ and $T_{\varepsilon 2}(x,t)$ represent temperature stress coefficients, $W_\varepsilon(x,t)$ is the coefficient of water stress, and ε_{max} is the maximum light use efficiency under ideal conditions [22]. Comparison between the calculated NPP and the reported study conducted in Northeast China and the western part of Jilin Province [23,24] as the cross checking and validation of the analysis. Mao et al. [23] verified the NPP by comparing the simulated value with the flux observation data. The simulated value was close to the measured value, and the error was within 25%.

2.4. Calculation of CUE

The CUE of ecosystem describes the relationship between photosynthesis and respiration, which is an important indicator of the ability of plants to transfer carbon [16]. As one of the key controlling factors of ecosystem carbon storage, CUE is defined as follows [19]:

$$CUE = \frac{NPP}{GPP} \quad (2)$$

where GPP represents the ability to capture energy and carbon through photosynthesis plants and the total amount of carbon assimilation. NPP reveals the energy of plants stored after losing carbon from GPP through autotrophic respiration [17]. The higher CUE means the greater proportion of

GPP kept by ecosystems after self-consumption. However, uncertainty issues have been recognized by studies using public-domain data, e.g., with respect to water use efficiency (WUE) [25].

2.5. Extraction of Land Surface Phenology Metrics

We used the dynamic threshold method to extract metrics of LSP. The polynomial method was used to fit and reconstruct the NDVI time series data from 2000 to 2016. The software TIMESAT with a seasonal parameter of 0.5, an adaptation strength of 2.0, a Savitzky–Golay window size of 2, and an amplitude of 20% was run in MATLAB R2015b (The Mathworks, Inc., Natick, MA, USA). The parameters were set according to Qi et al. [18]. The start of growing season (SOS) and the end of season (EOS) for each year were calculated, and the length of season (LOS) was obtained as the difference between SOS and EOS in each grid.

2.6. Statistical Analysis

Spatial trend of CUE was examined by applying a linear regression model with time as the independent variable and CUE as the dependent variables, respectively. The trend analysis method was used to analyze trend in seasonal CUE changes for the period 2001–2015. The outputs of the trend analysis are the maps of regression slope values, expressed by the following formula [19]:

$$\text{Slope} = \frac{n \times \sum_{i=1}^{n} i \times A_i - \sum_{i=1}^{n} i \sum_{i=1}^{n} A_i}{n \times \sum_{i=1}^{n} i^2 - \left(\sum_{i=1}^{n} i\right)^2} \quad (3)$$

where Slope is the slope of the fitted regression line at each pixel. n represents year range. i is 1 for the first year, 2 for the second year, and so on. A_i represents the CUE of the year i. A negative regression coefficient (Slope < 0) indicates a decline of CUE, whereas a positive value (Slope > 0) depicts an increase trend. F test was used to determine the significance of change trend.

To investigate the role of climate drivers and phenological factors affecting CUE, we analyzed the correlation between three phenological parameters (i.e., SOS, EOS and LOS) and CUE. In addition, Spearman partial correlation between CUE and two climate factors (i.e., precipitation and temperature) was calculated. The correlation coefficient and partial correlation coefficient were computed as follows [18]:

$$r_{BC} = \frac{\sum_{i=1}^{n}(B_i - \overline{B})(C_i - \overline{C})}{\sqrt{\sum_{i=1}^{n}(B_i - \overline{B})^2}\sqrt{\sum_{i=1}^{n}(C_i - \overline{C})^2}} \quad (4)$$

$$\overline{B} = \frac{1}{n}\sum_{i=1}^{n} B_i, \overline{C} = \frac{1}{n}\sum_{i=1}^{n} C_i \quad (5)$$

$$r_{BC,D} = \frac{r_{BC} - r_{BD}r_{CD}}{\sqrt{1 - r_{BD}^2}\sqrt{1 - r_{CD}^2}} \quad (6)$$

where r_{BC} represents the correlation coefficient between B and C, its threshold ranges from −1 to 1, and $r_{BC,D}$ is the partial correlation coefficient between B and C when we controlled D values. If r < 0, B is negatively correlated with C. If r > 0, there is a positive correlation between B and C. Furthermore, $\overline{B}, \overline{C}$ represent the average values of B_i and C_i, respectively. The significance of the results was examined by t-test.

3. Results

This study explained spatial patterns of ecosystem CUE at different temporal scales in a semi-humid and semi-arid transitional area. We identified that the variations of CUE in SNP were obvious at both seasonal and monthly scales. The CUE of GRA in the southwest and DBF in the east showed an upward trend. Monthly and seasonal CUE varied with ecosystem types. The earlier SOS, later EOS and longer LOS might encourage higher CUE. Spatially, CUE changes were positively correlated with precipitation and temperature in most of the SNP.

3.1. Monthly Change of CUE

The CUE value of natural ecosystems in the SNP started from May and continued to November, the CUE changed significantly within the year (Figure 3). The lowest CUE values occurred in May. After July, CUE increased, and exceeded 0.8. The highest CUE (over 0.9) values were mainly observed in October. The growing season started in May in SNP with low NPP and carbon sequestration capacity. In contrast, the proportion of GPP increased in October after self-consumption through the growing season. The CUE of the natural ecosystems was higher after July, along with the accumulation of more NPP, which meant that the natural ecosystem protection capacity may be stronger.

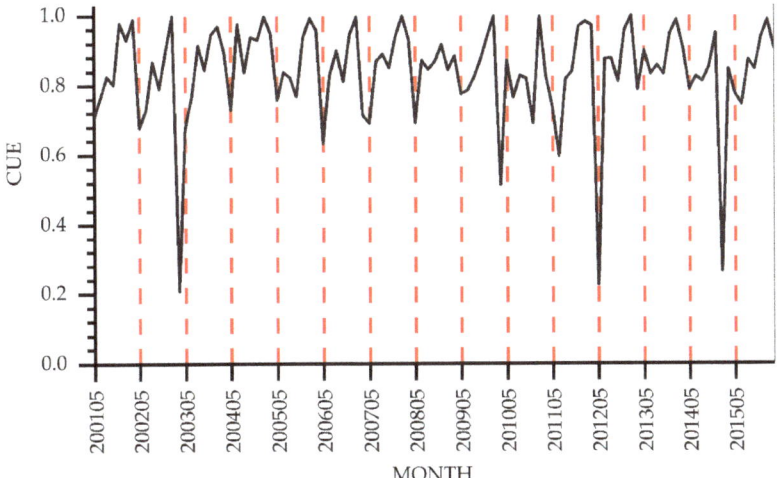

Figure 3. Regional average CUE at the monthly scale from 2001 to 2015.

There were three abnormally low CUE values in the SNP during the past 15 years (Figure 3). These were November 2002, May 2012 and October 2014. According to field-based meteorological measurements, the average temperature in November 2002 was the lowest in 15 years, which might be the reason that led to decrease in both GPP and NPP. In May 2012 and October 2014, the low CUE values might be associated with lower GPP and NPP due to reduced rainfall in those months.

The monthly CUE varied among different ecosystems in the SNP (Figure 4). Except for May and October, the CUE of DBF was generally lower than the regional monthly average, while the GRA was the opposite. GRA started with a green-up in May and gradually entered a senescence period in October. GRA might produce more net productivity in those two months. The CUE of MF was always higher than the regional CUE average. WET CUE in May, June and August were greater than the CUE mean value. WET CUE was the highest in August, indicating that the proportion of GPP kept by WET ecosystems after self-consumption was the greatest.

The CUE of MF was the highest from May to November except for August. Compared to other ecosystems, MF may have stronger ecological protection effects. Due to higher temperature in August, CUE of GRA may be restricted, while the area covered by humid MF could produce a higher CUE. From June to August, CUE of DBF was the lowest, because the accumulated NPP was relatively lower than other types.

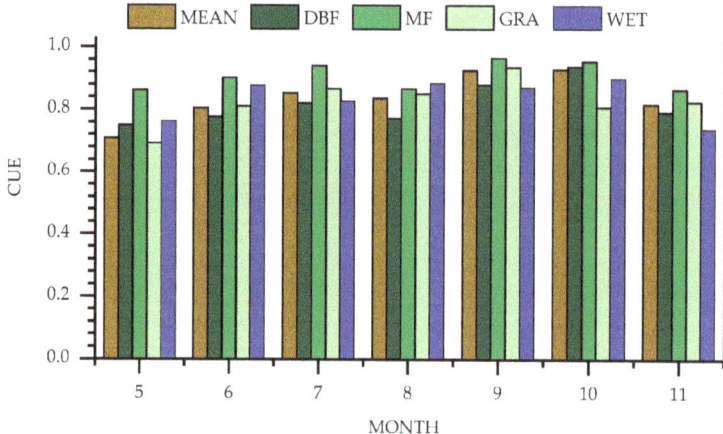

Figure 4. Monthly mean CUE of different ecosystems from 2001 to 2015.

3.2. Seasonal Changes in CUE

Considering the dormancy of vegetation in the SNP in the winter months from December to the following February, we calculated CUE in spring (March to May), summer (June to August) and autumn (September to November), respectively. Figure 5 shows the seasonal variation of ecosystem CUE. The regional average CUE was 0.236, 0.835 and 0.854 in spring, summer and autumn, respectively. The highest and lowest CUEs in spring were observed in the year of 2013 (0.299) and 2012 (0.075), respectively. The lowest value of CUE in spring 2012 may be due to the drought of that year [26]. The maximum summer CUE (0.916) was observed in 2004, whereas the minimum value (0.75) occurred in 2011. CUE in autumn reached the peak (0.974) in 2011, while the lowest value was found in 2014 (0.686).

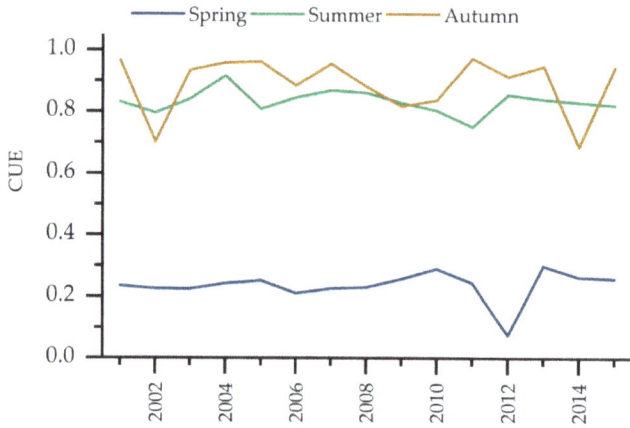

Figure 5. Average CUE variations in spring, summer and autumn from 2001 to 2015.

In most years, average CUE was the lowest in spring. The average CUE values in summer of 2002, 2009, and 2014 were greater than those in autumn, which was related to successive drought from summer to autumn. It was found that the degree of CUE decrease depends not only on the intensity of the drought, but also the duration of the drought intensity and the time of occurrence [27].

Spatially, spring CUE in the southwest of the SNP was higher than the east during the 15 years (Figure 6). It ranged from 0 to 0.4, with an average value of 0.24 (Figure 6a). In summer, vegetation grew vigorously, and the carbon sequestration capacity of vegetation increased (Figure 6b). Similar spatial distribution pattern was observed in summer with an average CUE of 0.83. With the arrival of autumn, the CUE in most regions also increased and ranged from 0.8 to 1.0 with an average value of 0.88 (Figure 6c). In summer and autumn, the carbon sequestration capacity of natural ecosystems was better, indicating the relatively stronger ecological protection function.

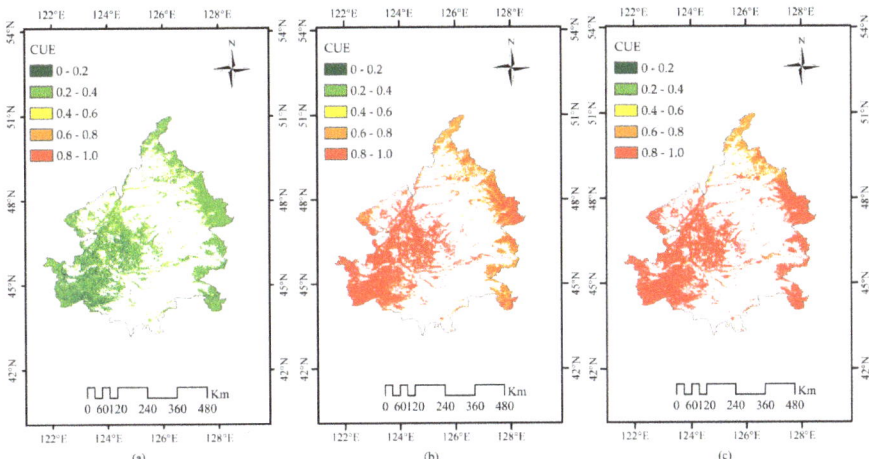

Figure 6. Spatial distribution of seasonal average CUE of the SNP from 2001 to 2015. (**a**) Spring; (**b**) summer; (**c**) autumn.

In terms of spatial distribution, the pixels showing an upward trend in three seasons were mainly found in grasslands in the southwest and deciduous broadleaf forest in the eastern fringe. According to the slope analysis, about two-thirds of the study area showed an upward trend of CUE in spring (Figure 7a). CUE in summer tended to increase in 53.7% of the study area (Figure 7b), while the CUE showed increasing trend in the area of 56.7% in autumn (Figure 7c). This increasing trend suggested that the carbon sequestration capacity of natural ecosystem in the SNP could be improving. More NPP accumulated in natural ecosystem may make their ecological protection function stronger. The change trends of CUE (over 90% pixels) passed the significance level test at $p < 0.05$.

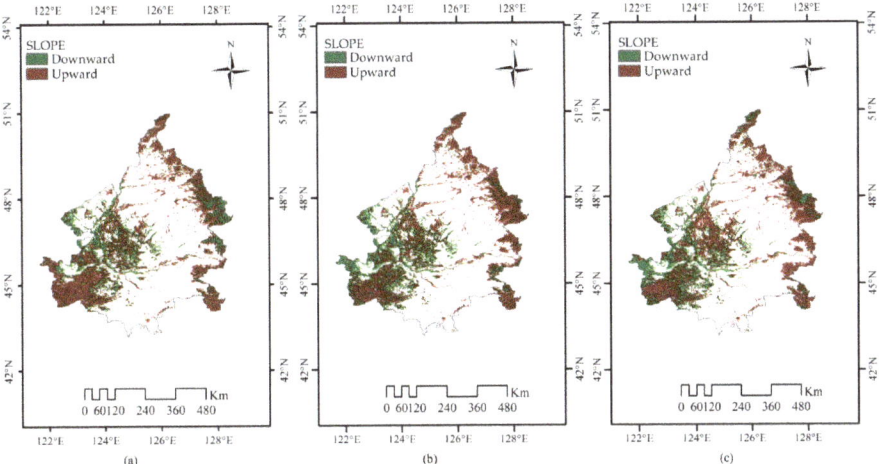

Figure 7. Spatial trend of average CUE in each season of SNP from 2001 to 2015. (**a**) Spring; (**b**) summer; (**c**) autumn.

The seasonal changes of CUE for different ecosystems CUE were also obviously changing in the SNP (Figure 8). In spring, all types of vegetation had low CUE values (Figure 8a). Relatively good hydrothermal conditions in summer were more favorable to vegetation growth and the CUE of each vegetation type was generally increasing (Figure 8b). In autumn, CUE of MF, DBF and GRA continued to rise, whereas WET CUE declined slightly (Figure 8c). Among different ecosystems, CUE of MF had been the highest (spring: 0.288; summer: 0.902; autumn: 0.928).

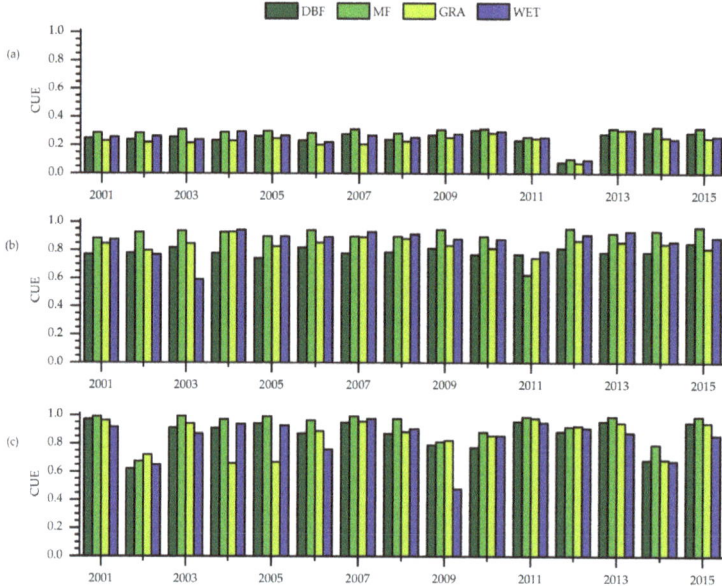

Figure 8. CUE of different ecosystems and seasons from 2001 to 2015. (**a**) Spring; (**b**) summer; (**c**) autumn.

3.3. The Mean Spatial Distribution of LSP

Spatial distributions of LSP parameters, i.e., SOS, LOS and EOS, in the SNP from 2001 to 2015 are illustrated in Figure 9. The SOS of the natural ecosystem mainly occurred at day of year (DOY) between 100 and 150. The earliest SOS was found in the eastern parts of the SNP region, while the southwestern region had the latest SOS (Figure 9a). The growing season of DBF and MF started from mid-March, and GRA and WET had later start of the growing season (early April). SOS began in March and April, the vegetation began to accumulate GPP, but the CUE value was in a very small range, almost neglected, so we began to record CUE from May.

The distribution of EOS dates showed similar pattern to that of SOS, gradually increasing from west to east, mainly in late October and November (290–330 DOY) (Figure 9b). The end dates of the growing season of DBF and MF occurred in early November. GRA and WET ended their growing seasons about ten days earlier than the forestland. During 2001–2015, the average LOS of natural ecosystems in the SNP was about 192 days, showing similar spatial distribution to SOS and EOS (Figure 9c). The average LOS of MF and GRA was 213 days and 176 days, respectively. LOS dates of DBF were about 5 days shorter than those of MF, and the growing season of WET was about 4 days longer than that of GRA.

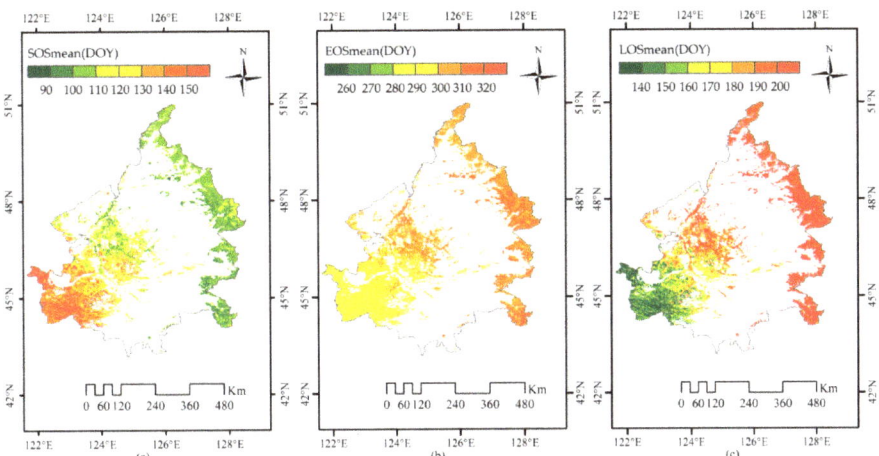

Figure 9. Spatial distribution of land surface phenology metrics in the SNP from 2001 to 2015: (**a**) average SOS; (**b**) average EOS; (**c**) average LOS.

3.4. Response of CUE to LSP Variation

After analyzing the correlation between CUE and LSP in the growing seasons, it was found that CUE was negatively correlated with SOS in about 70% of the study area (Figures 10a and 11). This indicated that earlier SOS would encourage higher CUE. In 72% of areas covered by GRA, CUE was negatively correlated with SOS. In 67% of the SNP, the later EOS would result in higher CUE. In 80% of areas covered by DBF, late EOS dates might have the positive effect on the increase of CUE (Figure 10b). CUE was positively correlated with LOS in more than 70% of areas covered by DBF, GRA and WET. The average CUE of MF with the longest growing season was highest (0.529). GRA with the second longest growing season (0.482). Although the LOS of GRA was the shortest, its average CUE (0.482) was greater than that of DBF (0.477). This would be because GRA in cold and dry regions consumed less energy to maintain growth. The area proportions of correlation coefficients after significant test for all the pixels were obtained (Figure 11).

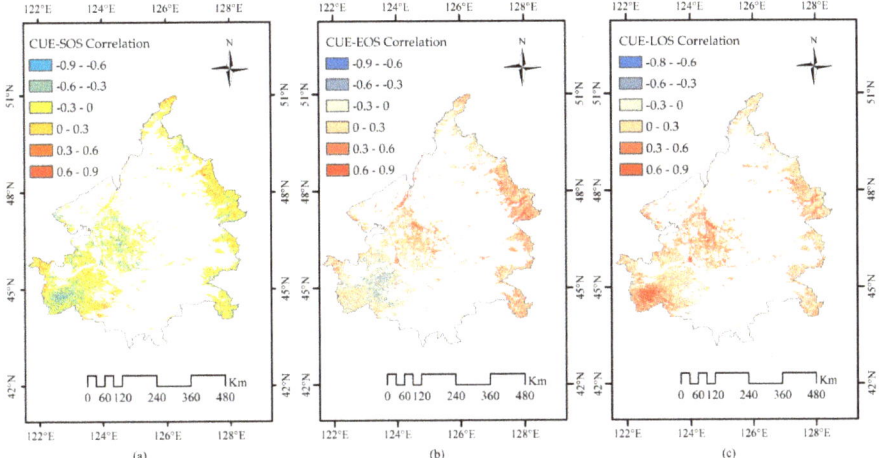

Figure 10. Spatial distribution of correlation coefficients (R) between CUE, SOS, EOS and LOS in the SNP during 2001–2015. (**a**) CUE and SOS; (**b**) CUE and EOS; (**c**) CUE and LOS.

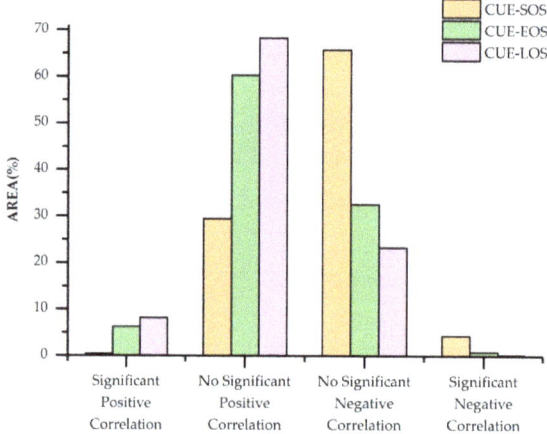

Figure 11. The area percentages of correlation coefficients between CUE, SOS, EOS and LOS. (Significant Positive Correlation ($r > 0$, $p < 0.05$): No Significant Positive Correlation ($r > 0$, $p > 0.05$), No Significant Negative Correlation ($r < 0$, $p > 0.05$), Significant Negative Correlation ($r < 0$, $p < 0.05$)).

3.5. Direct Effects of Local Climate Factors on CUE Change

This study revealed that a partial correlation existed between mean temperature and total precipitation and CUE in the growing season. CUE was negatively correlated with precipitation accounting for about 46.8% of the total pixels (Figure 12a). Among those, 0.98% had significant negative correlation, mainly distributed in the eastern and southwestern fringe areas of SNP. The area showing positive correlation between CUE of DBF and precipitation occupied 61.8% of the total area. About 60% of CUE values of GRA and WET were positively related to precipitation. CUE was positively affected by temperature in more than 90% of the region, of which 14.85% showed a significant positive correlation. Only in the northern and southern margins, CUE decreased with increasing temperature (Figure 12b).

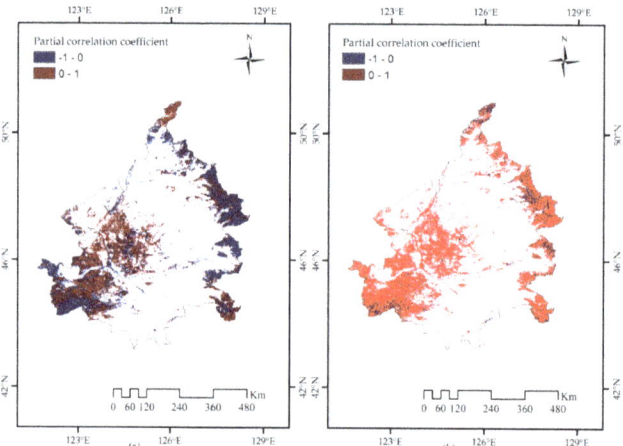

Figure 12. Partial correlation coefficients between CUE and major climatic factors in the growing season: (**a**) precipitation; (**b**) temperature.

At monthly scale, the responses of ecosystem CUE to climate drivers were also significantly different. Figures 13 and 14 showed the spatial pattern of correlation coefficients between monthly CUE, precipitation and temperature from 2001 to 2015. Overall, the pixels with a positive correlation coefficient took up higher area proportions of the study area. Except for November, increased precipitation could contribute to higher CUE for the corresponding months in more than 60% of naturally vegetated area in the SNP (Figure 13 and Table 2). From June to August, CUE in more than 50% of pixels in the natural ecosystem had a positive correlation with temperature. On the other hand, as temperature increased, plant ecosystem might suffer higher ecosystem respiration cost and lower net productivity. In May and September, the pixels showing negative correlation coefficient between temperature and CUE occupied most of the SNP (Figure 14 and Table 3).

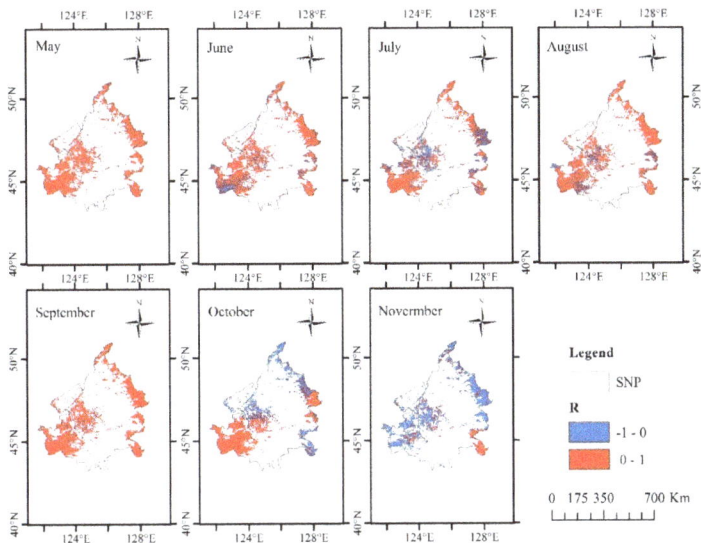

Figure 13. Correlation coefficients between CUE and monthly precipitation.

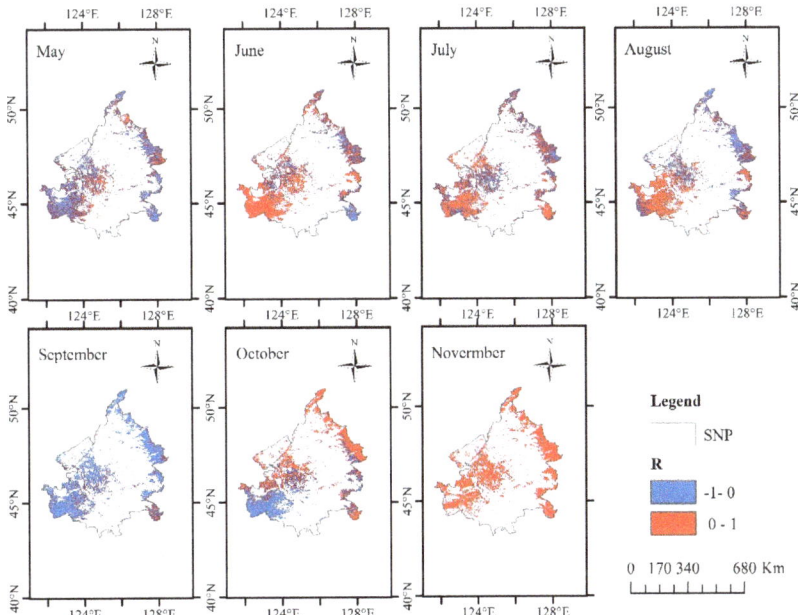

Figure 14. Correlation coefficients between CUE and temperature at monthly scale during 2001–2015.

Table 2. The number of pixels and their proportions of correlation coefficients between monthly precipitation and CUE.

Month	Positive Pixels (%)	Negative Pixels (%)
May	67,293 (91.60%)	6170 (8.40%)
June	58,717 (79.93%)	14,746 (20.07%)
July	49,180 (66.95%)	24,283 (33.05%)
August	61,691 (83.98%)	11,772 (16.02%)
September	69,366 (94.42%)	4097 (5.58%)
October	44,115 (60.05%)	29,348 (39.95%)
November	20,398 (27.77%)	53,065 (72.23%)

Table 3. The number of pixels and their proportions of the correlation coefficients between monthly temperature and CUE.

Month	Positive Pixels (%)	Negative Pixels (%)
May	25,181 (34.28%)	48,282 (65.72%)
June	48,525 (66.05%)	24,938 (33.95%)
July	43,585 (59.33%)	29,878 (40.67%)
August	39,588 (53.89%)	33,875 (46.11%)
September	7667 (10.44%)	65,796 (89.56%)
October	38,603 (52.55%)	34,860 (47.45%)
November	72,078 (98.11%)	1385 (1.89%)

4. Discussion

Previous studies on CUE using remote sensing methods mainly focus on changes at annual scale. In this paper, CUE at the seasonal and monthly scales were investigated. Thus, the change trends of CUE and the climate factors affecting CUE in different growth stages could be explained. CUE was considered to be a constant value regardless of ecosystem types or species [28,29]. However, this assumption at a global scale might be controversial, because it ignores the influence of environmental factors [30,31]. Tang et al. [3] estimated global average CUE using site data, which varied widely between 0.201 and 0.822. In this study, the estimated monthly CUE from satellite observations ranged from 0.021 to 0.999 in the SNP. The results suggested that CUE among ecosystems could not be a constant. The assumption of a constant CUE of 0.5 might lead to biased estimates for carbon cycling modelling across temporal-spatial scales.

We compared the CUE calculated by the same model of different ecosystems at the annual scale from other reported studies (Table 4). The order of annual CUE of different ecosystems in SNP was as follows: GRA (0.567) > WET (0.542) > MF (0.480) > DBF (0.479) [19]. Tang et al. [3] found the largest CUE for WET on a global scale. Khalifa et al. [32] estimated the CUE of different vegetation in sub Saharan area and found that the annual average CUE deceased in the following sequence: WET > GRA > MF > DBF. However, in our study, the order of average CUE of the growing season in the SNP was: MF > WET > GRA > DBF. This difference may be due to the different time scales and regions with of the studies.

Previous studies indicated that plant CUE might demonstrate a significant seasonal variation. In the short term, such as over one year, the dynamic patterns of carbohydrate storage and plant carbon allocation may lead to great changes in CUE [33]. Campioli et al. [7], using biometric methods and vortex correlation techniques, evaluated temporal and spatial variation of CUE in Fagus sylvatica forest and found that CUE in spring was the highest. Artificially grown apples have higher CUE in summer, which may be consistent with the higher accumulation of biomass and the lower respiratory consumption [34]. In contrast, as SNP is at mid-high latitudes, vegetation in the SNP may reduce the consumption of respiration and increase the carbon sequestration capacity in autumn, leading to the highest CUE.

Table 4. Comparison of estimated CUE at different time scales in different researches.

Ecosystem	Time Scale	CUE	Scale	Type of Data	Data Source
WET	Annual	0.607 ± 0.133	Global	Site data	Tang [3]
	Annual	0.550–0.60	Sudan and Ethiopia	Remote sensing data	Khalifa [32]
	Annual	0.542	SNP	Remote sensing data	Li [19]
	Growing season	0.488	SNP	Remote sensing data	Our article
GRA	Annual	0.457 ± 0.109	Global	Site data	Tang
	Annual	0.220–0.560	Sudan and Ethiopia	Remote sensing data	Khalifa
	Annual	0.567	SNP	Remote sensing data	Li
	Growing season	0.482	SNP	Remote sensing data	Our article
MF	Annual	0.464 ± 0.127	Global	Site data	Tang
	Annual	0.350–0.480	Sudan and Ethiopia	Remote sensing data	Khalifa
	Annual	0.480	SNP	Remote sensing data	Li
	Growing season	0.530	SNP	Remote sensing data	Our article
DBF	Annual	0.464 ± 0.127	Global	Site data	Tang
	Annual	0.340–0.420	Sudan and Ethiopia	Remote sensing data	Khalifa
	Annual	0.479	SNP	Remote sensing data	Li
	Growing season	0.477	SNP	Remote sensing data	Our article

CUE is regarded as a dynamic parameter, and differs among species of the same biome [35]. In this study, we found that the CUE of MF ecosystems in the SNP had great potential for carbon sequestration in different seasons. GPP and NPP of GRA were very small in spring, resulting in the lowest CUE. In summer and autumn, the CUE of GRA gradually increased. This study found that CUE of GRA in summer was higher than that of DBF, possibly because GRA had less investment in plant tissue respiration than that of broad-leaved forest, as reported by Law et al. [36]. Forest types showed high CUE in autumn, because trees with higher carbon storage might be more beneficial to the growth in the next year. After analyzing the abnormal values of different vegetation in different years (Figure 8), this study found that in the spring of 2012, CUE of all types of vegetation decreased to the lowest level, which would be related to different degrees of spring drought occurring in the western part of the SNP region [26]. With lower average temperature in the autumn of 2002, CUE decreased in the SNP as the temperature decreased, along with the CUE value. In the autumn of 2014, the CUE of vegetation decreased significantly, which was associated with moderate drought in the south-central Northeast China [37].

Phenological records can not only directly reveal the changes of natural seasons, but also illustrate the response and adaptation of ecosystem process and results to global environmental changes. Few previous studies have discussed the relationship between phenology and CUE. The phenological metrics that we extracted were similar to the study of Huang et al. [2]. Most of the existing literatures have focused on the relationship between NPP, GPP and phenology. Earlier SOS may extend the growing season longer and lead to an increase in GPP [38]. Similarly, the delay in EOS may also prolong the growth season, causing increases in GPP and NPP [18]; therefore, the CUE value of the vegetation will increase. Vegetation requires relatively less energy to maintain living tissues in lower temperature conditions, resulting in less respiration costs and higher CUE [39]. On the other hand, vegetation growth is generally constrained by the short growing season. Rising temperature could extend the growing season length and significantly increase GPP [12]. The sensitivity of CUE to temperature under lower-temperature conditions is lower because the temperature sensitivities of GPP and autotrophic respiration are of comparable size. In warm regions, especially in the tropics where the growing season is long, by contrast, the respiration consumption of vegetation are higher, leading to a lower CUE [1].

CUE is sensitive to environmental conditions and climate change [40]. Previous studies found that net productivity would increase linearly with higher average annual precipitation and temperature in cold and dry ecosystem [1]. As a function of GPP, NPP and respiration, CUE of vegetation (for instance, forest) may be affected by temperature and precipitation [41]. One reported study suggested that CUE exhibited a decreasing trend with the increase of precipitation when precipitation was less than 2300 mm year^{-1}. CUE showed an increasing trend along temperature when it was between −10 °C and 20 °C, as well as an increasing trend with rising temperature [1]. In this study, CUE showed an increasing trend from May to July and from August to October, respectively, possibly because the hydrothermal condition was more suitable during those two time periods. Increased precipitation may lead to a higher NPP/GPP ratio [6]. The variations of temperature affect both the photosynthesis and Ra rates, resulting in the changes of vegetation CUE [42]. The ratio of NPP to GPP might increase as the annual temperature increased between −10 and 20 °C [1], which was partially explained by the findings of this study.

In addition, in recent decades, to improve the local ecological environment and enhance the ecological protection barrier function, the Chinese government and local citizens have taken multiple measures and implemented actions for ecological and environmental protection [2]. Ecological and environmental restoration projects such as the "Three-North Shelterbelt Project" and the "Grain for Green Project" have achieved some positive effects [43,44]. We used the same method to calculate the CUE of farmland. By comparison, we found that the average CUE of the natural ecosystem in SNP showed a similar variation as that of the internal farmland from 2001 to 2015 (Figure 15). The CUE of farmland and natural ecosystem increased simultaneously. The respiration consumption of vegetation decreased. This also showed that the ecological protection function of natural ecosystem may have been strengthened during the past 15 years.

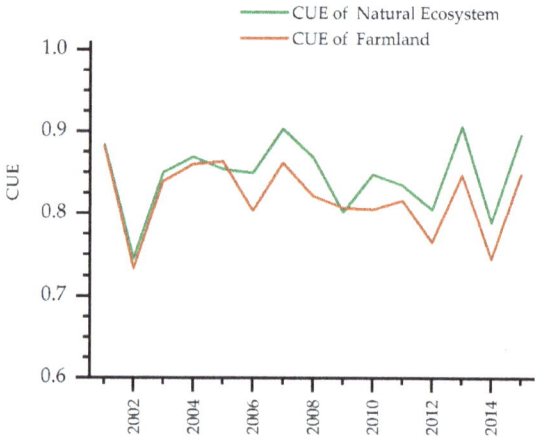

Figure 15. Average CUE variations of natural ecosystems and farmland in the SNP from 2001 to 2015.

5. Conclusions

Quantifying the variations of interannual CUE among ecosystems has proved to be a useful tool when calculating interannual carbon budgets. However, the intraannual change of CUE may present different characteristics. The assessment of temporal and spatial variations of CUE at shorter time scales and the impact on them of phenological and climatic factors are still poorly understood. This study attempted to reveal spatial patterns of CUE of natural ecosystems at different temporal scales in the SNP, China. The differences of CUE between months and seasons were significant. Monthly average CUE showed the highest in October and lowest in May. Average CUE was the highest in autumn, followed by summer. The variability of NPP accumulation in different seasons was significant. The highest CUE values were observed in MF in the growing season, indicating better ecological protection effects. The spatial variations of CUE were different. The pixels with rising CUE were mainly concentrated in southwest GRA and eastern DBF. The SOS was generally observed in March and April, while EOS dates were found in October and November. The earlier SOS and later EOS exerted a positive influence on CUE in the SNP, especially in 80% of the areas covered by broad-leaved forest. Longer LOS might cause the increase in CUE. In addition, CUE was positively correlated with precipitation and temperature in most areas of the SNP. Increasing trend of CUE in the SNP suggested a protective barrier function of natural ecosystems in the protected farmland region.

Author Contributions: B.L. and F.H. had the original idea for the study and designed the research. B.L. analyzed the data and wrote the paper. F.H. supervised the research and provided significant suggestions. L.Q., H.Q. and N.S. was involved in the data processing and the manuscript reviewing.

Funding: This research was supported by the National Natural Science Foundation of China (Grant No. 41571405 and 41571115) and the key project of the National Natural Science Foundation of China (Grant No. 41630749).

Acknowledgments: The authors would like to thank the reviewers and editors for their valuable comments and suggestions.

Conflicts of Interest: The authors declare no conflict of interest.

References

1. Zhang, Y.; Xu, M.; Chen, H.; Adams, J. Global pattern of NPP to GPP ratio derived from MODIS data: Effects of ecosystem type, geographical location and climate. *Glob. Ecol. Biogeogr.* **2009**, *18*, 280–290. [CrossRef]
2. Huang, F.; Wang, P.; Chang, S.; Li, B. Rain use efficiency changes and its effects on land surface phenology in the Songnen Plain, Northeast China. In Proceedings of the SPIE Remote Sensing 2018, Berlin, Germany, 11–12 September 2018. [CrossRef]
3. Tang, X.; Carvalhais, N.; Moura, C.; Ahrens, B.; Koirala, S.; Fan, S.; Guan, F.; Zhang, W.; Gao, S.; Magliulo, V.; et al. Global variability of carbon use efficiency in terrestrial ecosystems. *Biogeosci. Discuss.* **2019**, *2019*. [CrossRef]
4. Delucia, E.; Drake, J.; Thomas, R.; Gonzalez, M. Forest carbon use efficiency: Is respiration a constant fraction of gross primary production? *Glob. Chang. Biol.* **2007**, *13*, 1157–1167. [CrossRef]
5. Luyssaert, S.; Inglima, I.; Jung, M.; Richardson, A.; Reichsteins, M.; Papale, D.; Piao, S.; Schulzes, E.; Wingate, L.; Matteucci, G. CO_2 balance of boreal, temperate, and tropical forests derived from a global database. *Glob. Chang. Biol.* **2007**, *13*, 2509–2537. [CrossRef]
6. Zhang, Y.; Yu, G.; Jian, Y.; Wimberly, M.; Zhang, X.; Jian, T.; Jiang, Y.; Zhu, J. Climate-driven global changes in carbon use efficiency. *Glob. Ecol. Biogeogr.* **2014**, *23*, 144–155. [CrossRef]
7. Campioli, M.; Gielen, B.; Gockede, M.; Papale, D.; Bouriaud, O.; Granier, A. Temporal variability of the NPP-GPP ratio at seasonal and interannual time scales in a temperate beech forest. *Biogeosciences* **2011**, *8*, 2481–2492. [CrossRef]
8. Chambers, J.; Tribuzy, E.; Toledo, L. Rrepiration from a tropical forest ecosystem: Partitioning of sources and low carbon use efficiency. *Ecol. Appl.* **2004**, *14*, S72–S88. [CrossRef]
9. Amthor, J. The McCree–de Wit–Penning de Vries–Thornley Respiration Paradigms: 30 Years Later. *Ann. Bot.* **2000**, *86*, 1–20. [CrossRef]
10. Li, Y.; Fan, J.; Hu, Z.; Shao, Q.; Harris, W. Comparison of evapotranspiration components and water-use efficiency among different land use patterns of temperate steppe in the Northern China pastoral-farming ecotone. *Int. J. Biometeorol.* **2015**, *60*, 827–841. [CrossRef]
11. Yu, G.; Wang, Q.; Zhuang, J. Modeling the water use efficiency of soybean and maize plants under environmental stresses: Application of a synthetic model of photosynthesis-transpiration based on stomatal behavior. *J. Plant Physiology* **2004**, *161*, 303–318. [CrossRef]
12. Piao, S.; Wang, S. Forest annual carbon cost: A global-scale analysis of autotrophic respiration. *Ecology* **2010**, *91*, 652–661. [CrossRef] [PubMed]
13. He, Y.; Piao, S.; Li, X.; Chen, A.; Qin, D. Global patterns of vegetation carbon use efficiency and their climate drivers deduced from MODIS satellite data and process-based models. *Agric. For. Meteorol.* **2018**, *256*, 150–158. [CrossRef]
14. Noormets, A. (Ed.) *Phenology of Ecosystem Processes*; Springer: New York, NY, USA, 2009. [CrossRef]
15. Richardson, A.; Keenan, T.; Migliavacca, M.; Ryu, Y.; Sonnentag, O.; Toomey, M. Climate change, phenology, and phenological control of vegetation feedbacks to the climate system. *Agric. For. Meteorol.* **2013**, *169*, 156–173. [CrossRef]
16. Jin, J.; Ying, W.; Zhen, Z.; Magliulo, V.; Min, C. Phenology plays an important role in the regulation of terrestrial ecosystem water-use efficiency in the Northern Hemisphere. *Remote Sens.* **2017**, *9*, 664. [CrossRef]
17. Min, M.; Zhu, W.; Wang, W.; Xu, Y.; Liu, J. Evaluation of vegetation phenology remote sensing identification method based on carbon exchange data of flux tower net ecosystem. *Chin. J. Appl. Ecol.* **2012**, *23*, 319–327.
18. Qi, H.; Huang, F.; Zhai, H. Monitoring spatio-temporal changes of terrestrial ecosystem soil water use efficiency in Northeast China using time series remote sensing data. *Sensors* **2019**, *19*, 1481. [CrossRef]

19. Li, B.; Huang, F.; Chang, S.; Sun, N. The variations of satellite-based ecosystem water use and carbon use efficiency and their linkages with climate and human drivers in the Songnen Plain, China. *Adv. Meteorol.* **2019**, *2019*, 8659138. [CrossRef]
20. Huang, F.; Liu, X.; Wang, P.; Zhang, S.; Zhang, Y. Land use/cover change and its driving forces of west of Songnen Plain. *J. Soil Water Conserv.* **2003**, *17*, 14–17. [CrossRef]
21. Piao, S.; Fang, J.; Guo, Q. Application of CASA model to the estimation of Chinese terrestrial net primary productivity. *Acta Phytoecol. Sin.* **2001**, *25*, 603–608. [CrossRef]
22. Zhu, W.; Pan, Y.; He, H.; Yu, D.; Hu, H. Simulation of maximum light utilization rate of typical vegetation in China. *Chin. Sci. Bull.* **2006**, *51*, 700–706. [CrossRef]
23. Mao, D.; Wang, Z.; Han, J.; Ren, C. Temporal and spatial patterns and driving factors of vegetation NPP in Northeast China from 1982 to 2010. *Sci. Geogr. Sin.* **2012**, *32*, 1106–1111.
24. Tang, J.; Jiang, Y.; Zhang, N.; Hu, M. Estimation of vegetation net primary productivity and carbon sink in western jilin province based on CASA model. *J. Arid Land Resour. Environ.* **2013**, *27*, 1–7.
25. Tang, X.; Li, H.; Desai, A.; Nagy, Z.; Luo, J.; Kolb, T.; Olioso, A.; Xu, X.; Li, Y.; Kutsch, W. How is water-use efficiency of terrestrial ecosystems distributed and changing on Earth? *Sci. Rep.* **2014**, *4*, 7483. [CrossRef] [PubMed]
26. Wang, S.; Duan, H.; Feng, J. Drought events and its influence in spring of 2012 in China. *J. Arid Meteorol.* **2012**, *30*, 298–304. [CrossRef]
27. George, L. Carbon-use efficiency of terrestrial ecosystems under stress conditions in Southeast Europe (MODIS, NASA). *Proceedings* **2018**, *2*, 363. [CrossRef]
28. Running, S. A general model of forest ecosystem processes for regional applications I. Hydrologic balance, canopy gas exchange and primary production processes. *Ecol. Model.* **1988**, *42*, 125–154. [CrossRef]
29. Gifford, R. The global carbon cycle: A viewpoint on the missing sink. *Funct. Plant Biol.* **1994**, *21*, 1–15. [CrossRef]
30. Dewar, R.; Medlyn, B.; Mcmurtrie, R. Acclimation of the respiration/photosynthesis ratio to temperature: Insights from a model. *Glob. Chang. Biol.* **2010**, *5*, 615–622. [CrossRef]
31. Xiao, C.; Yuste, J.; Janssens, I.; Roskams, P.; Nachtergale, L.; Carrara, A.; Sanchez, B.Y.; Ceulemans, R. Above- and belowground biomass and net primary production in a 73-year-old scots pine forest. *Tree Physiol.* **2003**, *23*, 505–516. [CrossRef]
32. Khalifa, M.; Elagib, N.; Ribbe, L.; Schneider, K. Spatio-temporal variations in climate, primary productivity and efficiency of water and carbon use of the land cover types in Sudan and Ethiopia. *Sci. Total Environ.* **2018**, *624*, 790–806. [CrossRef]
33. Arneth, A.; Kelliher, F.; McSeveny, T.; Byers, J. Net ecosystem productivity, net primary productivity and ecosystem carbon sequestration in a pinus radiata plantation subject to soil water deficit. *Tree Physiol.* **1998**, *18*, 785. [CrossRef] [PubMed]
34. Zanotelli, D.; Montagnani, L.; Manca, G.; Tagliavini, M. Net primary productivity, allocation pattern and carbon use efficiency in an apple orchard assessed by integrating eddy-covariance, biometric and continuous soil chamber measurements. *Biogeosciences* **2013**, *10*, 3089–3108. [CrossRef]
35. Street, L.E.; Subke, J.A.; Sommerkorn, M.; Sloan, V.; Ducrotoy, H.; Phoenix, G.K.; Williams, M. The role of mosses in carbon uptake and partitioning in arctic vegetation. *New Phytol.* **2013**, *199*, 163–175. [CrossRef] [PubMed]
36. Law, B.; Falge, E.; Gu, L.; Baldocchi, D.; Bakwin, P.; Berbigier, P.; Davis, K.; Dolman, A.; Falk, M.; Fuentes, J. Environmental controls over carbon dioxide and water vapor exchange of terrestrial vegetation. *Agric. For. Meteorol.* **2015**, *113*, 97–120. [CrossRef]
37. Wang, S.; Duan, H.; Feng, J. Drought events and its influence in autumn of 2014 in China. *J. Arid Meteorol.* **2014**, *32*, 1031–1039. [CrossRef]
38. Keenan, T.; Gray, J.; Friedl, M.; Toomey, M.; Bohrer, G. Net carbon uptake has increased through warming-induced changes in temperate forest phenology. *Nat. Clim. Chang.* **2014**, *4*, 598–604. [CrossRef]
39. Ryan, M.; Linder, S.; Vose, J.; Hubbard, R. Dark respiration of pines. *Ecol. Bull* **1994**, *43*, 50–63. [CrossRef]
40. Bradford, M.; Crowther, T. Carbon use efficiency and storage in terrestrial ecosystems. *New Phytol.* **2013**, *199*, 7–9. [CrossRef]
41. Cox, P. Description of the "Triffid" Dynamic Global Vegetation Model. Hadley Centre Technical Note. 2001. Available online: https://www.researchgate.net/publication/245877262_Description_of_the_TRIFFID_dynamic_global_vegetation_model/related (accessed on 26 October 2019).

42. Giardina, C.P.; Ryan, M.G.; Binkley, D.; Fownes, J.H. Giardina, Primary production and carbon allocation in relation to nutrient supply in a tropical experimental forest. *Glob. Chang. Biol.* **2003**, *9*, 1438–1450. [CrossRef]
43. Li, A.; Han, Z.; Huang, C.; Tan, Z. Remote sensing monitoring on dynamic of sandy desertification degree in Horqin sandy land at the beginning of 21st century. *J. Desert Res.* **2007**, *27*, 546–551.
44. Du, Z.; Zhan, Y.; Wang, C.; Song, G. The dynamic monitoring of desertification in Horqin sandy land on the basis of MODIS NDVI. *Remote Sens. Land Resour.* **2009**, *21*, 14–18. [CrossRef]

© 2019 by the authors. Licensee MDPI, Basel, Switzerland. This article is an open access article distributed under the terms and conditions of the Creative Commons Attribution (CC BY) license (http://creativecommons.org/licenses/by/4.0/).

Article

Mapping Spatial Variations of Structure and Function Parameters for Forest Condition Assessment of the Changbai Mountain National Nature Reserve

Lin Chen [1,2,3], Chunying Ren [1,*], Bai Zhang [1], Zongming Wang [1,4] and Yeqiao Wang [3,*]

1. Key Laboratory of Wetland Ecology and Environment, Northeast Institute of Geography and Agroecology, Chinese Academy of Sciences, Changchun 130102, China; chenlin@iga.ac.cn (L.C.); zhangbai@iga.ac.cn (B.Z.); zongmingwang@iga.ac.cn (Z.W.)
2. University of Chinese Academy of Sciences, Beijing 100049, China
3. Department of Natural Resources Science, University of Rhode Island, Kingston, RI 02881, USA
4. National Earth System Science Data Center, Beijing 100101, China
* Correspondence: renchy@iga.ac.cn (C.R.); yqwang@uri.edu (Y.W.); Tel.: +86-431-8554-2297 (C.R.)

Received: 19 October 2019; Accepted: 9 December 2019; Published: 13 December 2019

Abstract: Forest condition is the baseline information for ecological evaluation and management. The National Forest Inventory of China contains structural parameters, such as canopy closure, stand density and forest age, and functional parameters, such as stand volume and soil fertility. Conventionally forest conditions are assessed through parameters collected from field observations, which could be costly and spatially limited. It is crucial to develop modeling approaches in mapping forest assessment parameters from satellite remote sensing. This study mapped structure and function parameters for forest condition assessment in the Changbai Mountain National Nature Reserve (CMNNR). The mapping algorithms, including statistical regression, random forests, and random forest kriging, were employed with predictors from Advanced Land Observing Satellite (ALOS)-2, Sentinel-1, Sentinel-2 satellite sensors, digital surface model of ALOS, and 1803 field sampled forest plots. Combined predicted parameters and weights from principal component analysis, forest conditions were assessed. The models explained spatial dynamics and characteristics of forest parameters based on an independent validation with all *r* values above 0.75. The root mean square error (RMSE) values of canopy closure, stand density, stand volume, forest age and soil fertility were 4.6%, 33.8%, 29.4%, 20.5%, and 14.3%, respectively. The mean assessment score suggested that forest conditions in the CMNNR are mainly resulted from spatial variations of function parameters such as stand volume and soil fertility. This study provides a methodology on forest condition assessment at regional scales, as well as the up-to-date information for the forest ecosystem in the CMNNR.

Keywords: forest parameter mapping; forest condition assessment; sentinel series; ALOS series; changbai mountain national nature reserve

1. Introduction

Forests occupy almost one third of the Earth's land area [1], playing a major role in sustaining global material and energy cycles [2]. Forests provide a variety of ecosystem services, which are important for human well-being and the overall health of the planet Earth [3,4]. Forest condition is an essential component of both forest management and ecological evaluations. It reflects the stability, resilience, and capability of carbon sequestration, timber production, as well as other services [5,6]. Current forest condition assessments are mainly based on the structure and function investigated in the field, which is costly and spatially limited [7]. It is essential to assess forest condition based on modeling structural and functional parameters. The condition assessment based on remote sensing usually contains indicators of community structure and productivity [6–8]. The sub-compartment measurements of the National Forest Inventory in China contain the information about structure, including canopy closure, stand density and forest age, and function, including stand volume and soil condition [9,10].

The explicit mapping of spatial variations of forest structure and function parameters has been an essential effort in ecological analysis [11–14]. Remote sensing modeling combined sample plot data has become a well adopted method to generate spatially explicit estimates of forest parameters [15,16]. The selection of predictor variables from various sensors and algorithms can affect the results considerably [17,18]. Variables from optical sensors are commonly applied to predict horizontal forest structure such as canopy closure and density [19,20]. This is due to the close relationship between horizontal forest structure and aggregate spectral signatures, i.e., reflectance or vegetation indices, with global coverage, repetitiveness, and cost-effectiveness [21,22]. However, synthetic aperture radar (SAR) and light detection and ranging (LiDAR) sensors are capable of penetrating cloud and canopies and are suitable for mapping vertical forest parameters such as tree height and stand volume [23–25]. Whereas, complex forest parameters such as biomass, soil fertility, and forest age are generally estimated by multi-sensor data [26–29].

Modeling vegetation parameters based on remote sensing can be divided into physically based models and empirical regression algorithms [18,21,30]. Physically based models depend on numerous factors to simulate canopy reflectance, such as leaf geometry, chlorophyll concentration, water and matter contents, soil reflectance, and bidirectional reflectance distribution function, which may not be readily available [31,32]. Those are built conventionally as semi-physical models by simplifying factors based on prerequisite assumptions and using machine learning or regression methods trained with radiative transfer, which achieve robust performance [33,34]. The biophysical products, such as leaf area index (LAI) and fraction of vegetation cover (FVC) from Sentinel-2, are generated by a physically based model, which has been implemented to Moderate Resolution Imaging Spectrometer (MODIS) and Landsat sensors [35,36]. Empirical regressions require support from abundant ground measurements, and depend on the modeling relationship between remote sensing-derived predictors and field-measured samples, including parametric and non-parametric algorithms [37,38]. The former refers to statistical regression methods, by which the expression relating to the dependent variable, i.e., forest parameters, and the independent variables are estimated [39,40]. These regressions are suitable to model explicit relationships and are easily applied to a large scale [12,41,42]. As for the complex forest parameters such as stand volume, forest age, and soil fertility, it is a challenge to formulate their relationships with remote sensing data because of many affecting factors [43–45] which require non-parametric algorithms. Among the various non-parametric techniques, random forests (RF) has been recognized to be efficient and accurate in modeling complex relationships between remote sensing data and forest parameters [7,17,46,47].

Forest parameter modeling based on satellite data has advantages such as repetition rate enabling long-term monitoring [48,49]. Sentinel-1 C band SAR and Sentinel-2 multispectral instrument (MSI) have the global coverage [50]. Those publicly accessible data have been applied in vegetation studies and provided capabilities for forest parameter modeling using both active and passive remote sensing techniques [51,52]. The Advanced Land Observing Satellite (ALOS/ALOS-2) Phased Array type L band SAR (PALSAR/PALSAR-2) images from L band SAR contain comprehensive information on the orientation and structure of tree canopy and stems within the pixel [53–55]. It makes the yearly mosaic ALOS/ALOS-2 images with global and free-access observations particularly useful for forest parameter mapping [56,57]. The digital surface model (DSM) from ALOS L band interferometric SAR (InSAR) had greater accuracy and can provide useful topographic indices to estimate forest parameters [58,59]. Although estimates of forest parameters from moderate resolution satellite images and abovementioned algorithms have achieved varying success [12,13,49], inventory and application of efficient algorithms and predictors from open-access remote sensing data on forest structure, function, and condition assessment continuously deserve exploration.

The Changbai Mountain National Nature Reserve (CMNNR) in Northeast China is covered with large areas of old-growth forests, which are under strict protection [60,61]. It has been regarded as the most typical natural composite body on the Eurasia Continent with a complex biota composition and abundant flora and fauna [62,63]. Due to its ecological importance, substantial researches on landscape structure, function, and productivity in the Changbai Mountain region have been reported since the early 1980s [64–69]. However, there is a lack of systematic maps of forest parameters and conditions in this vital ecosystem site.

In this study, we developed an effective methodology for evaluating forest conditions by mapping canopy closure, stand density, volume, age, and soil fertility in the CMNNR. The specific objectives were to: (1) model forest structure and function parameters by determining their relationships with predictors from satellite data, including L and C band SAR, topographical indices from L band InSAR, and Sentinel-2 MSI variables; (2) map five parameters using efficient algorithms with remote sensing data; and (3) assess forest conditions based on structural and functional parameters, which can provide baseline information for forest management.

2. Materials and Methods

2.1. Study Area

The CMNNR (41°42′–42°25′N, 127°42′–128°17′E) was established in 1961. It is located in Jilin Province of northeastern China (Figure 1). The top of the reserve is a volcanic summit cupping a crater lake named Tianchi at 2693 m above sea level on the China-North Korea border. The reserve occupies an area of 195,852 ha in the Chinese side. It has the largest protected temperate forests which supports a significant species gene base and biodiversity in Northeast Asia. The CMNNR was admitted into the United Nations Educational, Scientific, and Cultural Organization's (UNESCO's) Man and Biosphere Program in 1979 [70]. This site has a continental temperate climate characterized by a long cold winter and short cool summers with recognizable vertical climate and vegetation zones. The annual average precipitation and temperature range from 700 to 1400 mm and −7 °C to 3 °C, respectively. This area has dense forests covered 177,082 ha (90.4%). The forests in the reserve are divided into three functional management zones including the core, buffer, and transition areas [71]. Harvesting and poaching are prohibited in the core area. The human disturbances are prevented from the core area by the buffer zone. The endemic species, ecotourism, and bases for reproduction of natural resources are established in the transition area. From the foot of the mountain to the peak, vegetation changes in distinguishable vertical zones with elevation. The distributions include mixed coniferous and broad-leaved forest (<1100 m), dark-coniferous spruce-fir forest (1100–1700 m), Ermans birch forest (1700–2000 m), and alpine tundra (>2000 m) [72]. Soils also differ in each vertical zone, typically marked by dark-brown and brown earths, meadow, volcanic, bog, and bleached baijiang soil.

2.2. Data

2.2.1. Field Data

The field campaign was carried out from the end of August to the beginning of October in 2017. The stratified sampling design was adopted. Non-forest areas were masked out and the distribution of sampling plots was randomly generated in forest areas, while the plot sites which were impossible to access were replaced by the nearest sites. Following the national guidelines for forest resource survey [73], ten teams took part in the field campaign and collected measured data under the same protocol. A total of 1803 30 m by 30 m plots were located and sampled (Figure 1). At each sample site, tree species, diameter at breast height (DBH, the diameter at 1.3 m from the ground), tree height, soil depth and types, as well as the number of trees, were measured and recorded. Overhead photos of the canopy by fisheye lens were taken at the center of each sample site. The measured parameters were acquired based on field sample sites and processes shown in Table 1. Stand volume of each sample was estimated by DBH and tree height according to the National Standard of China: Tree volume tables (LY/T 1353–1999) [74]. Canopy closure was estimated from fisheye photos after processing of circle clipping, binary conversion, and raster calculation [75]. Soil fertility was assessed by the depth weighted according to soil types as shown in Table 1. The six types of soils in the study area were assigned three different weights based on previous studies of their organic carbon density [76,77]. The descriptive statistics of field-based forest parameters were shown in Table 2. The 1803 sample sites were randomly split into training (n = 1202) and validation (n = 601) sets (Figure 1) for modeling and validating the spatial variation of forest parameters, respectively.

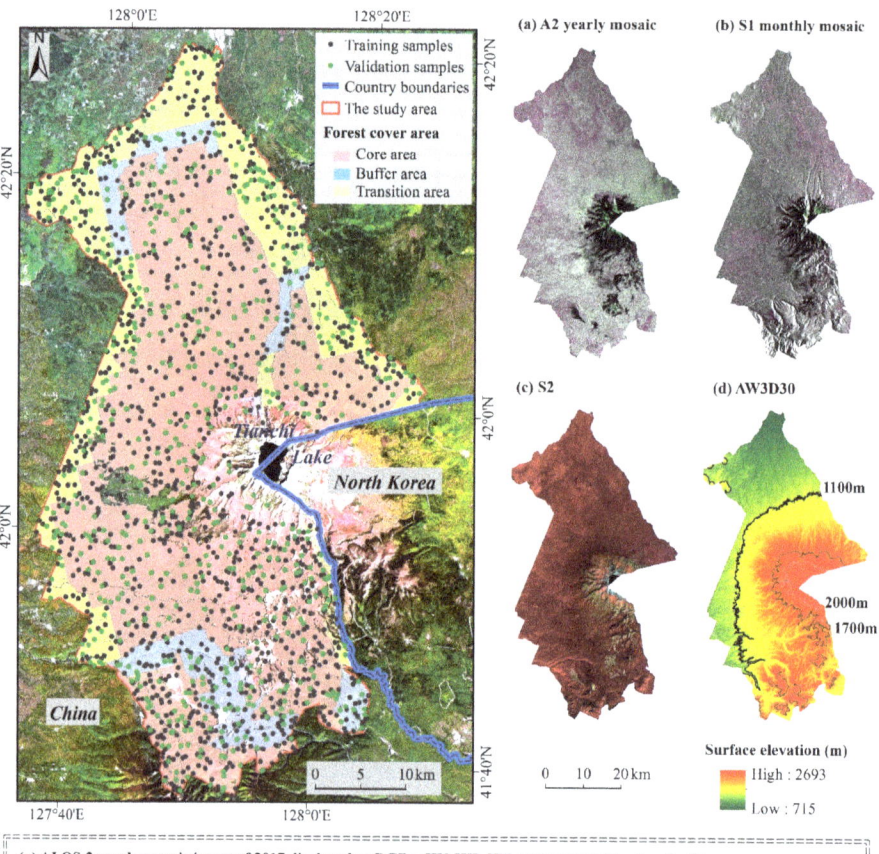

(a) ALOS-2 yearly mosaic image of 2017 displayed as RGB = HV, HH, HV, gamma naught values in dB.
(b) Sentinel-1 Level-1 GRD monthly (Sep., 2017) mosaic image displayed as RGB = VV, VH, VV, gamma naught values in dB.
(c) Sentinel-2B Level-1C image acquired in 25th Sep., 2017 displayed as RGB = Band 8, 4, 3.
(d) ALOS Digital Surface Model "ALOS World 3D-30m" (AW3D30).

Figure 1. The figure illustrates the shape of the Changbai Mountain National Nature Reserve (CMNNR), and locations of field sample sites, and employed satellite remote sensing data derived from Advanced Land Observing Satellite (ALOS), ALOS-2, Sentinel-1, and Sentinel-2 series.

Table 1. Field measurements and processing to acquire measured forest parameters.

Measurements	Parameters	Processing
Diameter at breast height (D_t, cm) Tree height (H_t, m)	Stand volume (m³/ha)	$a \cdot D_t^b \cdot H_t^c$, a–c are the species specific constants, as provided by Tree volume tables (LY/T 1353-1999)
Fisheye photos	Canopy closure (%)	Canopy area/total area times 100
Soil types	Soil fertility (no unit)	Dark-brown earths or Bog soil:1 × D_s, Meadow soil or Volcanic soil: 0.8 × D_s, Brown earths or Bleached baijiang soil: 0.6 × D_s
Soil depth (D_s, cm)		
Forest age	Forest age (no unit)	Classes from one to five meaning young to over-mature forests were acquired from the forest manager's archives at the local forestry bureau
Tree number	Stand density (tree/ha)	Number/area = number/(0.09 ha)

Table 2. Descriptive statistics of field-measured forest parameters.

Parameters	Minimum	Maximum	Mean	Median	Standard Deviation	Coefficient of Variation (%)
Canopy closure (%)	20	90	78.89	80	9.02	11.43
Stand density (tree/ha)	100	8000	619	500	602.26	97.30
Stand volume (m³/ha)	5	553	227	240	99.70	43.92
Forest age	1	5	3.32	4	1.02	30.72
Soil fertility	15	70	38.68	32	15.70	40.59

2.2.2. Remote Sensing Data

In this study, 18 predictor variables related to forest parameters were selected and extracted from multi-sensors imagery (Tables 3 and 4) [12,78–80]. ALOS-2 PALSAR-2 yearly mosaic image of 2017 was masked and converted to gamma naught values in decibel unit (dB) from 16-bit digital number (DN) (Table 4) using the following equation [81]:

$$\gamma^0 = 10 \log_{10}(DN^2) - 83 \tag{1}$$

where γ^0 is gamma naught backscatter coefficient of horizontal transmit-horizontal channel (HH) or horizontal transmit-vertical channel (HV); DN is the polarization data in HH or HV.

Monthly mosaic predictors from C band SAR were generated from seven Sentinel-1 Ground Range Detected images by masking and mosaic using Google Earth Engine (GEE). Those data were pre-processed by thermal noise removal, radiometric calibration, and terrain correction stored in dB via log scaling [82]. The cloud-free Sentinel-2B Level-1C images acquired on 25 September 2017 were downloaded from the Copernicus Sentinel Scientific Data Hub (https://scihub.copernicus.eu/) to extract vegetation and soil indices, as well as biophysical variables. Previous studies explored numerous Sentinel-2 spectral indices. They found that four vegetation indices and two soil indicators were useful in modelling forest age and soil fertility, respectively [78,80,83–85]. The MSI data had 13 spectral bands with 10 m (bands 2–4, 8), 20 m (band 5–7, 8a, 11–12), and 60 m (band 1, 9–10) spatial resolutions. Bottom-of-atmosphere-corrected reflectance (Level-2A) images were atmospherically corrected from the Level-1C data by SEN2COR atmospheric correction processor based on the radiative transfer model (version 2.5.5, European Space Agency, Paris, France). Then, eight predictors were acquired by resampling, band math, biophysical processor, and mosaic from Level-2A images based on SNAP software (version 6.0, European Space Agency).

The DSM data from ALOS as AW3D30 were download from Japan Aerospace Exploration Agency (https:ww.eroc.jaxa.jp/ALOS/en/aw3d30/data/index/htm) to extract topographic indices based on Spatial Analyst of ArcGIS software (version 10.0, ESRI, RedLands, CA, USA). All predictor data were re-projected into UTM Zone 52 WGS84, and then resampled to the 30 m pixel size by ArcGIS.

Table 3. The ALOS-2, Sentinel-1, Sentinel-2, and digital surface model (DSM) data used in this study.

Sensors	Elements	Time	Spatial Resolution (m)	Source
ALOS-2	1	2017	25	A2 mosaic
Sentinel-1	Two of Sentinel-1A five of Sentinel-1B	20170906/0918 20170903/0910/0915/0922/0927	10	S1 mosaic
Sentinel-2	Two of Sentinel-2B, T52TDM/T52TCM	20170925	10	S2
ALOS	N041E127/N041E128/ N042E127/N042E128	Derived from PALSAR data during 2006 to 2011	30	AW3D30

Table 4. Predictors from remote sensing data for spatial modeling of forest parameters.

Sources	Predictors	Description	Parameters	Processing
A2 mosaic	HH	Gamma naught backscatter coefficient of horizontal transmit-horizontal channel in dB	Stand volume, soil fertility, forest age	Masking, conversion to gamma naught values based on Google Earth Engine (GEE)
	HV	Gamma naught backscatter coefficient of horizontal transmit-vertical channel in dB		
S1 mosaic	VV	Gamma naught backscatter coefficient of vertical transmit-vertical channel in dB	Soil fertility, forest age	Masking and mosaic based on GEE
	VH	Gamma naught backscatter coefficient of vertical transmit-horizontal channel in dB		
S2	LAI	Leaf area index	Canopy closure, stand density	Atmosphere correction based on Sen2Cor, then resampling, biophysical processor, and mosaic based on SNAP
	FVC	Fraction of vegetation cover		
	NDVI	Normalized difference vegetation index, (B8 − B4)/(B8 + B4)	Forest age	Atmosphere correction based on Sen2Cor, then resampling, vegetation radiometric indices processing, and mosaic based on SNAP
	GEMI	Global environmental monitoring index, eta × (1 − 0.25 × eta) − (B4 − 0.125)/(1 − B4), where eta = [2 × (B8A − B4) + 1.5 × B8A + 0.5 × B4]/(B8A + B4 + 0.5)		
	GNDVI	Green normalized difference vegetation index, (B7 − B3)/(B7 + B3)		
	S2REP	Sentinel-2 red-edge position index, 705 + 35 × [(B4 + B7)/2 − B5] × (B6 − B5)		
	BI2	The second brightness index, sqrt ((B4 × B4 + B3 × B3 + B8 × B8)/3)	Soil fertility	Atmosphere correction based on Sen2Cor, then resampling, soil radiometric indices processing, and mosaic based on SNAP
	CI	The color index, (B4 − B3)/(B4 + B3)		
AW3D30	H	Surface elevation	Soil fertility, forest age	Spatial analysis based on ArcGIS
	Slope	Slope		
	Aspect	Aspect		
	C_v	Profile curvature		
	C_h	Plan curvature		
	TWI	Topographic wetness index, Ln[Ac/tanβ], Ac is the catchment area directed to the vertical flow		

2.3. Assessment of Forest Conditions

2.3.1. Spatial Modeling of Canopy Closure and Stand Density by Statistical Regressions

Stand density is a prominent component of forest structure, which governs elemental processing and retention, competition, and habitat suitability [86,87]. Canopy closure is the proportion of the sky hemisphere occupied by tree crowns when viewed from a single ground point [88,89]. It is closely associated with understory light and has wide-reaching effects for ecological processes in forests [90]. Whereas FVC is defined as the percentage of the forest area covered by the vertical projection of trees [91,92]. LAI is one half of the total green leaf area per unit ground surface area [93]. FVC and LAI are critical biodiversity variables as recognized by international organizations such as Global Climate Observation System and Global Terrestrial Observation System [35]. The generalized linear correlations were discovered between stand density and LAI, canopy closure and FVC, and LAI and FVC in previous studies [42,94,95]. Thus, it was assumed in this study that canopy closure and stand density could be modeled by generalized linear regressions based on FVC and LAI from Sentinel-2 images. In this study, five types of regression models (linear, quadratic polynomial, power, exponential, and logarithmic) were built. LAI or FVC, and canopy closure or stand density, were used as the input variables to derive the empirical parameters for the models. This study selected the model with the largest value of coefficient of determination (R^2) to map the canopy closure and stand density.

2.3.2. Spatial Modeling of Stand Volume and Forest Age by Random Forests

Firstly, a semi-physical simple water cloud model (WCM) was used for the investigation of the relationship between stand volume and backscatters (HH and HV) derived from ALOS-2 data. The prerequisite assumption of WCM was that the dielectric constant of dry vegetation matter was much smaller than that of the water content of vegetation, and almost all volume backscatters were composed of air in the vegetation canopy [96]. Therefore, WCM was developed assuming that the canopy "cloud", called the water cloud, contained identical water droplets showed the random distribution within the canopy [55]. In this study, the WCM was adopted for the initial exploration [97], which was written as Equation (2):

$$\gamma_f^0 = \gamma_g^0 e^{-\delta SV} + \gamma_v^0 (1 - e^{-\delta SV}) \tag{2}$$

where γ_f^0 is the backscatter from the forest, as the gamma naught value of HH or HV (dB); γ_g^0 is the direct backscatter from the forest floor through gaps in the canopy (dB); γ_v^0 is the volume backscatter from an opaque canopy without gaps (dB); SV is stand volume (m^3/ha); and δ is the extinction coefficient.

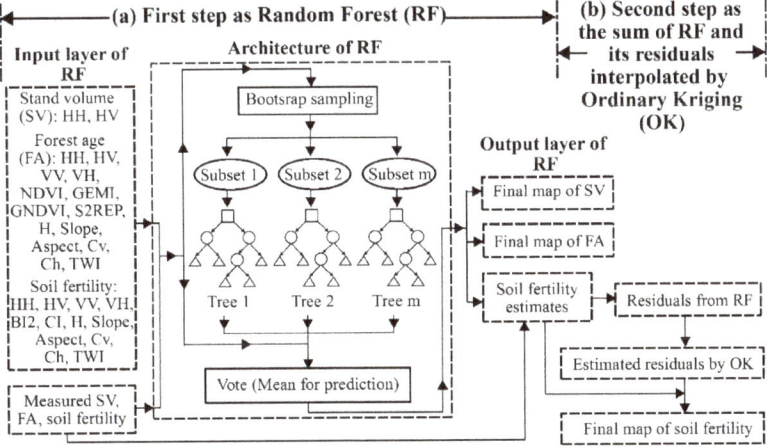

Figure 2. Steps of spatial modeling on stand volume and forest age by random forests and that on soil fertility by random forest kriging.

Then, RF was used to model the spatial distribution of stand volume with predictors of HH and HV. The RF was an ensemble of decision trees, which was created by a subset of training sample through replacement as a bagging approach [98]. Each decision tree was independently developed without any pruning and each node was divided using a user-defined number of features selected at random [99]. By producing the forest up to a user-defined number of trees, RF creates trees with large variance and small bias [98,99]. The abovementioned two user-defined parameters, i.e., numFeatures and numIterations, were selected by the smallest root mean square error (RMSE) in WEKA software (version 3.8, The University of Waikato, Hamilton, New Zealand), and the attribute importance was also estimated [100]. The new unlabeled data were input to evaluate and vote, and the finial prediction was the average of the membership (Figure 2a). Additionally, forest age, as indirect and complex retrieved parameters for remote sensing techniques, was also modeled by RF with multi-sensor predictors (Table 4 and Figure 3).

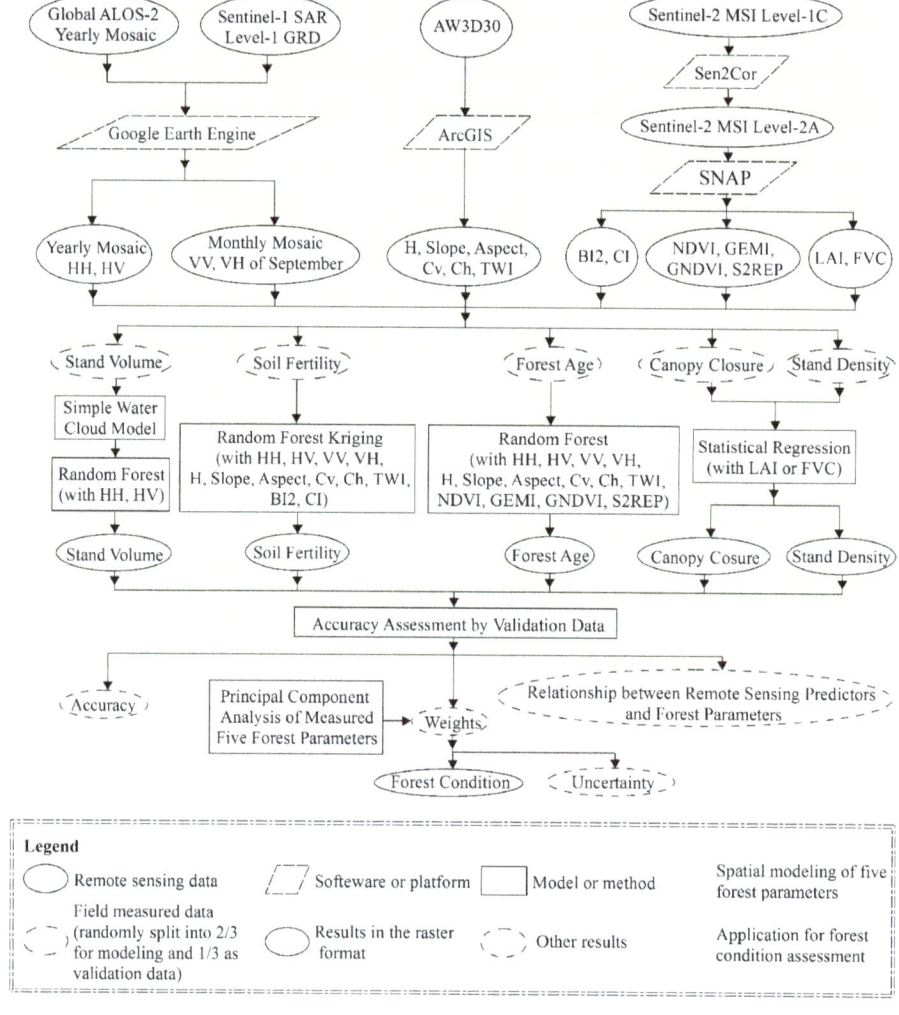

Figure 3. The flowchart for spatial modeling of parameters and application on forest condition assessment.

2.3.3. Spatial Modeling of Soil Fertility by Random Forest Kriging

Random forest kriging (RFK) was the extension of RF, which integrated RF prediction values and estimation of the residuals by ordinary kriging (OK) using Equation (3) [101]. It considered spatial parametric non-stationarity with the effects of environmental variables derived from the benefits of

RF [102,103]. RFK also added the spatial dependence of the residuals interpolated through OK to the estimated trend, as part of the spatial autocorrelation. RFK has been conducted in soil attribute mapping and had a greater accuracy than RF [104,105]. Its implementation included two steps (Figure 2). First, RF was used to model the relationship between soil fertility and multi-sensor predictors. Second, the result of RFK was predicted as the sum of the RF result and its residuals interpolated by the OK approach using Equation (4).

$$SF_{RFK} = SF_{RF} + R_{OK} \tag{3}$$

$$\begin{cases} R_{OK} = \sum_{i=1}^{n} \omega_i R_i \\ \sum_{i=1}^{n} \omega_i = 1 \end{cases} \tag{4}$$

where SF_{RFK}, SF_{RF} are predication of soil fertility based on RFK and RF, respectively; R_{OK} is the estimated residuals of soil fertility from RF models; R_i is the residuals of soil fertility from RF models at a measured sample i; ω_i is the weight estimated by the stationary OK system as an error variance model to minimize the error from the semivariogram modeling [106]; and n is the number of measured values within a neighborhood.

2.3.4. Model Evaluation and Forest Condition Assessment

The validation set (Figure 1) was used to test performances of spatial modeling of forest parameters based on the root mean squared error (Equation (5)), mean absolute error (MAE, Equation (6)), mean error (ME, Equation (7)), and correlation coefficient between the measured and predicted parameters (r, Equation (8)). In order to better estimate accuracy, the mean measured value of the parameter ($\overline{y}$) was applied to divide the RMSE, MAE, and ME (Equations (5)–(7)).

$$\text{RMSE} = \frac{1}{\overline{y}} \sqrt{\sum_{1}^{n} \frac{(y_i - \hat{y}_i)^2}{n}} \times 100\% \tag{5}$$

$$\text{MAE} = \frac{1}{\overline{y}} \sum_{1}^{n} \frac{|y_i - \hat{y}_i|}{n} \times 100\% \tag{6}$$

$$\text{ME} = \frac{1}{\overline{y}} \sum_{1}^{n} \frac{y_i - \hat{y}_i}{n} \times 100\% \tag{7}$$

$$r = \frac{\sum_{1}^{n}(y_i - \overline{y})(\hat{y}_i - \overline{\hat{y}})}{\sqrt{\sum_{1}^{n}(y_i - \overline{y})} \sqrt{\sum_{1}^{n}(\hat{y}_i - \overline{\hat{y}})}} \tag{8}$$

where y_i is the measured parameter value; $\hat{y}_i$ is the predicted parameter value; $\overline{y}$ and $\overline{\hat{y}}$ are the average of measured and predicted values of the parameter, respectively; and n is 601 in this study. The RMSE and MAE should be as small as possible. The ME should be close to zero, while r should be larger.

After that, each map of forest parameters was transformed into the spatial distribution of the parameter score as Equation (9). All parameters were positive indicators for forest condition, except that stand density was considered as a complex indicator. Indeed, excessive or insufficient stand density was harmful to forest conditions [107]. In this study, optimum stand density for forest condition was assigned as the median of measured values (500 tree/ha in Table 2). In other words, stand density below 500 tree/ha was regarded as a positive parameter along with canopy closure, stand volume, forest age, and soil fertility. While stand density above 500 tree/ha was regarded as a negative parameter.

$$Score_j = \begin{cases} \frac{P_i - P_{min}}{P_{max} - P_{min}} \times 100, & P \text{ is, CC, SV, FA, SF or SD} \leq 500 \\ \frac{P_{max} - P_i}{P_{max} - P_{min}} \times 100, & P \text{ is SD} > 500 \end{cases} \tag{9}$$

where $Score_j$ is the score of parameter j; P_i, P_{min}, and P_{max} are raw data, minimum, and maximum values of spatial modeled parameters, respectively; CC, SV, FA, SF, and SD are canopy closure, stand volume, forest age, soil fertility, and stand density, respectively.

To estimate quantitatively the weight of each parameters, the principal component analysis (PCA) is a common method to use [108,109]. PCA was performed under the factor analysis in SPSS (version 21.0, IBM, Armonk, NY, USA) using Equation (10) by three elements, i.e., coefficients of parameters in linear combinations of different principal components, variance contribution rate of principal components, and normalization of weights. Finally, the forest condition assessment map was generated by the score and weight of each parameter according to Equation (11).

$$\begin{cases} w_j = \dfrac{\dfrac{\sum_{k=1}^{q} \dfrac{C_{jk}}{\sqrt{E_k}} V_k}{\sum_{k=1}^{q} V_k}}{\sum_{j=1}^{5} \dfrac{\sum_{k=1}^{q} \dfrac{C_{jk}}{\sqrt{E_k}} V_k}{\sum_{k=1}^{q} V_k}} \\ \sum_{j=1}^{5} w_j = 1 \\ \sum_{j=1}^{5} V_k \geq 8 \end{cases} \quad (10)$$

where w_j is the weight of parameter j; C_{jk} is the component matrix value of parameter j in component k; E_k is eigenvalue of component k; V_k is variation contribution rate of component k; and q is principal component number.

$$Condition_{score} = \sum_{j=1}^{5} Score_j \times w_j \quad (11)$$

where $Condition_{score}$ is the score of forest condition s; $Score_j$ is the score of parameter j; and w_j is the weight of parameter j.

3. Results

3.1. Canopy Closure and Stand Density

The five types of statistical regression models were built as illustrated in Figure 4. Among five models, logarithmic, and quadratic power regressions with the largest values of R^2 were the best at explaining relationship of canopy closure with LAI and FVC, respectively. Considered the much larger R^2 value of a LAI-based model, LAI derived from Sentinel-2 was selected to map canopy closure based on the logarithmic regression model. Likewise, the exponential regression model with FVC was selected to map stand density. For comparison with field measured values, the modeled output of canopy closure and stand density were divided into several levels for displaying (Figure 5). Specifically, each level had an equal number of measured sample sites. The better performance of spatial modeling of canopy closure and stand density can be indicated by the agreement pattern at each level. Generally, predicted canopy closure and stand density were close to field measured values (Table 2). The large values of canopy closure and stand density were distributed in lower altitude regions. There is no forest in the high elevation alpine tundra and the volcanic summit of the study site. The southern slope of the Changbai Mountain showed less canopy closure and stand density than the north, as affected by historical volcanic damages (Figure 5).

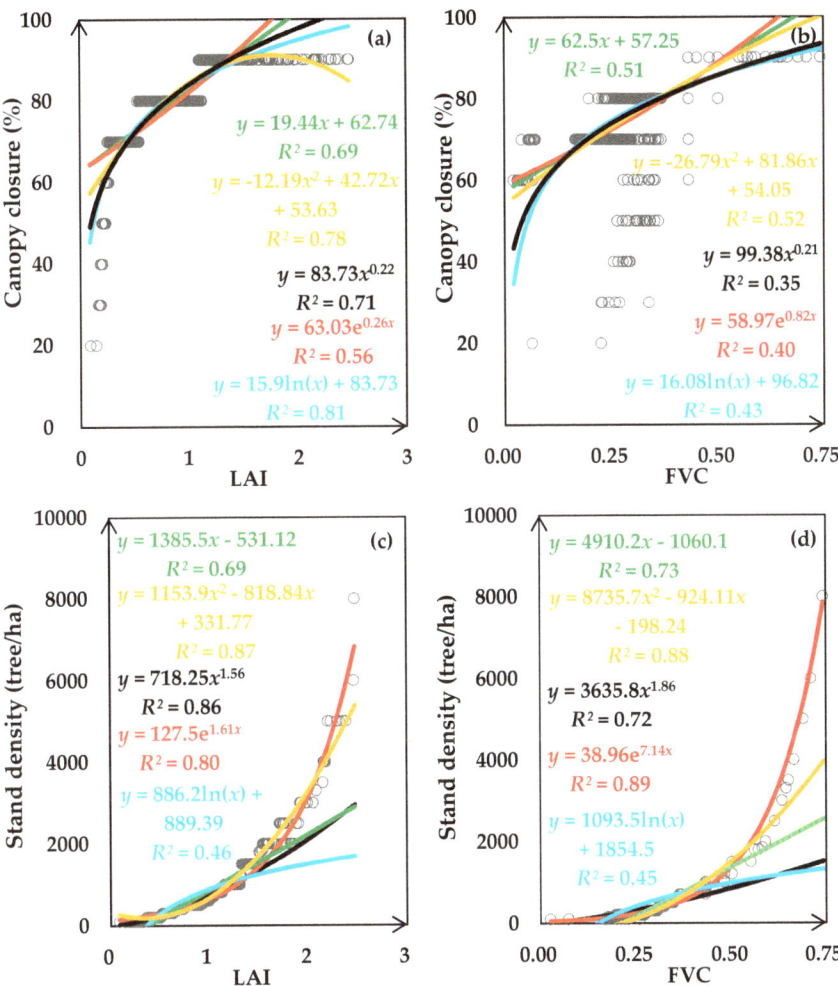

Figure 4. Statistical regressions of canopy closure and stand density based on Sentinel-2 leaf area index (LAI) and fraction of vegetation cover (FVC). Regressions of canopy closure by LAI and FVC were illustrated as (**a**,**b**), respectively. Models of stand density by LAI and FVC were shown in (**c**,**d**).

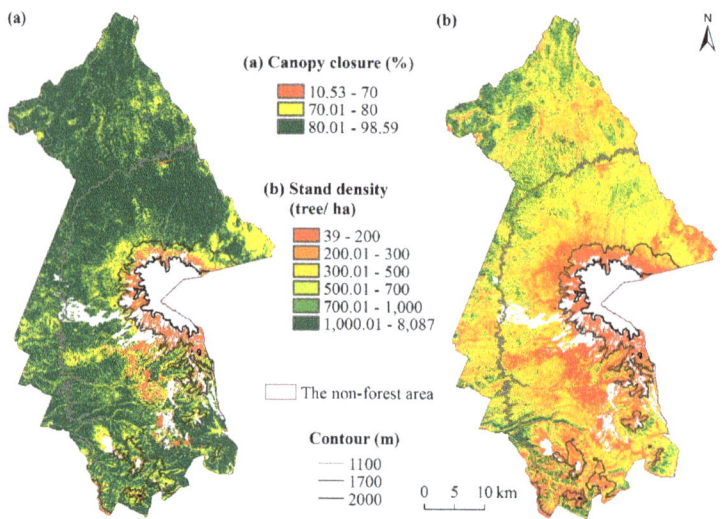

Figure 5. Canopy closure (**a**) and stand density (**b**) in the CMNNR.

3.2. Stand Volume and Forest Age

The fitting line of HV showed a much flatter range than that of HH (Figure 6), indicating that HV was much more sensitive to stand volume than HH. It was also shown that the L band ALOS-2 backscatters reached saturation at around 400 m^3/ha, which was greater than 97.4% of measured stand volume (Table 2). Thus, ALOS-2 data were considered suitable for spatial modeling of stand volume in the study area. Based on 1000 decision trees and one feature, the RF model was built to predict stand volume with more contribution from HV than HH. The result was depicted in Figure 7a and different levels were divided against measured values. It was delineated that the northeastern part of the study area was a large valued region.

The RF model with 1000 decision trees and six features was trained to predict forest age. The attribute importance ranking in decreasing order was H, Slope, Global environmental monitoring index (GEMI), Aspect, normalized difference vegetation index (NDVI), green normalized difference vegetation index (GNDVI), topographic wetness index (TWI), Sentinel-2 red-edge position index (S2REP), HV, C_v, C_h, VV, HH, and VH. It was indicated that topographic and vegetation indices contributed more than SAR backscatters. According to measured age, the majority of forests were mature or over-mature with a median value of four (Table 2). The predicted forest age, as Figure 7b, was consistent with the measured values. Greater values of forest age were found in the western part of the study area. It was also shown that small values of forest age and stand density in higher altitude regions where forest distribution was limited.

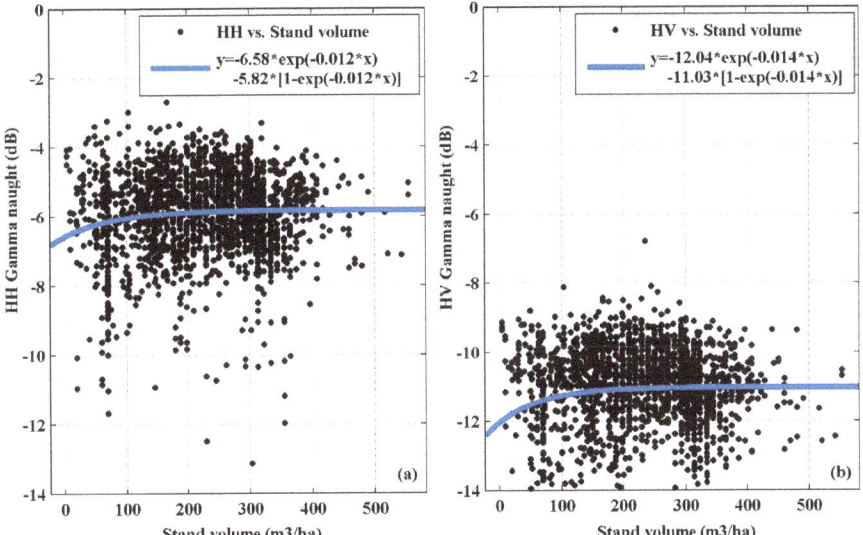

Figure 6. Relationships between backscatters and stand volume using the simple water cloud model. (**a**) horizontal transmit-horizontal channel (HH) backscatters from pure forest canopy is −5.82 dB, from bare soil is −6.58 dB, and the extinction coefficient is 0.012. (**b**) horizontal transmit-vertical channel (HV) backscatters from pure forest canopy is −11.03 dB, from bare soil is −12.04, and the extinction coefficient is 0.014.

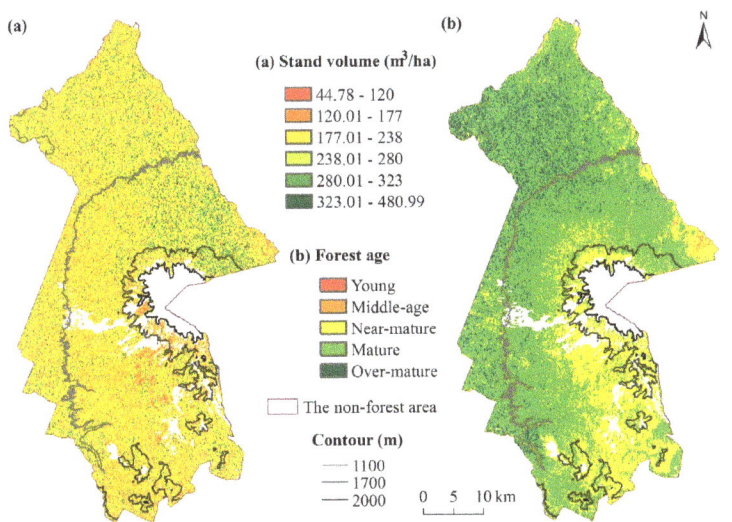

Figure 7. Stand volume (**a**) and forest age (**b**) in the CMNNR.

3.3. Soil Fertility

The RF model with 1000 decision trees and seven features was trained to predict soil fertility at the first step. The ranking of attribute importance in decreasing order was H, Slope, Aspect, C_v, C_h, TWI, second brightness index (BI2), color index (CI), HV, HH, VV, and VH. It was revealed that topographic and soil indices contributed more than SAR backscatters. By dividing into three levels against measured values, the RF-based soil fertility displayed zonal distribution (Figure 8a). The small and large values were mixed in distribution. The RF model overestimated small values while underestimating large values in comparison to measured soil fertility (Table 2). With range, nugget, and sill values of 14.5 km, 48.6, and 230.5, respectively, semivariogram in OK interpolation was modeled as an exponential type and used to predict soil fertility residuals from RF at the second step. The small value of basal effect (nugget/sill = 0.21) showed a strong autocorrelation of residuals. It was suggested that soil fertility prediction be overestimated by RF in the high-altitude area (Figure 8b). Final predicted soil fertility by RFK with more equal area of each level than RF prediction, which means closer to the measured soil fertility, was shown in Figure 8c. Soil fertility decreased as altitude increased, which agreed with vertical zonal patterns affected by the volcanic eruption hundreds of years ago.

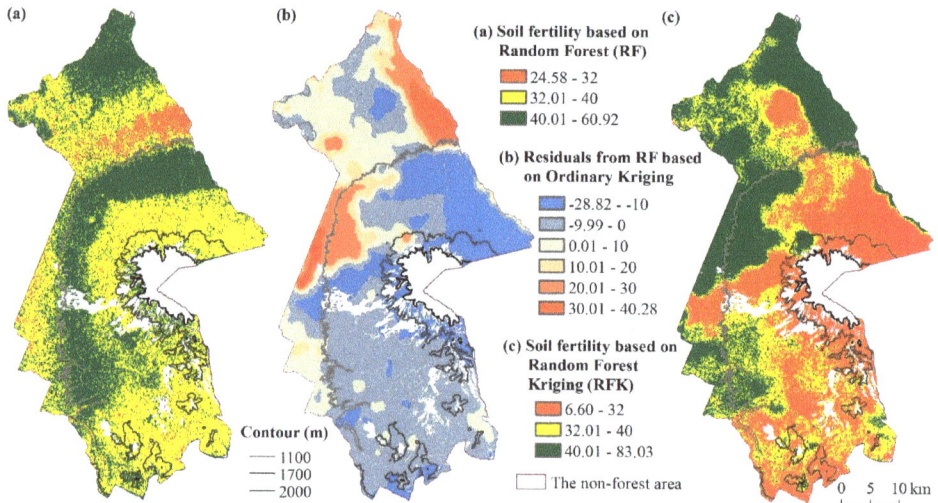

Figure 8. Soil fertility in the CMNNR. Soil fertility predicted by random forest was shown in (**a**), and its residuals interpolated by ordinary kriging was (**b**). Final map of soil fertility was (**c**).

3.4. Assessment of Modeling Accuracy and Forest Condition

The spatial modeling accuracy of five forest parameters was estimated by independent data (n = 601). Canopy closure was modeled with the greatest accuracy among the five parameters, while modeling stand density performed the worst with the least accuracy (Table 5). With large values of r and R^2 ($r \geq 0.75$ and $R^2 \geq 0.6$), it was revealed that all models explained spatial dynamics and characteristics of parameters to a good extent (Figure 9). The modeled parameters were credible ($r \geq 0.75$) for application in forest condition assessment.

Table 5. Accuracy assessment of forest parameter modeling based on independent validation data.

Parameters	ME (%)	MAE (%)	RMSE (%)	r
Canopy closure	−0.15	3.65	4.62	0.91
Stand density	−3.27	17.29	33.80	0.96
Stand volume	−0.60	17.42	29.41	0.75
Forest age	0.51	11.77	20.50	0.76
Soil fertility	0.13	9.45	14.31	0.94

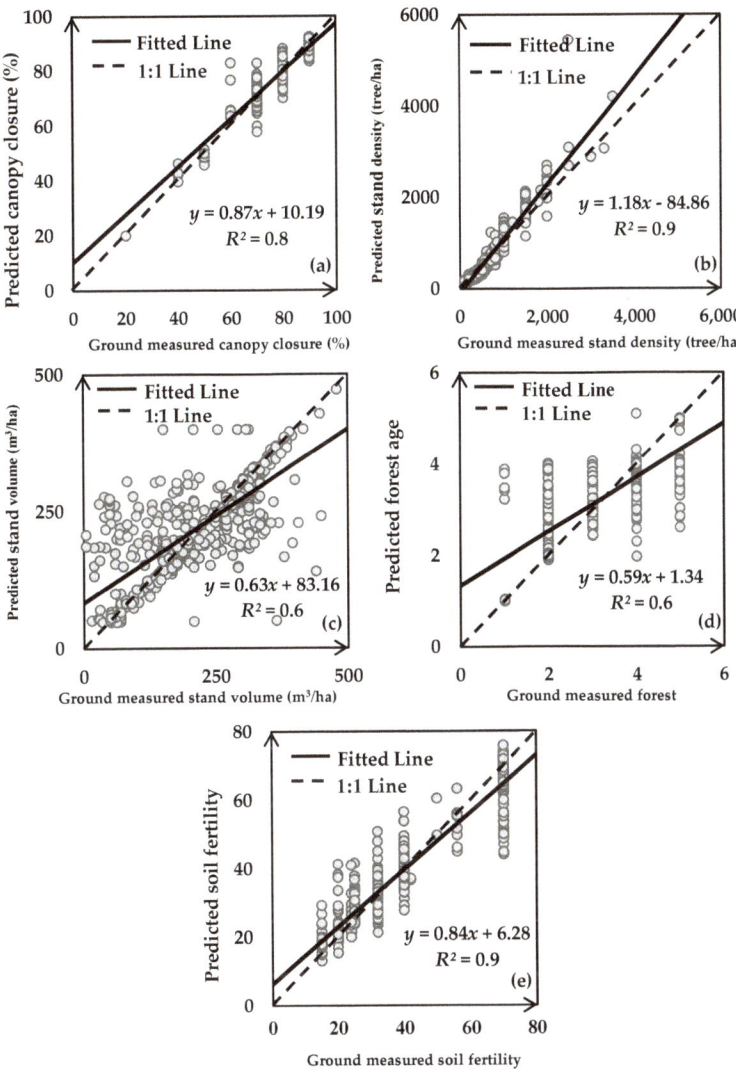

Figure 9. Scatter plots of predicted versus ground measured parameters from validation data including canopy closure (**a**), stand density (**b**), stand volume (**c**), forest age (**d**), and soil fertility (**e**). As for forest age, one to five represented young to over-mature.

According to PCA of measured parameters, two components were acquired and shown as Table 6. Forest age contributed the most among the five parameters with a weight of 0.23, followed by stand volume and canopy closure. Stand density had the weakest influence with a weight of 0.12. The normalized parameters scores were shown in Figure 10a–e. The largest score in canopy closure was distributed homogeneously, with almost all above 60. However, the score of soil fertility showed the strongest spatial variations, followed by that of stand density. Weighted by five parameters, forest condition score was mapped as Figure 10f. The forest of the study area in 2017 had major scores were between 50 to 70 and coefficient of variation (CV) as 14.79% (Figure 10g).

Table 6. Component matrix and weights of five parameters.

Parameters	Component 1 with Contribution Rate of 44.73% and Eigenvalue of 2.69	Component 2 with Contribution Rate of 36.19% and Eigenvalue of 1.31	Weight
Canopy closure	0.89	−0.05	0.21
Stand density	0.80	−0.29	0.12
Stand volume	0.07	0.75	0.23
Forest age	0.14	0.79	0.26
Soil fertility	0.47	0.23	0.18

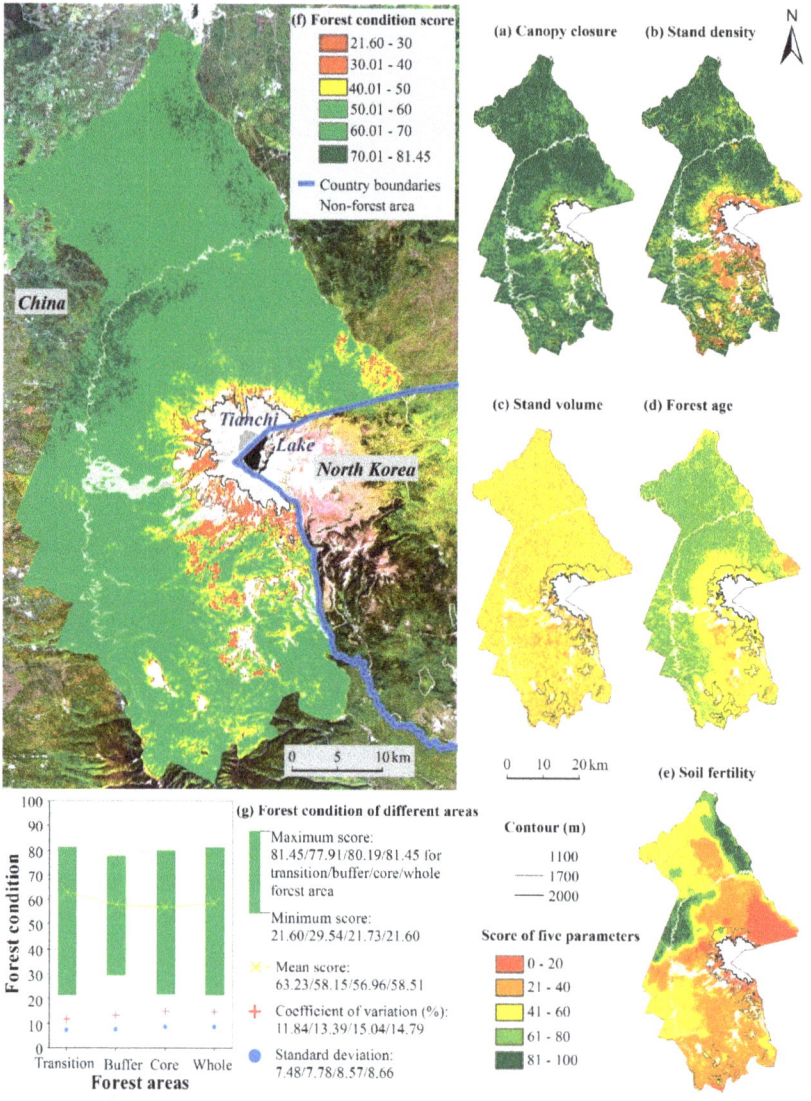

Figure 10. Forest parameters and condition in the CMNNR. Forest parameters included canopy closure (**a**), stand density (**b**), stand volume (**c**), forest age (**d**) and soil fertility (**e**). Distribution of condition was in (**f**). Forest condition of different areas was summarized in (**g**).

4. Discussion

4.1. Understanding Forest Parameters with Remote Sensing Predictors

Close relationships were found for canopy closure and stand density with LAI and FVC from Sentinel-2 (Figure 4). LAI explained more canopy closure ($R^2 = 0.81$) than FVC ($R^2 = 0.52$). However, the performances of LAI ($R^2 = 0.87$) and FVC ($R^2 = 0.89$) on modeling stand density were similar. The assessment by independent sample sites (Table 5 and Figure 9) showed that LAI and FVC revealed spatial variations of canopy closure and stand density ($r \geq 0.9$ and $R^2 \geq 0.8$). While RMSE values showed that stand density was estimated with the largest error among the five parameters (Table 5). This may have resulted from the saturation problems from Sentinel-2. It was indicated that biophysical products of Sentinel-2, especially LAI, had good abilities to delineate the spatial variation of simple horizontal structure, such as canopy closure and stand density, in the study area.

The backscatter from HV was more sensitive to stand volume than HH based on the WCM models (Figure 6) and the attribute importance in the RF model. This revealed that HV backscatter was more helpful than HH to model forest productivity, which was consistent with previous findings of aboveground biomass [110,111]. The extinction coefficients modeled by WCM models in this study were much larger than previous studies in modeling aboveground biomass with backscatters and without mosaic [55,97]. This resulted in a relatively less accuracy of stand volume among five forest parameters.

The attribute importance in RF models demonstrated that topographic and spectral indices from L band InSAR and MSI contributed more than backscatters from L and C band SAR in modeling forest age and soil fertility. Additionally, backscatters from HV and VV had influenced more on forest parameter modeling than HH and VH, respectively. However, the ranking of predictor importance was different between forest age and soil fertility. The L band InSAR predictors showed the absolute dominance in soil fertility modeling, followed by variables from MSI, L band SAR, and C band SAR. This was on account of the good penetrability of L band and sensitivity to vegetation (CI) and soil humidity (BI2) of indices from MSI. As for forest age modeling, HV backscatter was much more important than second-derivative topographic micro indices (C_v and C_h). Additionally, vegetation indices from near-infrared (Band 8 and 8A), red (Band 4), vegetation red edge (Band 7), and green band (Band 3) had a greater effect than the complex macro topographic index (TWI) on forest age modeling. Also, the backscatter from VV was much more significant than that from HH. It was denoted that forest age had a more complex relationship than soil fertility with SAR and MSI data, which contained multisource influences form basic vertical and horizontal forest parameters. Moreover, the multi-sensor modeling of soil fertility based on RF algorithms showed certain limits in predicting minimum and maximum values (Figure 8a) with strongly auto-correlated residuals (nugget/sill = 0.21). It was also revealed that soil fertility had the heterogeneity conveyed by backscatters and reflectance, and the spatial autocorrelation dependence on own attributes.

4.2. Uncertainty of Spatial Modeling

The uncertainty analysis is crucial for understanding the quality of remote sensing-based forest parameters. RMSE used in this study is the common statistic to characterize the uncertainty [112,113]. Overall, the uncertainty of forest parameter modeling was acceptable with all r values above 0.75 and RMSE below 35% based on the independent validation data (Table 5). The uncertainties were from three aspects in this study as field-measurements, predictor variables and modeling. In order to get representative sample sites, a total of 1803 plots covering forests in the study area were measured (Figure 1). To match the remote sensing data, the plot size was set as 30 m by 30 m. Then, open-access remote sensing data from four different sources were selected to match the field campaign time. Predictor variables were derived from the monthly mosaic of filtered Sentinel-1 images. Limited by the cloud cover, only one cloud-free image from Sentinel-2 acquired on 25 September 2017 was used.

The forest parameters in this study were modeled with efficient predictor variables from minimum data sources based on previous findings. Specifically, canopy closure, and stand density were modeled by Sentinel-2 and stand volume was modeled by L band SAR. It was accorded from previous studies that SAR data were sensitive to vertical structure and function while MSI were primary in horizontal canopy modeling [18,21]. L band SAR penetrated into the canopy and scatters back from leaves, branches, and stems [114]. Hence, L band SAR was used to model stand volume, and DSM from L band InSAR was chosen to extract topographic indices in this study, rather than SRTM DEM from C band as most researches used. Nevertheless, as complex parameters, forest age and soil fertility were modeled by multi-sensor data to reflect the information on basic structure and function.

The uncertainty of modeling was reduced by using efficient algorithms combined with remote sensing predictors based on existing researches. First, the physically based models were considered to acquire the basic variables which were directly related to remote sensing data, such as LAI and FVC. Then, basic structure parameters such as canopy closure and stand density were modeled by parametric algorithms to show the explicit relationships with biophysical variables. The physically based model was also used to test the suitability of L band SAR for stand volume modeling. However, the function and comprehensive parameters had complex relationships with remote sensing-derived variables. Therefore, recognized nonparametric algorithms with great accuracy, such as RF and RFK, were selected to model stand volume, forest age, and soil fertility.

4.3. Forest Condition from Structure and Function

Forest parameters and condition showed variations along the elevation gradient (Figure 10). Among four vertical vegetation zones, the mixed coniferous and broad-leaved forest had the highest scores, followed by dark-coniferous spruce-fir and Ermans birch forest. While the northern slope area within dark-coniferous spruce-fir forest had large values of stand volume (Figure 7a). This was mainly due to taller and matured trees are distributed in this region [115]. The intensity of soil fauna activities, moisture, temperature, and plant diversity in lower altitudes were more favorable than those at higher elevations in the Changbai Mountain [72,116–119], so that forest parameters and conditions generally decreased with increasing altitude.

Forest conditions in the CMNNR showed spatial variations, which were assessed by the weighted structural and functional parameters (Figure 10). The forests with higher condition scores were located in the area with lager values of soil fertility. While low values of forest condition were mainly consistent with smaller scores of stand volume. It was demonstrated that function parameters were primary in assessment of forest conditions in the CMNNR. Among three functional zones, forests in the core area showed the largest variation and were vital for improving forest conditions.

5. Conclusions

Most of current forest condition assessments are mainly based on structural and functional parameters investigated in the field. To evaluate forest conditions in a comprehensive and comparable manner, this study developed a methodology on forest condition assessment based on explicit modeling and mapping of forest parameters from satellite images. Efficient predictors and algorithms were implemented to map structure and function parameters in the CMNNR of 2017 based on ALOS-2, Sentinel-1, Sentinel-2, and DSM from ALOS. With parameter modeling, this study assessed forest conditions to provide a foundation of methodology and up-to-date information of the CMNNR.

The results included performances of predictor variables and models on spatial modeling of the structure and function, maps of forest parameters, and conditions. First, explicit relationships between Sentinl-2-derived biophysical variables and simple forest structure parameters such as canopy closure and stand density were discovered. Topographic and spectral indices from L band InSAR and MSI contributed more than L and C band SAR in RF modeling of complex forest parameters such as forest age and soil fertility. While backscatters of HV were more important in the RF modeling of stand volume, forest age, and soil fertility than those of HH. Meanwhile, backscatters of VV were more sensitive to forest age and soil fertility than those of VH. Models explained spatial dynamics and characteristics of forest parameters to a good extent based on the independent validation set ($r \geq 0.75$). Second, all maps of forest parameters showed that the lower altitude northern slope had larger values than the south. Third, the mean score of forest conditions in the CMNNR was 58.51, with the smallest in the core zone (56.96) and the largest in the transition area (63.23). The assessment illustrated that the distribution of forest conditions in the CMNNR mainly resulted from spatial variations of function parameters including stand volume and soil fertility.

Author Contributions: L.C., C.R., and B.Z. designed this research. L.C. conducted field sampling, performed the experiments, conducted the analysis and drafted the manuscript. Y.W. supervised preparation of the manuscript. L.C., Y.W., C.R., B.Z., and Z.W. revised and finalized the manuscript.

Funding: This study is supported by the National Key Research and Development Project of China (No. 2016YFC0500300), the Jilin Scientific and Technological Development Program (No. 20170301001NY), the funding from Youth Innovation Promotion Association of Chinese Academy of Sciences (No. 2017277, 2012178) and National Earth System Science Data Center of China. The principal author appreciates the scholarship provided by the China Scholarship Council (CSC) (No. 201804910492) for her study in the University of Rhode Island.

Acknowledgments: We appreciate critical and constructive comments and suggestion from the reviewers that helped improve the quality of this manuscript. The authors are grateful to the support from colleagues and local forestry bureau who participated in the field surveys and data collection. We thank the National Earth System Science Data Center (http://www..geodata.cn) for providing geographic information data. This study is supported by the National Key Research and Development Project of China (No. 2016YFC0500300, the Jilin Scientific and Technological Development Program (No. 20170301001NY), the funding from Youth Innovation Promotion Association of Chinese Academy of Sciences (No. 2017277, 2012178) and National Earth System Science Data Center of China. The principal author appreciates the scholarship provided by the China Scholarship Council (CSC) (No. 201804910492) for her study in the University of Rhode Island.

Conflicts of Interest: The authors declare no conflict of interest.

References

1. FAO. *FAO Global Forest Resources Assessment 2015*; UN Food and Agriculture Organization: Rome, Italy, 2015.
2. UNFCCC. Report of the Conference of the Parties on its Twenty-First Session, Held in Paris from 30 November to 13 December 2015. Addendum. Part Two: Action Taken by the Conference of the Parties at Its Twenty-First Session. Available online: http://unfccc.int/resource/docs/2015/cop21/eng/10a01.pdf (accessed on 29 January 2016).
3. Binder, S.; Haight, R.G.; Polasky, S.; Warziniack, T.; Mockrin, M.H.; Deal, R.L.; Arthaud, G. *Assessment and Valuation of Forest Ecosystem Services: State of the Science Review*; U.S. Department of Agriculture, Forest Service, Northern Research Station: Newtown, PA, USA, 2017.
4. Brockerhoff, E.G.; Barbaro, L.; Castagneyrol, B.; Forrester, D.I.; Gardiner, B.; González-Olabarria, J.R.; Lyver, P.O.B.; Meurisse, N.; Oxbrough, A.; Taki, H.; et al. Forest biodiversity, ecosystem functioning and the provision of ecosystem services. *Biodivers. Conserv.* **2017**, *26*, 3005–3035. [CrossRef]
5. Sugden, A.; Fahrenkamp-Uppenbrink, J.; Malakoff, D.; Vignieri, S. Forest health in a changing world. *Science* **2015**, *349*, 800–801. [CrossRef] [PubMed]
6. Trumbore, S.; Brando, P.; Hartmann, H. Forest health and global change. *Science* **2015**, *349*, 814–818. [CrossRef] [PubMed]
7. Mao, D.H.; Wang, Z.M.; Wu, B.F.; Zeng, Y.; Luo, L.; Zhang, B. Land degradation and restoration in the arid and semiarid zones of China: Quantified evidence and implications from satellites. *Land Degrad. Dev.* **2018**, *29*, 3841–3851. [CrossRef]
8. Zhao, Q.X.; Yu, S.C.; Zhao, F.; Tian, L.H.; Zhao, Z. Comparison of machine learning algorithms for forest parameter estimations and application for forest quality assessments. *Forest Ecol. Manag.* **2019**, *434*, 224–234. [CrossRef]
9. Fang, J.Y.; Brown, S.; Tang, Y.H.; Nabuurs, G.-J.; Wang, X.P.; Shen, H.H. Overestimated biomass carbon pools of the northern mid—And high latitude forests. *Clim. Chang.* **2006**, *74*, 355–368. [CrossRef]
10. Shen, W.J.; Li, M.S.; Huang, C.Q.; Tao, X.; Wei, A.S. Annual forest aboveground biomass changes mapped using ICESat/GLAS measurements, historical inventory data, and time-series optical and radar imagery for Guangdong province, China. *Agric. For. Meteorol.* **2018**, *259*, 23–38. [CrossRef]

11. Moeser, D.; Roubinek, J.; Schleppi, P.; Morsdorf, F.; Jonas, T. Canopy closure, LAI and radiation transfer from airborne LiDAR synthetic images. *Agrc. For. Meteorol.* **2014**, *197*, 158–168. [CrossRef]
12. Crowther, T.W.; Glick, H.B.; Covey, K.R.; Bettigole, C.; Maynard, D.S.; Thomas, S.M.; Smith, J.R.; Hintler, G.; Duguid, M.C.; Amatulli, G.; et al. Mapping tree density at a global scale. *Nature* **2015**, *525*, 201–205. [CrossRef]
13. Sanderman, J.; Hengl, T.; Fiske, G.; Solvik, K.; Adame, M.F.; Benson, L.; Bukoski, J.J.; Carnell, P.; Cifuentes-Jara, M.; Donato, D. A global map of mangrove forest soil carbon at 30 m spatial resolution. *Environ. Res. Lett.* **2018**, *13*, 055002. [CrossRef]
14. Xu, Y.; Chao, L.; Sun, Z.; Jiang, L.; Fang, J. Tree height explains stand volume of closed-canopy stands: Evidence from forest inventory data of China. *For. Ecol. Manag.* **2019**, *438*, 51–56. [CrossRef]
15. Miettinen, J.; Stibig, H.-J.; Achard, F. Remote sensing of forest degradation in Southeast Asia—Aiming for a regional view through 5–30 m satellite data. *Glob. Ecol. Conserv.* **2014**, *2*, 24–36. [CrossRef]
16. Wittke, S.; Yu, X.W.; Karjalainen, M.; Hyyppä, J.; Puttonen, E. Comparison of two-dimensional multitemporal Sentinel-2 data with three-dimensional remote sensing data sources for forest inventory parameter estimation over a boreal forest. *Int. J. Appl. Earth Obs.* **2019**, *76*, 167–178. [CrossRef]
17. Fassnacht, F.E.; Hartig, F.; Latifi, H.; Berger, C.; Hernández, J.; Corvalán, P.; Koch, B. Importance of sample size, data type and prediction method for remote sensing-based estimations of aboveground forest biomass. *Remote Sens. Environ.* **2014**, *154*, 102–114. [CrossRef]
18. Lausch, A.; Erasmi, S.; King, D.J.; Magdon, P.; Heurich, M. Understanding forest health with remote sensing-part II—A review of approaches and data models. *Remote Sens.* **2017**, *9*, 129. [CrossRef]
19. Vicente-Serrano, S.M.; Camarero, J.J.; Olano, J.M.; Martín-Hernández, N.; Peña-Gallardo, M.; Tomás-Burguera, M.; Gazol, A.; Azorin-Molina, C.; Bhuyan, U.; EI Kenawy, A. Diverse relationships between forest growth and the Normalized Difference Vegetation Index at a global scale. *Remote Sens. Environ.* **2016**, *187*, 14–29. [CrossRef]
20. Landry, S.; St-Laurent, M.-H.; Nelson, P.R.; Pelletier, G.; Villard, M.-A. Canopy cover estimation from Landsat images: Understory impact on top-of-canopy reflectance in a northern Hardwood forest. *Can. J. Remote Sens.* **2018**, *44*, 435–446. [CrossRef]
21. Lu, D.S.; Chen, Q.; Wang, G.X.; Liu, L.J.; Li, G.Y.; Moran, E. A survey of remote sensing-based aboveground biomass estimation methods in forest ecosystems. *Int. J. Digit. Earth* **2016**, *9*, 63–105. [CrossRef]
22. Li, W.; Cao, S.; Campos-Vargas, C.; Sanchez-Azofeifa, A. Identifying tropical dry forests extent and succession via the use of machine learning techniques. *Int. J. Appl. Earth Obs.* **2017**, *63*, 196–205. [CrossRef]
23. Abdullahi, S.; Kugler, F.; Pretzsch, H. Prediction of stem volume in complex temperate forest stands using TanDEM-X SAR data. *Remote Sens. Environ.* **2016**, *174*, 197–211. [CrossRef]
24. Cazcarra-Bes, V.; Tello-Alonso, M.; Fischer, R.; Heym, M.; Papathanassiou, K. Monitoring of forest structure dynamics by means of L-band SAR tomography. *Remote Sens.* **2017**, *9*, 1229. [CrossRef]
25. Mauya, E.W.; Koskinen, J.; Tegel, K.; Hämäläinen, J.; Kauranen, T.; Käyhkö, N. Modelling and predicting the growing stock volume in small-scale plantation forests of Tanzania using multi-sensor image synergy. *Forests* **2019**, *10*, 279. [CrossRef]
26. Koch, B. Status and future of laser scanning, synthetic aperture radar and hyperspectral remote sensing data for forest biomass assessment. *ISPRS J. Photogramm.* **2010**, *65*, 581–590. [CrossRef]
27. Mulder, V.L.; De Bruin, S.; Schaepman, M.E.; Mayr, T.R. The use of remote sensing in soil and terrain mapping—A review. *Geoderma* **2011**, *162*, 1–2, 1–19. [CrossRef]
28. Lizuka, K.; Tateishi, R. Estimation of CO_2 sequestration by the forests in Japan by discriminating precise tree age category using remote sensing techniques. *Remote Sens.* **2015**, *7*, 15082–15113.
29. Hribljan, J.A.; Suarez, E.; Bourgeau-Chavez, L.; Endres, S.; Lilleskov, E.A.; Chimbolema, S.; Wayson, C.; Serocki, E.; Chimner, R.A. Multidate, multisensor remote sensing reveals high density of carbon-rich mountain peatlands in the páramo of Ecuador. *Glob. Chang. Biol.* **2017**, *23*, 5412–5425. [CrossRef]
30. Ganguly, S.; Nemani, R.R.; Zhang, G.; Hashimoto, H.; Milesi, C.; Michaelis, A.; Wang, W.; Votava, P.; Samanta, A.; Melton, F.; et al. Generating global Leaf Area Index from Landsat: Algorithm formulation and demonstration. *Remote Sens. Environ.* **2012**, *122*, 185–202. [CrossRef]
31. Atzberger, C. Object-based retrieval of biophysical canopy variables using artificial neural nets and radiative transfer models. *Remote Sens. Environ.* **2004**, *93*, 53–67. [CrossRef]

32. Yue, J.B.; Feng, H.K.; Yang, G.J.; Li, Z.H. A comparison of regression techniques for estimation of above-ground winter wheat biomass using near-surface spectroscopy. *Remote Sens.* **2018**, *10*, 66. [CrossRef]
33. Tang, H.; Brolly, M.; Zhao, F.; Strahler, A.H.; Schaaf, C.L.; Ganguly, S.; Zhang, G.; Dubayah, R. Deriving and validating Leaf Area Index (LAI) at multiple spatial scales through lidar remote sensing: A case study in Sierra National Forest, CA. *Remote Sens. Environ.* **2014**, *143*, 131–141. [CrossRef]
34. Wolanin, A.; Camps-Valls, G.; Gómez-Chova, L.; Mateo-García, G.; van der Tol, C.; Zhang, Y.G.; Guanter, L. Estimating crop primary productivity with Sentinel-2 and Landsat 8 using machine learning methods trained with radiative transfer simulations. *Remote Sens. Environ.* **2019**, *225*, 441–457. [CrossRef]
35. Weiss, M.; Baret, F. *Sentinel 2 Toolbox Level 2 Products: LAI, FAPAR, FCOVER*; INRA: Paris, France, 2016.
36. Djamai, N.; Fernandes, R.; Weiss, M.; McNairn, H.; Goïta, K. Validation of the Sentinel Simplified Level 2 Product Prototype Processor (SL2P) for mapping cropland biophysical variables using Sentinel-2/MSI and Landsat-8/OLI data. *Remote Sens. Environ.* **2019**, *225*, 416–430. [CrossRef]
37. Ahmed, O.S.; Franklin, S.E.; Wulder, M.A.; White, J.C. Characterizing stand-level forest canopy cover and height using Landsat time series, samples of airborne LiDAR, and the Random Forest algorithm. *ISPRS J. Photogramm.* **2015**, *101*, 89–101. [CrossRef]
38. Watt, M.S.; Dash, J.P.; Bhandari, S.; Watt, P. Comparing parametric and non-parametric methods of predicting Site Index for radiata pine using combinations of data derived from environmental surfaces, satellite imagery and airborne laser scanning. *For. Ecol. Manag.* **2015**, *357*, 1–9. [CrossRef]
39. Ozdemir, I.; Karnieli, A. Predicting forest structural parameters using the image texture derived from WorldView-2 multispectral imagery in a dryland forest, Israel. *Int. J. Appl. Earth Obs.* **2011**, *13*, 701–710. [CrossRef]
40. Wang, V.; Gao, J. Importance of structural and spectral parameters in modelling the aboveground carbon stock of urban vegetation. *Int. J. Appl. Earth Obs.* **2019**, *78*, 93–101. [CrossRef]
41. Mohammadi, J.; Joibary, S.S.; Yaghmaee, F.; Mahiny, A.S. Modelling forest stand volume and tree density using Landsat ETM+ data. *Int. J. Remote Sens.* **2010**, *31*, 2959–2975. [CrossRef]
42. Taureau, F.; Robin, M.; Proisy, C.; Fromard, F.; Imbert, D.; Debaine, F. Mapping the mangrove forest canopy using spectral unmixing of very high spatial resolution satellite Images. *Remote Sens.* **2019**, *11*, 367. [CrossRef]
43. Latifi, H.; Nothdurft, A.; Koch, B. Non-parametric prediction and mapping of standing timber volume and biomass in a temperate forest: Application of multiple optical/LiDAR-derived predictors. *Forestry* **2010**, *83*, 395–407. [CrossRef]
44. Tan, K.P.; Kanniah, K.D.; Cracknell, A.P. Use of UK-DMC 2 and ALOS PALSAR for studying the age of oil palm trees in southern peninsular Malaysia. *Int. J. Remote Sens.* **2013**, *34*, 7424–7446. [CrossRef]
45. Beguin, J.; Fuglstad, G.-A.; Mansuy, N.; Paré, D. Predicting soil properties in the Canadian boreal forest with limited data: Comparison of spatial and non-spatial statistical approaches. *Geoderma* **2017**, *306*, 195–205. [CrossRef]
46. Abdollahnejad, A.; Panagiotidis, D.; Joybari, S.S.; Surový, P. Prediction of dominant forest tree species using QuickBird and environmental data. *Forests* **2017**, *8*, 42. [CrossRef]
47. Lu, W.; Lu, D.S.; Wang, G.X.; Wu, J.S.; Huang, J.Q.; Li, G.Y. Examining soil organic carbon distribution and dynamic change in a hickory plantation region with Landsat and ancillary data. *Catena* **2018**, *165*, 576–589. [CrossRef]
48. Popkin, G. US government considers charging for popular Earth-observing data. *Nature* **2018**, *556*, 417–418. [CrossRef] [PubMed]
49. Wallis, C.I.B.; Homeier, J.; Peña, J.; Brandl, R.; Farwig, N.; Bendix, J. Modeling tropical montane forest biomass, productivity and canopy traits with multispectral remote sensing data. *Remote Sens. Environ.* **2019**, *225*, 77–92. [CrossRef]
50. Malenovsky, Z.; Rott, H.; Cihlar, J.; Schaepman, M.E.; Garcia-Santos, G.; Fernandes, R.; Berger, M. Sentinels for science: Potential of Sentinel-1, -2, and -3 missions for scientific observations of ocean, cryosphere, and land. *Remote Sens. Environ.* **2012**, *120*, 91–101. [CrossRef]
51. Laurin, G.V.; Balling, J.; Corona, P.; Mattioli, W.; Papale, D.; Puletti, N.; Rizzo, M.; Truckenbrodt, J.; Urban, M. Above-ground biomass prediction by Sentinel-1 multitemporal data in central Italy with integration of ALOS2 and Sentinel-2 data. *J. Appl. Remote Sens.* **2018**, *12*, 016008. [CrossRef]
52. Jia, M.M.; Wang, Z.M.; Wang, C.; Mao, D.H.; Zhang, Y.Z. A new vegetation index to detect periodically submerged mangrove forest using single-tide Sentinle-2 imagery. *Remote Sens.* **2019**, *11*, 2043. [CrossRef]

53. Takada, M.; Mishima, Y.; Natusume, S. Estimation of surface soil properties in peatland using ALOS/PALSAR. *Landsc. Ecol. Eng.* **2009**, *5*, 45–58. [CrossRef]
54. Thiel, C.; Schmullius, C. The potential of ALOS PALSAR backscatter and InSAR coherence for forest growing stock volume estimation in Central Siberia. *Remote Sens. Environ.* **2016**, *173*, 258–273. [CrossRef]
55. Huang, X.D.; Ziniti, B.; Torbick, N.; Ducey, M.J. Assessment of forest above ground biomass estimation using multi-temporal C-band Sentinel-1 and polarimetric L-band PALSAR-2 data. *Remote Sens.* **2018**, *10*, 1424. [CrossRef]
56. Ma, J.; Xiao, X.M.; Qin, Y.W.; Chen, B.Q.; Hu, Y.M.; Li, X.P.; Zhao, B. Estimating aboveground biomass of broadleaf, needleleaf, and mixed forests in Northeastern China through analysis of 25-m ALOS/PALSAR mosaic data. *For. Ecol. Manag.* **2017**, *389*, 199–210. [CrossRef]
57. Bouvet, A.; Mermoz, S.; Le Toan, T.; Villard, L.; Mathieu, R.; Naidoo, L.; Asner, G.P. An above-ground biomass map of African savannahs and woodlands at 25 m resolution derived from ALOS PALSAR. *Remote Sens. Environ.* **2018**, *206*, 156–173. [CrossRef]
58. Aslan, A.; Rahman, A.F.; Warren, M.W.; Robeson, S.M. Mapping spatial distribution and biomass of coastal wetland vegetation in Indonesian Papua by combining active and passive remotely sensed data. *Remote Sens. Environ.* **2016**, *183*, 65–81. [CrossRef]
59. Florinsky, I.V.; Skrypitsyna, T.N.; Luschikova, O.S. Comparative accuracy of the AW3D30 DSM, ASTER GDEM, and SRTM1 DEM: A case study on the Zaoksky testing ground, Central European Russia. *Remote Sens. Lett.* **2018**, *9*, 706–714. [CrossRef]
60. Tang, L.N.; Li, A.X.; Shao, G.F. Landscape-level forest ecosystem conservation on Changbai Mountain, China and North Korea (DPRK). *BioOne* **2011**, *31*, 169–175. [CrossRef]
61. Zhang, J.L.; Liu, F.Z.; Cui, G.F. The efficacy of landscape-level conservation in Changbai Mountain Biosphere Reserve, China. *PLoS ONE* **2014**, *9*, e95081. [CrossRef]
62. Yu, D.D.; Han, S.J. Ecosystem service status and changes of degraded natural reserves—A study from the Changbai Mountain Natural Reserve, China. *Ecosyst. Serv.* **2016**, *20*, 56–65. [CrossRef]
63. Gu, X.P.; Lewis, B.J.; Niu, L.J.; Yu, D.P.; Zhou, L.; Zhou, W.M.; Gong, Z.C.; Tai, Z.J.; Dai, L.M. Segmentation by domestic visitor motivation: Changbai Mountain Biosphere Reserve, China. *J. Mt. Sci.* **2018**, *15*, 1711–1727. [CrossRef]
64. Zheng, D.L.; Wallin, D.O.; Hao, Z.Q. Rates and patterns of landscape change between 1972 and 1988 in the Changbai Mountain area of China and North Korea. *Landsc. Ecol.* **1997**, *12*, 241–254. [CrossRef]
65. Stone, R. A threatened nature reserve breaks down Asian borders. *Science* **2006**, *313*, 1379–1380. [CrossRef] [PubMed]
66. Zhou, L.; Dai, L.M.; Wang, S.X.; Huang, X.T.; Wang, X.C.; Qi, L.; Wang, Q.W.; Li, G.W.; Wei, Y.W.; Shao, G.F. Changes in carbon density for three old-growth forests on Changbai Mountain, Northeast China: 1981–2010. *Ann. For. Sci.* **2011**, *68*, 953–958. [CrossRef]
67. Shen, C.C.; Xiong, J.B.; Zhang, H.Y.; Feng, Y.Z.; Lin, X.G.; Li, X.Y.; Liang, W.J.; Chun, H.Y. Soil pH drives the spatial distribution of bacterial communities along elevation on Changbai Mountain. *Soil Biol. Biochem.* **2013**, *57*, 204–211. [CrossRef]
68. Chi, H.; Sun, G.Q.; Huang, J.L.; Li, R.D.; Ni, W.J.; Fu, A.M. Estimation of forest aboveground biomass in Changbai Mountain region using ICESat/GLAS and Landsat/TM data. *Remote Sens.* **2017**, *9*, 707. [CrossRef]
69. Du, H.B.; Liu, J.; Li, M.H.; Büntgen, U.; Yang, Y.; Wang, L.; Wu, Z.F.; He, H.S. Warming-induced upward migration of the alpine treeline in the Changbai Mountains, northeast China. *Glob. Chang. Biol.* **2018**, *24*, 1256–1266. [CrossRef]
70. Wang, Y.Q.; Wu, Z.F.; Yuan, X.; Zhang, H.Y.; Zhang, J.Q.; Xu, J.W.; Lu, Z.; Zhou, Y.Y.; Feng, J. Resources and ecological security of the Changbai Mountain region in Northeast Asia. In *Remote Sensing of Protected Lands*; Wang, Y.Q., Ed.; CRC Press: Boca Raton, FL, USA, 2011; pp. 203–232.
71. World Resources Institute; International Union of Conservation of Nature; United National Environment Programme. *Global Biodiversity Strategy*; World Resources Institute: Washington WA, USA; New York, NY, USA, 1992.
72. Xu, Z.W.; Yu, G.R.; Zhang, X.Y.; Ge, J.P.; He, N.P.; Wang, Q.F.; Wang, D. The variations in soil microbial communities, enzyme activities and their relationships with soil organic matter decomposition along the northern slope of Changbai Mountain. *Appl. Soil Ecol.* **2015**, *86*, 19–29. [CrossRef]

73. MOF (Ministry of Forestry). *Standards for Forestry Resource Survey*; China Forestry Publisher: Beijing, China, 1982.
74. Forestry Administration of China. *Tree Volume Tables (National standard # LY/T 1353-1999)*; Forestry Administration of China: Beijing, China, 1999.
75. Tang, X.G. Estimation of Forest Aboveground Biomass by Integrating ICESat/GLAS Waveform and TM Data. Ph.D. Thesis, University of Chinese Academy of Sciences, Beijing, China, 2013.
76. Wang, S.Q.; Zhou, C.H.; Liu, J.Y.; Tian, H.Q.; Li, K.R.; Yang, X.M. Carbon storage in northeast China as estimated from vegetation and soil inventories. *Environ. Pollut.* **2002**, *116*, S157–S165. [CrossRef]
77. Wu, H.B.; Guo, Z.T.; Peng, C.H. Distribution and storage of soil organic carbon in China. *Glob. Biogechem. Cycles* **2003**, *17*, 1048. [CrossRef]
78. SNAP. *Sentinels Application Platform Software ver. 4.0.0*; European Space Agency: Paris, France, 2016.
79. Guo, T.; Zhu, J.J.; Yan, Q.L.; Deng, S.Q.; Zheng, X.; Zhng, J.X.; Shang, G.D. Mapping growing stock volume and biomass carbon storage of larch plantations in Northeast China with L-band ALOS PALSAR backscatter mosaics. *Int. J. Remote Sens.* **2018**, *39*, 7978–7997. [CrossRef]
80. Morin, D.; Planells, M.; Guyon, D.; Villard, L.; Mermoz, S.; Bouvet, A.; Thevenon, H.; Dejoux, J.F.; Toan, T.L.; Dedieu, G. Estimation and mapping of forest structure parameters from open access satellite images: Development of a generic method with a study case on coniferous plantation. *Remote Sens.* **2019**, *11*, 1275. [CrossRef]
81. Shimada, M.; Isoguchi, O.; Tadono, T.; Isono, K. PALSAR radiometric and geometric calibration. *IEEE Trans. Geosci. Remote Sens.* **2009**, *47*, 3915–3932. [CrossRef]
82. Hird, J.N.; DeLancey, E.R.; McDermid, G.J.; Kariyeva, J. Google Earth Engine, open-access satellite data, and machine learning in support of large-area probabilistic wetland mapping. *Remote Sens.* **2017**, *9*, 1315. [CrossRef]
83. Carreiras, J.M.B.; Jones, J.; Lucas, R.M.; Shimabukuro, Y.E. Mapping major land cover types and retrieving the age of secondary forests in the Brazilian Amazon by combining single-date optical and radar remote sensing data. *Remote Sens. Environ.* **2017**, *194*, 16–32. [CrossRef]
84. Ceddia, M.B.; Gomes, A.S.; Vasques, G.M.; Pinheiro, E.F.M. Soil carbon stock and particle size fractions in the central Amazon predicted from remotely sensed relief, multispectral and radar data. *Remote Sens.* **2017**, *9*, 124. [CrossRef]
85. Hallik, L.; Kuusk, A.; Lang, M.; Kuusk, J. Reflectance properties of hemiboreal mixed forest canopies with focus on red edge and near infrared apectral regions. *Remote Sens.* **2019**, *11*, 1717. [CrossRef]
86. Leathwick, J.R.; Austin, M.P. Competitive interactions between tree species in New Zealand old-growth indigenous forests. *Ecology* **2001**, *82*, 2560–2573. [CrossRef]
87. Walker, A.P.; Zaehle, S.; Medlyn, B.E.; De Kauwe, M.G.; Asao, S.; Hickler, T.; Parton, W.; Ricciuto, D.M.; Wang, Y.P.; Wårlind, D.; et al. Predicting long-term carbon sequestration in response to CO_2 enrichment: How and why do current ecosystem models differ? *Glob. Biogeochem. Cycles* **2015**, *29*, 476–495. [CrossRef]
88. Jennings, S.B.; Brown, N.D.; Sheil, D. Assessing forest canopies and understorey illumination: Canopy closure, canopy cover and other measures. *Forestry* **1999**, *72*, 59–74. [CrossRef]
89. Mon, M.S.; Mizoue, N.; Htun, N.Z.; Kajisa, T.; Yoshida, S. Estimating forest canopy density of tropical mixed deciduous vegetation using Landsat data: A comparison of three classification approaches. *Int. J. Remote Sens.* **2012**, *33*, 1042–1057. [CrossRef]
90. Smith, A.M.; Ramsay, P.M. A comparison of ground-based methods for estimating canopy closure for use in phenology research. *Agrc. For. Meteorol.* **2018**, *252*, 18–26. [CrossRef]
91. Korhonen, L.; Korhonen, K.T.; Rautiainen, M.; Stenberg, P. Estimation of forest canopy cover: A comparison of field measurement techniques. *Silva Fenn.* **2006**, *40*, 577–588. [CrossRef]
92. Paletto, A.; Tosi, V. Forest canopy cover and canopy closure: Comparison of assessment techniques. *Eur. J. For. Res.* **2009**, *128*, 265–272. [CrossRef]
93. Chen, J.M.; Black, T.A. Defining leaf-area index for non-flat leaves. *Plant. Cell Environ.* **1992**, *15*, 421–429. [CrossRef]
94. Sprintsin, M.; Karnieli, A.; Berliner, P.; Rotenberg, E.; Yakir, D.; Cohen, S. The effect of spatial resolution on the accuracy of leaf area index estimation for a forest planted in the desert transition zone. *Remote Sens. Environ.* **2007**, *109*, 416–428. [CrossRef]

95. Jump, A.S.; Ruiz-Benito, P.; Greenwood, S.; Allen, C.; Kitzberger, T.; Fensham, R.; Martinez-vilalta, J.; Lloret, F. Structural overshoot of tree growth with climate variability and the global spectrum of drought-induced forest dieback. *Glob. Chang. Biol.* **2017**, *23*, 3742–3757. [CrossRef] [PubMed]
96. Attema, E.P.W.; Ulaby, F.T. Vegetation modeled as a water cloud. *Radio Sci.* **1978**, *13*, 357–364. [CrossRef]
97. Cartus, O.; Santoro, M.; Kellndorfer, J. Mapping forest aboveground biomass in the Northeastern United States with ALOS PALSAR dual-polarization L-band. *Remote Sens. Environ.* **2012**, *124*, 466–478. [CrossRef]
98. Breiman, L. Random forests. *Mach. Learn.* **2001**, *45*, 5–32. [CrossRef]
99. Belgiu, M.; Drăguţ, L. Random forest in remote sensing: A review of applications and future directions. *ISPRS J. Photogramm.* **2016**, *114*, 24–31. [CrossRef]
100. Chen, L.; Wang, Y.Q.; Ren, C.Y.; Zhang, B.; Wang, Z.M. Optimal combination of predictors and algorithms for forest above-ground biomass mapping from Sentinel and SRTM data. *Remote Sens.* **2019**, *11*, 414. [CrossRef]
101. Fayad, I.; Baghdadi, N.; Bailly, J.S.; Barbier, N.; Gond, V.; Hérault, B.; Hajj, M.E.; Fabre, F.; Perrin, J. Regional scale rain-forest height mapping using regression-kriging of spaceborne and airborne LiDAR data: Application on French Guiana. *Remote Sens.* **2016**, *8*, 240. [CrossRef]
102. Viscarra Rossel, R.A.; Webster, R.; Kidd, D. Mapping gamma radiation and its uncertainty from weathering products in a Tasmanian landscape with a proximal sensor and random forest kriging. *Earth Surf. Proc. Land.* **2014**, *39*, 735–748. [CrossRef]
103. Liu, Y.; Cao, G.F.; Zhao, N.Z.; Mulligan, K.; Ye, X.Y. Improve ground-level PM2.5 concentration mapping using a random forests-based geostatistical approach. *Envron. Pollut.* **2018**, *235*, 272–282. [CrossRef] [PubMed]
104. Kidd, D.B.; Malone, B.P.; McBratney, A.B.; Minasny, B.; Webb, M.A. Digital mapping of a soil drainage index for irrigated enterprise suitability in Tasmania, Australia. *Soil Res.* **2014**, *52*, 107–119. [CrossRef]
105. Guo, P.T.; Li, M.F.; Luo, W.; Tang, Q.F.; Liu, Z.W.; Lin, Z.M. Digital mapping of soil organic matter for rubber plantation at regional scale: An application of random forest plus residuals kriging approach. *Geoderma* **2015**, *237–238*, 49–59. [CrossRef]
106. Isaaks, E.H.; Srivastava, R.M. *An Introduction to Applied Geostatistics*; Oxford University Press: Oxford, UK, 1989.
107. Moreno, G.; Cubera, E. Impact of stand density on water status and leaf gas exchange in Quercus ilex. *For. Ecol. Manag.* **2008**, *254*, 74–84. [CrossRef]
108. Luke, S.H.; Barclay, H.; Bidin, K.; Chey, V.K.; Ewers, R.M.; Foster, W.A.; Nainar, A.; Pfeifer, M.; Reynolds, G.; Turner, E.C.; et al. The effects of catchment and riparian forest quality on stream environmental conditions across a tropical rainforest and oil palm landscape in Malaysian Borneo. *Ecohydrology* **2017**, *10*, e1827. [CrossRef]
109. Wu, L.Y.; You, W.B.; Ji, Z.R.; Xiao, S.H.; He, D.J. Ecosystem health assessment of Dongshan Island based on its ability to provide ecological services that regulate heavy rainfall. *Ecol. Indic.* **2018**, *84*, 393–403.
110. Sinha, S.; Santra, A.; Sharma, L.; Jeganathan, C.; Nathawat, M.S.; Das, A.K.; Mohan, S. Multi-polarized Radarsat-2 satellite sensor in assessing forest vigor from above ground biomass. *J. For. Res.* **2018**, *29*, 1139–1145. [CrossRef]
111. Vafaei, S.; Soosani, J.; Adeli, K.; Fadaei, H.; Naghavi, H.; Pham, T.D.; Bui, D.T. Improving accuracy estimation of Forest Aboveground Biomass Based on Incorporation of ALOS-2 PALSAR-2 and Sentinel-2A Imagery and Machine Learning: A Case Study of the Hyrcanian Forest Area (Iran). *Remote Sens.* **2018**, *10*, 172. [CrossRef]
112. Zolkos, S.G.; Goetz, S.J.; Dubayah, R. A meta-analysis of terrestrial aboveground biomass estimation using lidar remote sensing. *Remote Sens. Environ.* **2013**, *128*, 289–298. [CrossRef]
113. Huang, H.B.; Liu, C.X.; Wang, X.Y.; Zhou, X.L.; Gong, P. Integration of multi-resource remotely sensed data and allometric models for forest aboveground biomass estimation in China. *Remote Sens. Environ.* **2019**, *221*, 225–234. [CrossRef]
114. Rosenqvist, A.; Shimada, M.; Suzuki, S.; Ohgushi, F.; Tadono, T.; Watanabe, M.; Tsuzuku, K.; Watanabe, T.; Kamijo, S.; Aoki, E. Operational performance of the ALOS global systematic acquisition strategy and observation plans for ALOS-2 PALSAR-2. *Remote Sens. Environ.* **2014**, *155*, 3–12. [CrossRef]
115. Yu, D.P.; Wang, Q.W.; Liu, J.Q.; Zhou, W.M.; Qi, L.; Wang, X.U.Y.; Zhou, L.; Dai, L.L. Formation mechanisms of the alpine Erman's birch (Betula ermanii) treeline on Changbai Mountain in Northeast China. *Trees Struct. Funct.* **2014**, *28*, 935–947. [CrossRef]

116. Guo, D.; Zhang, H.Y.; Hou, G.L.; Zhao, J.J.; Liu, D.Y.; Guo, X.Y. Topographic controls on alpine treeline patterns on Changbai Mountain, China. *J. Mt. Sci.* **2014**, *11*, 429–441. [CrossRef]
117. Shen, C.C.; Liang, W.J.; Shi, Y.; Lin, X.G.; Zhang, H.Y.; Wu, X.; Xie, G.; Chain, P.; Grogan, P.; Chu, H.Y. Contrasting elevational diversity patterns between eukaryotic soil microbes and plants. *Ecology* **2014**, *95*, 3190–3202. [CrossRef]
118. Jiang, Y.F.; Yin, X.Q.; Wang, F.B. Composition and spatial distribution of soil mesofauna slong an elevation gradient on the north slope of the Changbai Mountains, China. *Pedosphere* **2015**, *25*, 811–824. [CrossRef]
119. Cong, Y.; Li, M.H.; Liu, K.; Dang, Y.C.; Han, H.D.; He, H.S. Decreased temperature with increasing elevation decreases the end-season leaf-to-wood reallocation of resources in deciduous Betula ermanii Cham. *Trees For.* **2019**, *10*, 166. [CrossRef]

© 2019 by the authors. Licensee MDPI, Basel, Switzerland. This article is an open access article distributed under the terms and conditions of the Creative Commons Attribution (CC BY) license (http://creativecommons.org/licenses/by/4.0/).

Letter

Monitoring Droughts in the Greater Changbai Mountains Using Multiple Remote Sensing-Based Drought Indices

Yang Han [1,†], Ziying Li [1,†], Chang Huang [2,*], Yuyu Zhou [3], Shengwei Zong [1], Tianyi Hao [1], Haofang Niu [1] and Haiyan Yao [1]

1. Key Laboratory of Geographical Processes and Ecological Security in Changbai Mountains, Ministry of Education, School of Geographical Sciences, Northeast Normal University, Changchun 130024, China
2. Institute of Earth Surface System and Hazards, Northwest University, Xi'an 710127, China
3. Department of Geological and Atmospheric Sciences, Iowa State University, Ames, IA 50011-1051, USA
* Correspondence: changh@nwu.edu.cn
† These authors contributed equally to this work.

Received: 26 December 2019; Accepted: 3 February 2020; Published: 6 February 2020

Abstract: Various drought indices have been developed to monitor drought conditions. Each index has typical characteristics that make it applicable to a specific environment. In this study, six popular drought indices, namely, precipitation condition index (PCI), temperature condition index (TCI), vegetation condition index (VCI), vegetation health index (VHI), scaled drought condition index (SDCI), and temperature–vegetation dryness index (TVDI), have been used to monitor droughts in the Greater Changbai Mountains(GCM) in recent years. The spatial pattern and temporal trend of droughts in this area in the period 2001–2018 were explored by calculating these indices from multi-source remote sensing data. Significant spatial–temporal variations were identified. The results of a slope analysis along with the F-statistic test showed that up to 20% of the study area showed a significant increasing or decreasing trend in drought. It was found that some drought indices cannot be explained by meteorological observations because of the time lag between meteorological drought and vegetation response. The drought condition and its changing pattern differ from various land cover types and indices, but the relative drought situation of different landforms is consistent among all indices. This work provides a basic reference for reasonably choosing drought indices for monitoring drought in the GCM to gain a better understanding of the ecosystem conditions and environment.

Keywords: multi-source remote sensing data; drought index; trend analysis; MODIS; TMPA

1. Introduction

Drought is considered an environmental disaster, and many researchers, including environmentalists, ecologists, hydrologists, meteorologists, geologists, and agricultural scientists have investigated droughts [1]. Drought causes soil degradation, desertification, water deficit, plant death, sandstorm, fire disaster, and other disaster phenomena [2]. Moreover, drought also affects crop growth, influences global food prices, and contributes to political unrest [3–6]. Therefore, monitoring drought and studying its spatiotemporal dynamics are important for improving agricultural production, protecting the environment, and promoting sustainable social economic development [7].

Traditional drought-monitoring methods are based on ground- or station-based meteorological and hydrological observations, such as precipitation, air temperature, soil moisture, evapotranspiration, and surface runoff. A series of meteorological drought indices, including the Standardized Precipitation Index (SPI) and Palmer Drought Severity Index (PDSI), were developed based on these observation data. However, it is difficult to ensure the reliability of such interpolation because of the limited spatial density and uneven distribution of the observation stations [7,8]. Therefore, there is increased focus on remote sensing for drought monitoring because of its comprehensive, fast, and dynamic features that can rapidly and accurately yield multiscale and multitemporal information [9–13].

Many remote sensing-based drought indices have been established to reflect drought conditions. One of the most extensively used one is the normalized difference vegetation index (NDVI). However, when conducting drought monitoring over nonhomogeneous areas, NDVI is less reliable because of the effects of geographical location, ecological systems, and soil conditions [9,10]. To overcome these problems, Kogan proposed the vegetation condition index (VCI) by normalizing NDVI values to the maximum range of a specific area [9]. The weather-related NDVI component is smaller than the one related to the ecosystem; therefore, normalization successfully minimizes the ecosystem component. The VCI has been widely applied to drought monitoring and analysis, and its reliability has been verified by many studies [14–21]. VCI can individually monitor the effect of drought on vegetation health but is insufficient because it indicates only one moisture condition [14]. Considering that temperature may also reflect drought conditions to some extent, Kogan further developed the temperature condition index (TCI) by normalizing land surface temperature (LST) values to the maximum range of a specific area as an indicator of drought [10]. The vegetation health index (VHI), which averages the sum of VCI and TCI, too was introduced by Kogan [22]. The VHI has also been frequently used for agricultural purposes, such as crop yield estimation [18,19,23]. The principle of using VHI for drought monitoring is that an assessment of temperature conditions helps identify subtle changes in vegetation health because the effect of drought is more drastic if shortage of moisture is accompanied by excessive temperatures. The feasibility of using VHI has been validated in all major agricultural countries [22]. Precipitation deficit is an important condition for drought formation; therefore, the precipitation condition index (PCI) can reflect drought conditions [24]. Since drought usually is induced by precipitation deficit and rise in temperature and poses a threat to vegetation health, the scaled drought condition index (SDCI), which combines the PCI, TCI, and VCI, was proposed [25]. The abovementioned indices can be calculated from easily available satellite remote sensing data. Other researchers proposed the temperature–vegetation dryness index (TVDI) using the spatial relationship between the LST and NDVI based on the spectral reflectance of near-infrared (NIR) and red channels to indicate drought [26] and soil moisture conditions [27]. These indices take advantage of one or more aspects of droughts to reflect drought conditions; as a result, these indices have distinct characteristics that makes them suitable for different scenarios.

Remote sensing can provide long-term series data with broad spatial coverage; hence, these data are a perfect source for Earth observation and land surface monitoring. Drought-monitoring research also benefits greatly from remote sensing techniques, which help track the long-term trend of drought condition easily. Liang et al. used the VCI to evaluate the spatiotemporal variations of drought in different regions in China based on a trend analysis of tendency rate (slope) [7]. The occurrence of drought events in Northeast China from 2001 to 2014 was also investigated using slope analysis [6]. Spatial and temporal variations of drought in Nepal were examined by trend analysis based on satellite-derived VCI [28]. The Greater Changbai Mountains (GCM) is extremely important from the ecological viewpoint for the entire Northeast Asia as well as the world owing to the well preserved and most abundant forests of different types. Therefore, understanding the drought conditions of the GCM is critical for understanding the ecosystem conditions and environment of this region.

The main objectives of this study are (1) to evaluate the six widely used drought indices (PCI, VCI, TCI, VHI, SDCI, and TVDI) for drought monitoring in the GCM, considering that they reflect three different aspects of drought, namely precipitation, temperature, and vegetation conditions, and also that their data are easily available; (2) to explore the spatiotemporal patterns and the changing trend of drought in the GCM during the period 2001–2018; and (3) to analyze the correlations of the drought indices with meteorological factors and land cover types.

2. Study Area and Data

2.1. Study Area

The GCM include a part of the northeast provinces of Heilongjiang, Jilin, and Liaoning in China and have the largest protected temperate forest in Northeast Asia with many rare animal and plant resources. This area has been a focal point of ecosystem and biodiversity research based on remote sensing applications [27]. The Greater Changbai Mountains (GCM) has a northeast–southwest orientation and extends in the region 38°46′–47°30′ N latitude and 121°08′–134° E longitude (Figure 1). It mainly includes parallel fault block mountain areas, such as Changbai, Laoyeling, Zhangguangcailing, and Hadaling. It extends 1300 km from the north to the south and stretches 400 km from the east to the west. The Changbai Mountain is somewhat spindle shaped, and it has a large elevation difference. Its highest peak is located in Jilin Province and is 2670 m high. The total area of the mountains is approximately 2.8×10^5 km^2. The Changbai Mountain is a treasure trove of world resources as it contains all types of vegetation from temperate to polar types. It covers climatic zones ranging from warm temperate to mid-temperate and from humid to semi humid zones. The mountain also has diverse soil types and complex landform types. The Changbai Mountain has diverse variety and is rich in species. The area mainly has coniferous and broad-leaved mixed forests dominated by Korean pine; coniferous forests dominated by fir, spruce, and larch; and broad-leaved forests in the temperate zone. The Changbai Mountain has a temperate humid monsoon climate and is located on the northeastern edge of the global monsoon climate regions. The region is also affected by the continental climate. Its climate is mainly characterized by long and cold winters and cool and short summers. The climate difference between the north and south is large as the region is spread across nearly 10° latitude, and the climate types vary greatly with the terrain because of the influence of altitude.

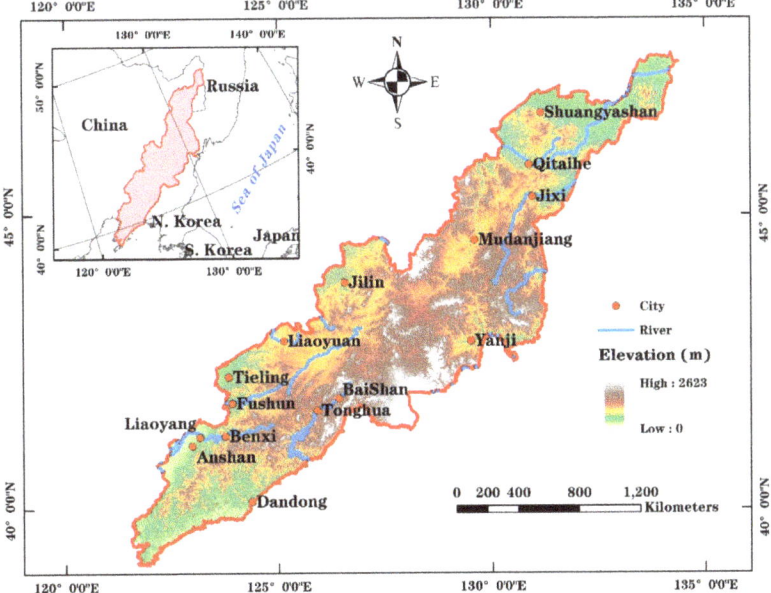

Figure 1. Location map of the Greater Changbai Mountains (GCM).

2.2. Data

Satellite remote sensing data from the Moderate Resolution Imaging Spectroradiometer (MODIS) and Tropical Rainfall Measurement Missions (TRMM) multi-satellite precipitation analyses (TMPA) were used as the inputs to generate different drought index images. Four products including the NDVI, LST, Land cover and precipitation were employed as listed in Table 1. All data have a monthly or yearly temporal resolution. The monthly data were then aggregated to yearly data by averaging. For Land cover product (MCD12Q1), Land Cover Type 3 layer with annual Leaf Area Index classification scheme, which has 10 land cover classes, was used here. All these products have enough time scale to cover the whole of 2001–2018.

Table 1. Satellite remote sensing products used in this study

Satellite Mission	Product	Data Type	Spatial Resolution (m)	Temporal Resolution	Source
MODIS	MOD13A3.006	NDVI	1,000	monthly	[29]
	MYD11C3.006	LST	500	monthly	[30]
	MCD12Q1.006	Land cover	500	yearly	[31]
TMPA	TRMM 3B43	Precipitation	~25,000	monthly	[32]

The annual average temperature and total precipitation data for 14 prefecture-level cities in the GCM area from 2001 to 2017 (data of 2018 were missing) were collected from local statistical yearbooks. They were used to explore the correlation between remotely sensed drought indices and meteorological observations.

3. Methodology

3.1. Drought Indices

These six widely used drought indices (PCI, VCI, TCI, VHI, SDCI, and TVDI) can be categorized into three types. One type comprises a single-factor index, including PCI, TCI, and VCI, calculated only from one of the three—precipitation, LST, and NDVI (Table 2). The second type is an index comprising a combination of factors; examples of this type are the VHI and SDCI, which are calculated from weighted combinations of multiple single-factor indices (Table 3). The other one is TVDI, which employs the spatial relationship between LST and NDVI to reflect drought information (Equation (1)).

$$\text{TVDI} = \frac{LST - LST_{min}}{LST_{max} - LST_{min}} \quad (1)$$

$$LST_{min} = a + b \times NDVI \quad (2)$$

$$LST_{max} = c + d \times NDVI \quad (3)$$

Table 2. Single-factor drought indices*.

Drought Index	Data Source	Formula	Reference
PCI	TRMM	$PCI = \frac{TRMM_i - TRMM_{min}}{TRMM_{max} - TRMM_{min}}$	[33]
TCI	MODIS	$TCI = \frac{LST_{max} - LST_i}{LST_{max} - LST_{min}}$	[10]
VCI	MODIS	$VCI = \frac{NDVI_i - NDVI_{min}}{NDVI_{max} - NDVI_{min}}$	[9]

* The minimum and maximum values in these formulas were selected for the whole 2001–2018 period over the whole study area

Table 3. Drought indices based on a combination of factors.

Drought Index	Formula	Weights	Weight Determination Method	Reference
VHI	VHI = αTCI + βVCI	α = 0.5, β = 0.5	Empirical weights	[11]
SDCI	SDCI = αTCI + βVCI + γPCI	α = 0.25, β = 0.25, γ = 0.5	Empirical weights	[25]

In Equation (1), *LST* is the observed surface temperature at a given pixel; LST_{min}, the minimum surface temperature in the triangle for a given *NDVI* defining the wet edge (Equation (2)); and LST_{max}, the maximum surface temperature in the triangle for a given *NDVI* defining the dry edge (Equation (3)) [27]. Their coefficients (a, b, c, and d) can be estimated by fitting the dry and wet edges of the triangle (Figure 2).

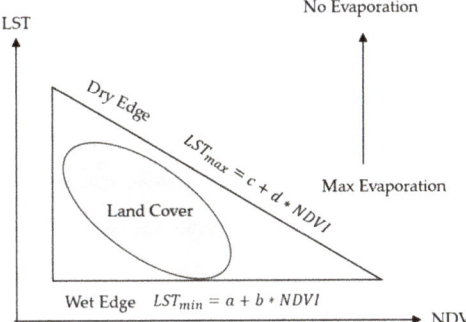

Figure 2. Simplified land surface temperature (LST)- normalized difference vegetation index (NDVI) triangle (Adopted from [27]).

MODIS and TRMM datasets were acquired and processed as the inputs to generate six drought indices. The values of each of these indices range from 0 to 1, with a higher value indicating less drought for most indices. However, for TVDI, a higher value indicates a more severe drought. Table 4 lists the classification scheme of these indices for drought levels according to [34].

Table 4. Classification scheme of the drought indices employed in this study [34].

Name of Class	PCI	TCI	VCI	SDCI	VHI	TVDI
Extreme drought	0–0.1	0–0.1	0–0.1	0–0.2	0–0.1	
Severe drought	0.1–0.2	0.1–0.2	0.1–0.2	0.2–0.3	0.1–0.2	0.8–1
Moderate drought	0.2–0.3	0.2–0.3	0.2–0.3	0.3–0.4	0.2–0.3	0.6–0.8
Mild drought	0.3–0.4	0.3–0.4	0.3–0.4	0.4–0.5	0.3–0.4	0.4–0.6
Abnormal drought	0.4–0.5	0.4–0.5	0.4–0.5			
No drought	0.5–1	0.5–1	0.5–1	0.5–1	0.4–1	0–0.4

3.2. Trend Analysis

To reveal the trend of the drought condition from 2001 to 2018, a temporal trend analysis based on the ordinary least squares (OLS) regression was conducted for each drought index (*DI*) pixel. Then, a linear equation of a *DI* was fit as a function of the variable "YEAR" to calculate the slope (Equation (4)). An image of the changing slope over the period 2001–2018 was thus obtained.

$$SLOPE = \frac{n \times \sum_{i=1}^{n} i \times DI_i - \sum_{i=1}^{n} i \sum_{i=1}^{n} DI_i}{n \times \sum_{i=1}^{n} i^2 - \left(\sum_{i=1}^{n} i\right)^2} \quad (4)$$

In Equation (4), *n* represents the total number of observation years ($n = 18$). DI_i represents the mean value of drought index for the *i*th year. *SLOPE* > 0 represents an increasing trend of *DI* from 2001 to 2018. Conversely, *SLOPE* < 0 represents a decreasing trend. F-statistics were conducted to determine the significance of the fitted linear regression model.

4. Results

4.1. Spatial Pattern

Figure 3 shows six *DI* maps that were averaged based on the annual draught indices from 2001 to 2018. Figure 3a–e presents those drought indices whose lower values indicate severe drought, while Figure 3f shows the TVDI map for which a higher value indicates severe drought.

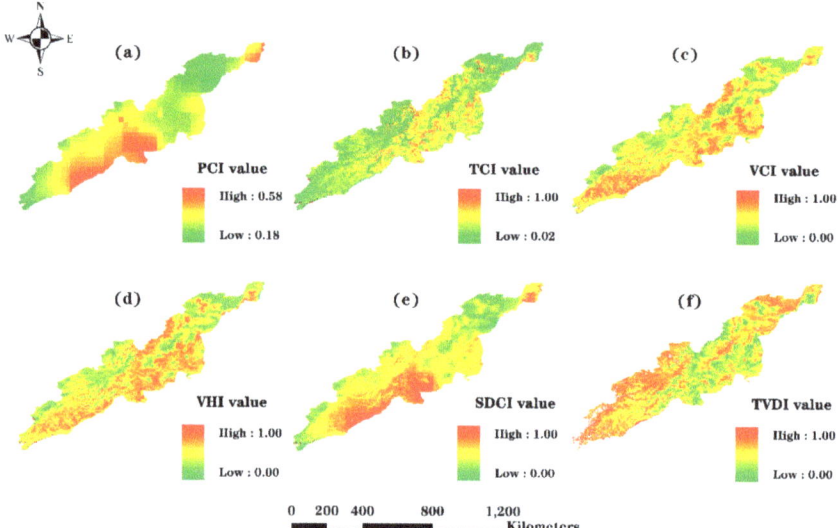

Figure 3. Six drought indices averaged over 18 years for the Greater Changbai Mountains (GCM). (**a**) Precipitation condition index (PCI), (**b**) temperature condition index (TCI), (**c**) vegetation condition index (VCI), (**d**) vegetation health index (VHI), (**e**) scaled drought condition index (SDCI), and (**f**) temperature–vegetation dryness index (TVDI).

According to the PCI shown in Figure 3a and the classification scheme presented in Table 4, the southeast area of the GCM region where Baishan, Tonghua, and Dandong Cities are located, exhibits no drought conditions, whereas the northeast area (ShuangyaShan) has a low PCI value, indicating obvious drought conditions. Most of the remaining areas had slight or no drought. The TCI and VCI maps show slightly different drought patterns. According to the TCI map, most of the GCM area experienced drought conditions, except for the mountains in the central region. The VCI map indicates that most of the area experienced slight or no drought, except for the low-lying areas in the southwest and northeast. Since the TCI is related to LST, and the VCI is related to NDVI, the conflict between the TCI and VCI suggests that vegetation flourished in the mountains despite the high LST, indicating that the vegetation in the GCM area has high drought endurance. The VHI is high (no drought/wet conditions) in the southeast area with high elevation, while low (drought conditions) in the low-lying southwest and northeast areas. The other area has moderate VHI values, indicating mild drought conditions according to the classification in Table 4. The SDCI map shows a pattern similar to that of the PCI map, largely because the PCI is an important factor in the SDCI. According to the TVDI map, no drought conditions occurred in the southeast area with high elevation. The remaining parts of the region exhibited slight to moderate drought conditions. The spatial pattern of the TVDI was similar to that of the VHI. Thus, these six indices exhibit different drought information for the GCM area. This is mainly because they are calculated from different combinations of precipitation, LST, and vegetation status. Thus far, we cannot conclude which index is more reliable because they all try to reflect one or more aspects of drought conditions. The differences among the index maps also suggests that the method for drought monitoring should be selected carefully.

4.2. Temporal Trend

Trend analyses were conducted for the annual index images using the aforementioned OLS regression and F-statistics. On the basis of the calculated slopes and significant levels, the trends of changes in the drought indices were classified into seven categories: (1) highly significant decrease ($SLOPE < 0$, $p \leq 0.01$), (2) significant decrease ($SLOPE < 0$, $0.01 < p \leq 0.05$), (3) moderately significant decrease ($SLOPE < 0$, $0.05 < p \leq 0.1$), (4) no significant change ($p > 0.1$), (5) moderately significant increase ($SLOPE < 0$, $0.05 < p \leq 0.1$), (6) significant increase ($SLOPE > 0$, $0.01 < p \leq 0.05$), and (7) highly significant increase ($SLOPE > 0$, $p \leq 0.01$) [35], as shown in Figure 4. Figure 4 shows a significant PCI decrease for 8.8% of the study area and a significant increase for 13.6% of the area. The remaining area exhibited no significant change during the last 18 years. The area with a significant increase in the PCI is mainly located near Mudanjiang, Jilin, and Yanji City, while the area with significantly decreasing PCI is mainly distributed near Liaoyang, Anshan, and Dandong City. For TCI, 3.3% of the area experienced a significant increase and 3.3% of the area witnessed a significant decrease. The area with a significant decrease was distributed near the north of Shuangyashan and Qitaihe City and south of Anshan City, suggesting that these areas experienced very severe droughts in the recent 18 years. The positive trend was mainly observed in Mudanjiang and Tonghua City. With regard to the VCI, a significant increase was seen for 5.3% of the study area, whereas a significant decrease was seen for 21.3%, indicating that a larger area tended to experience very severe droughts recently. The areas with a significant decrease in the VCI are mostly located in the southern coastal region, Yanji in Jilin, and the northern part of Shuangyashan. The rest of Shuangyashan and Baishan exhibited an increase in the VCI. VHI trend analysis showed that 4.9% of the GCM area experienced a significant drought alleviation, whereas 14.1% experienced a significant drought aggravation. The spatial pattern of the VHI trend is similar to that of the VCI trend. SDCI trend analysis revealed that about 11% of the central area of the GCM experienced a significant increase (i.e., drought condition is relieving); meanwhile, almost an identical extent in the southwest experienced drought aggravation. TVDI trend analysis showed that a sparse area of 9.0% of the GCM experienced a significant increase, whereas 3.5% of the area experienced a significant decrease. The region with the decreasing TVDI was mostly distributed near Mudanjiang, Liaoyuan, and Tonghua.

Thus, the trend analyses indicate different drought trends for different indices. For the GCM, both PCI and SDCI show similar patterns that the central area is getting wetter and the southwest area is getting drier. VCI and VHI exhibit similar patterns showing sparse areas with decrease (drying) and overwhelming increase (wetting).

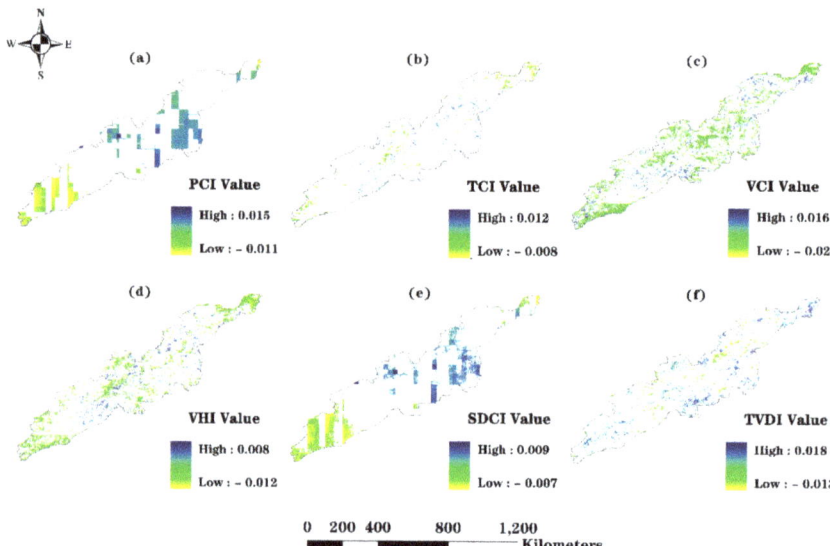

Figure 4. Slopes of drought indices during 2001–2018. (**a**) PCI, (**b**) TCI, (**c**) VCI, (**d**) VHI, (**e**) SDCI, and (**f**) TVDI.

4.3. Correlations between Drought Indices and Meterological Factors

The annual average precipitation and temperature were collected from 14 prefecture-level cities in the GCM. Six annual drought indices were plotted for these cities, as shown in Figures 5 and 6. As shown in Figure 5, PCI and SDCI both show similar annual patterns with precipitation. This is because precipitation is an important input for both these indices. The other four indices (VCI, TCI, TVDI, and VHI) do not show any clear relation with annual precipitation. TCI mainly reflects the surface temperature variation, while for the other three indices, NDVI is an important input.

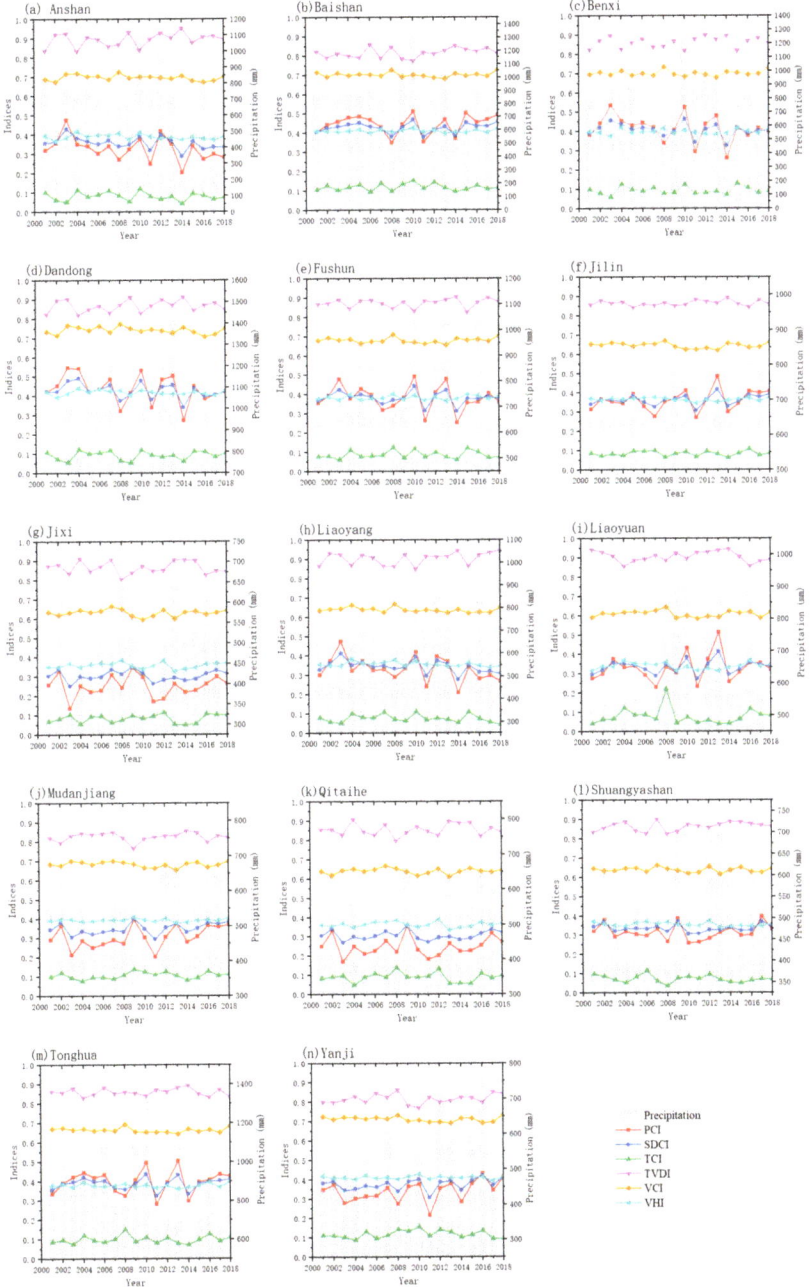

Figure 5. Relationship between annual drought indices and precipitation in the GCM. (**a**–**n**) are the 14 prefecture-level cities. (**a**) Anshan; (**b**) Baishan; (**c**) Benxi; (**d**) Dandong; (**e**) Fushun; (**f**) Jilin; (**g**) Jixi; (**h**) Liaoyang; (**i**) Liaoyuan; (**j**) Mudanjiang; (**k**) Qitaihe; (**l**) Shuangyashan; (**m**) Tonghua; (**n**) Yanji.

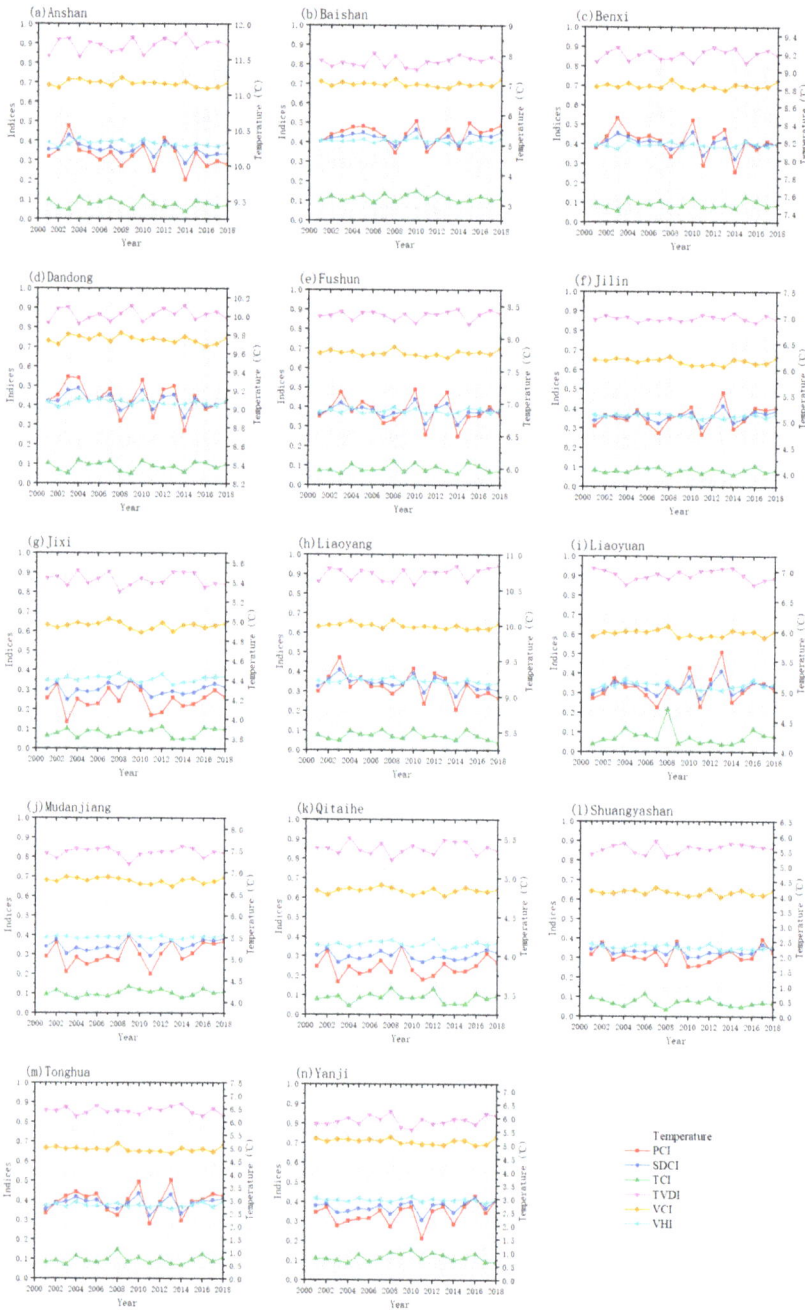

Figure 6. Relationship between annual drought indices and temperature in the GCM. (**a–n**) are the 14 prefecture-level cities. (**a**) Anshan; (**b**) Baishan; (**c**) Benxi; (**d**) Dandong; (**e**) Fushun; (**f**) Jilin; (**g**) Jixi; (**h**) Liaoyang; (**i**) Liaoyuan; (**j**) Mudanjiang; (**k**) Qitaihe; (**l**) Shuangyashan; (**m**) Tonghua; (**n**) Yanji.

Dandong has a relatively larger precipitation volume with high variations from 2001 to 2018 because of its special geolocation. Its forest coverage is as high as 65%, endowing the region with high water-holding capacity. Precipitation-based drought indices in this area indicate clearly humid characteristics. For example, the precipitation was very high in 2012 and 2013, and the average annual temperature was lower (Figure 6d). Therefore, most of the indices identified the lack of a drought condition in these 2 years. Benxi, which is a city neighboring Dandong, had slightly lower annual average temperature and precipitation. Its drought condition was similar to Dandong according to different indices. Anshan and Liaoyang are located inland and receive less precipitation than Dandong and Benxi. In particular, in 2014, the precipitation of Liaoyang was only 300 mm, and the average temperature was higher than 10 °C, which was significantly higher than that in the other years. PCI and SDCI clearly indicate the drought situation in Liaoyang. Anshan received precipitation as low as approximately 400 mm in 2014, and the average annual temperature was higher than 11 °C. PCI and SDCI also reflected the drought situation of Anshan correctly. Fushun received an annual precipitation exceeding 1,000 mm in 2010 and 2013, and its average annual temperature was between 5.5 °C and 6 °C; meanwhile, in 2011 and 2014, it received a very low precipitation of 500 mm, and the PCI and SDCI values clearly reflect these changes. Baishan and Tonghua are located close to the Changbai Mountain Nature Reserve. The annual precipitation and temperature were relatively stable from 2001 to 2018. The drought indices also tended to vary smoothly. Jilin and Liaoyuan experienced similar annual precipitation variations in the recent years. In 2011, their annual precipitation was the lowest, only about 500 mm, which was captured by PCI and SDCI. Yanji, Jixi, Mudanjiang, Qitaihe, and Shuangyashan are located inland north of the Changbai Mountain Nature Reserve and have high latitudes, low annual average temperatures, and annual precipitations less than 800 mm. Their PCI, TCI, VCI, VHI, and SDCI values are lower than those in other regions. In the GCM, the TVDI, VCI, TCI, and VHI values also exhibit distinct annual patterns that do not appear to be related to the annual precipitation and temperature. This may because these indices do not consider precipitation, and indicate droughts based on the LST anomalies and vegetation health. The annual average temperature is usually not sensitive enough to reflect LST anomalies; hence, LST-based indices fail to show consistency with annual average temperature variations. Further, vegetation health may be sometimes affected by factors besides drought, and there is usually a time lag before drought can cause deterioration of vegetation health. This may be the reason why vegetation-based indices also fail to show consistency with the precipitation and temperature patterns.

4.4. Correlations between Drought Indices and Land Cover Types

An examination of the annual land cover data from 2001 to 2018 revealed very little land cover change over these years. Therefore, in this study, we assumed that there was no land cover change, and we used the land cover map of 2018 as the current condition of land cover to investigate the correlations between drought indices and land cover types. According to the 2018 land cover map, in the GCM, deciduous broadleaf forests, grasslands, and savannas accounted for 46.8%, 30.4%, and 18.1%, respectively. The other seven land cover types account for less than 5%.

The distribution of vegetation types has a strong relationship with regional climatic factors, and surface temperature is an important climatic factor that affects the zonal distribution of vegetation types. Temperature is an important limiting factor for plant growth in the GCM; forest vegetation types, in particular, have distinct vertical zonal distribution characteristics. NDVI represents vegetation characteristics, and there exists a clear relationship between NDVI and LST. Le Page et al. [36] found that the negative correlation between NDVI and LST in agricultural area is due to drastic evaporation that decreases LST. However, there exists a positive correlation between NDVI and LST in the northeastern part of the study area, and this correlation is explained by the simultaneous forest leaf loss and fall in surface temperature (the coldest months). Figure 7a,b present the mean values of the six drought indices for different land cover types in 2018 and 2001. For the NDVI-based index (i.e., VCI) and the LST-based index (TCI), when the land cover changes from deciduous needleleaf forests to unvegetated lands, VCI decreases, while TCI increases. These two indices exhibit distinct patterns in different land cover types. The variability in the slope of the inverse LST–NDVI relationship in association with local topographic and environmental conditions has been assessed in previous studies [37]. The validity of VHI as a drought detection tool relies on the assumption that the NDVI and LST at a given pixel vary inversely over time with variations in VCI and TCI driven by local moisture conditions. However, over vast areas and long periods, the LST–NDVI relationship is nonunique and often nonnegative [37–39]. According to [37–39], NDVI and LST are positively related usually in energy-limited ecosystems, which implies that high temperature promotes the growth of vegetation. In this case, VHI and TVDI may not be appropriate for indicating drought conditions. However, for water-limited ecosystems, where high temperature may inhibit vegetation growth, NDVI has a negative relation with LST, conforming with the assumption of TVDI. In this case, TVDI and VHI are applicable for indicating drought conditions. Since the water-holding capacity of deciduous broadleaf forests is stronger than that of grasslands, deciduous broadleaf forests generally are more humid than grasslands. This is reflected by almost all the drought indices. Savannas comprise a mixed forest–grassland type of vegetation and have drought indices intermediate between those of the deciduous forests and grasslands.

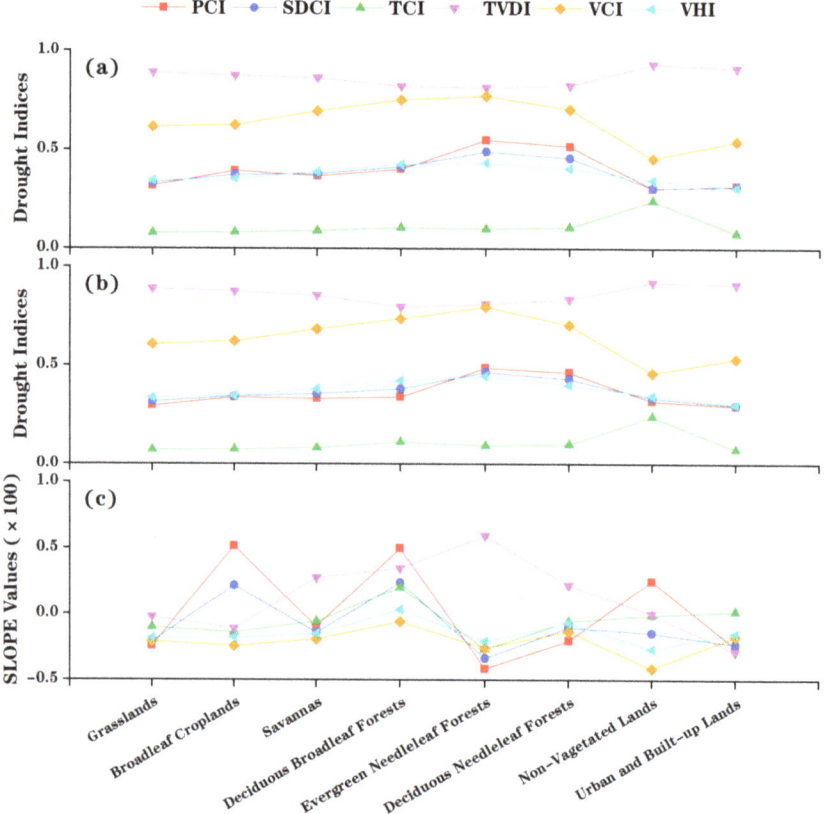

Figure 7. Drought indices and their changing slopes for different land cover types: (**a**) Drought indices in 2018; (**b**) drought indices in 2001; and (**c**) slopes (×100) of drought indices from 2001 to 2018.

We examined the zonal statistic to the slope value based on land cover types. Figure 7c shows the mean *SLOPE* values (scaled by 100) of drought indices from 2001 to 2018 for different land cover types. The TVDI *SLOPE* values of evergreen needleleaf forests is the highest among all land cover types; the *SLOPE* values of the other five indices are low and negative. This suggests that evergreen needleleaf forests in the GCM clearly have experienced severe drought conditions. Savannas and deciduous needleleaf forests also have similar but weaker changing patterns with positive slope for TVDI but negative slope for the other indices, which is similar to Zribi et al. [40], who analyzed drought affection on vegetation coverage based on time series Vegetation Anomaly Index (VAI). Our analysis of slope and land cover types showed that VCI and VHI have negative values with similar variation patterns across different land cover types. This means that the VCI and VHI decreased from 2001 to 2018, indicating an overall drying trend. The sharpest trend occurs in the case of unvegetated Lands. Since PCI is a precipitation-based index, when the precipitation increases, the slope of PCI is positive, indicating a wetting situation. Broadleaf croplands and deciduous broadleaf forests, which together account for a large proportion of the GCM, have high water-holding capacity. Figure 7c shows that the slope values for these types is positive, indicating a wetting situation.

5. Discussion

Spatiotemporal analyses of *DI* values showed that approximately 80% of the GCM experienced stable drought conditions without significant changes over the last 18 years. In general, precipitation-based indices (PCI and SDCI) showed similar spatiotemporal patterns, while two vegetation-based indices (VCI and VHI) exhibited similar patterns. The other two indices based on LST revealed vague changing patterns of drought, with pixels indicating significant changes distributed sparsely across the whole study area. This may be because the LST is extremely sensitive to location and land cover type and varies with time. By comparing the annual pattern of change of drought indices with the meteorological factors such as precipitation and temperature, we found that the LST-based indices (TCI and TVDI) were not closely related to either annual average precipitation or temperature. Analysis of the relationship of the land cover type with drought indices as well as their slope showed that grasslands are more easily affected by droughts than are deciduous broadleaf forests. The changing rates of the drought situation of three forests (i.e., evergreen needleleaf forests, deciduous broadleaf forests, and deciduous needleleaf forests) are clearly greater than those of the other land cover types.

As the GCM is a mountainous area, the terrain too may affect the drought situation. We extracted landforms of the GCM based on the terrain information extracted from Shuttle Radar Topography Mission digital elevation model according to [41]. Eight geomorphic units with different combinations of altitude and relief were determined. Relief is the difference between the maximum and minimum altitude in a landform unit. It is usually calculated in square or circular sampling units. These units include the low-altitude plain (LAP), low-altitude hill (LAH), low-relief and low-altitude mountain (LR-LAM), middle-relief and low-altitude mountain (MR-LAM), middle-altitude plain, middle-altitude hill (MAH), low-relief and middle-altitude mountain (LR-MAM), and middle-relief and middle-altitude mountain (MR-MAM) (Figure 8a). Zonal statistics of the six *DI* maps of 2018, shown in Figure 8, were obtained based on the geomorphic unit map in order to calculate the mean *DI* values in different units. The relative drought situation was consistent regardless of the index used. Overall, the MR-MAM and MAH have smaller TVDI values and higher values for the other five indices, indicating a mild drought condition; this may be a result of the higher altitude of the MR-MAM and MAH along with the lower temperature and the generally greater precipitation and better forest coverage in these regions. Drier areas generally correspond to the geomorphic units of LAP and LAH, which have low altitude and poorer vegetation coverage compared to high-altitude areas.

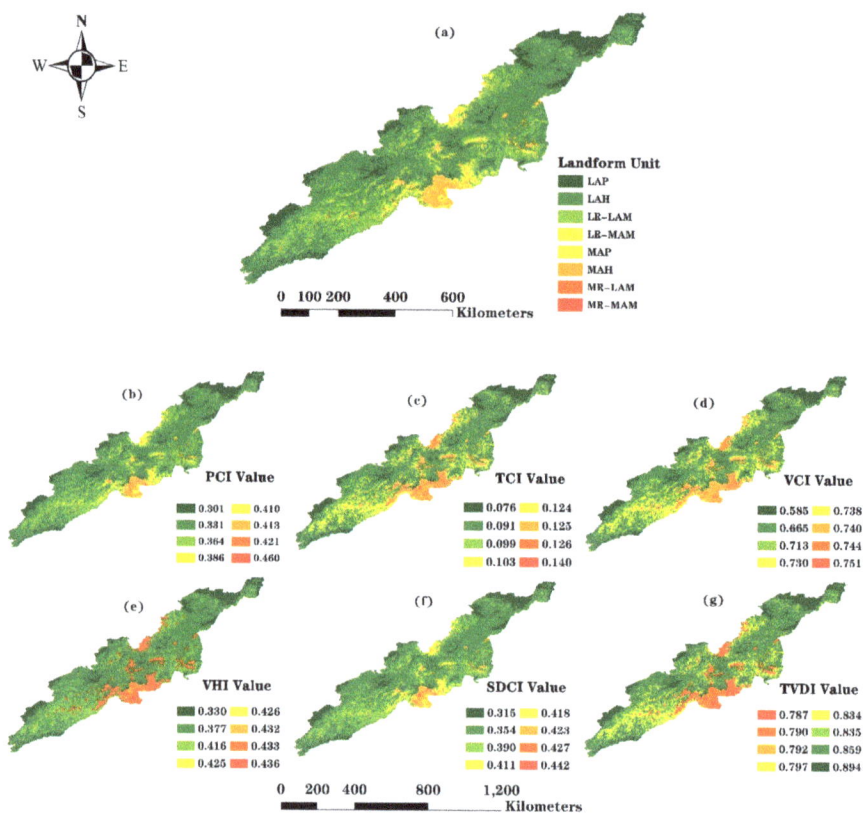

Figure 8. (a) Landforms of the GCM; (b–g) Mean values of drought indices in different geomorphic units (**b**: PCI; **c**: TCI; **d**: VCI; **e**: VHI; **f**: SDCI; and **g**: TVDI).

6. Conclusions

In this study, six widely used drought indices (PCI, VCI, TCI, VHI, SDCI, and TVDI) were used for monitoring drought in the GCM from the viewpoints of temperature, precipitation, and vegetation condition. Unlike the remote sensing products such as LST, precipitation, and vegetation indices, which provide absolute information, the aforementioned indices are computed for a certain area based on one or more of the remote sensing products, to reflect drought conditions. The spatiotemporal variations of drought were examined using the annual series of these indices from 2001 to 2018. The drought trends may be different for different indices. This indicates that the applicability of different indices differs with the location within the study area itself. Both PCI and SDCI show similar patterns, indicating that the central region of the GCM is getting wetter and the southwestern area is getting drier. VCI and VHI exhibit similar patterns with more drying trends. The correlations between these drought indices and meteorological factors were discussed to reveal that different indices are affected differently by precipitation and temperature variations. This is because these indices focus on different aspects of drought causes and symptoms, namely, precipitation, LST, and vegetation health. A comparison of the time series of the indices with precipitation and temperature showed that some drought indices cannot be explained by meteorological observations probably because of the time lag between meteorological drought and vegetation response. In particular, VHI and SDCI were generally employed for agricultural drought monitoring based on empirical weights. Note that the weights of these indices are adjustable. They may help us gain a better judgment about the drought conditions in different study areas. For this purpose, in situ observations can be employed in the future.

An examination of the slope of changes in drought for the different land cover types showed that the evergreen needleleaf forests in the GCM experienced increasingly severe drought conditions in recent years. Similar patterns with weaker changes were obtained for savannas and deciduous needleleaf forests. VCI and VHI exhibited similar variation patterns across different land cover types. The slopes of VCI and VHI for all land cover types are negative, indicating an overall drying trend. Note that the trend of vegetation-based indices may be affected by the long-term physical changes in vegetation.

The terrain is regarded an important factor for drought conditions. Based on the statistical analysis of drought patterns in different landforms, it was found that although different indices indicate distinct drought conditions, the relative drought situation of different landforms is consistent regardless of the index. This implies that the landform type may be important ancillary information for drought monitoring.

Note that many drought indices and drought-monitoring methods were not considered in this study. From the perspective of drought consequences, soil moisture is a direct indicator for drought and has been adopted in several drought indices. Our future work is to adopt the indices related to soil moisture for a more comprehensive study in order to propose the most appropriate method for monitoring drought in the GCM. Nevertheless, the results of this study so far have preliminarily demonstrated the convenience of using remote sensing product-based indices for drought monitoring in the GCM as well as the differences between these indices. The results are expected to provide guidance for drought monitoring to help understand and monitor the ecosystem conditions and the environment in this region.

Author Contributions: Conceptualization, Y.H. and Z.L.; methodology, Y.H., and Z.L.; software, Z.L.; validation, Z.L., Y.H. and T.H.; formal analysis, Z.L. and Y.H.; investigation, H.N. and H.Y.; resources, Z.L., Y.H. and S.Z.; data curation, Z.L., H.N. and H.Y.; writing—original draft preparation, Y.H. and Z.L.; writing—review and editing, Y.H., Z.L., C.H. and Y.Z.; visualization, Y.H. and Z.L.; supervision, Y.H., C.H. and Y.Z.; project administration, Y.H. and C.H.; funding acquisition, Y.H. All authors have read and agreed to the published version of the manuscript.

Funding: This research was supported by the Fundamental Research Funds for the Central Universities (2412018ZD012); National Natural Science Foundation of China (41301364; 41630749).

Acknowledgments: The authors are grateful to three anonymous reviewers for their helpful comments. We also want to thank Y. Wang, H. Du, S. Liu and Z. Liu for the constructive suggestions.

Conflicts of Interest: The authors declare no conflict of interest.

References

1. Mishra, A.K.; Singh, V.P. A review of drought concepts. *J. Hydrol.* **2010**, *391*, 204–216. [CrossRef]
2. Guo, E.L.; Liu, X.P.; Zhang, J.Q.; Wang, Y.F.; Wang, C.L.; Wang, R.; Li, D.J. Assessing spatiotemporal variation of drought and its impact on maize yield in Northeast China. *J. Hydrol.* **2017**, *553*, 231–247. [CrossRef]
3. Wu, J.J.; Zhou, L.; Liu, M.; Zhang, J.; Leng, S.; Diao, C.Y. Establishing and assessing the Integrated Surface Drought Index (ISDI) for agricultural drought monitoring in mid-eastern China. *Int. J. Appl. Earth Obs. Geoinf.* **2013**, *23*, 397–410. [CrossRef]
4. Kelley, C.P.; Mohtadi, S.; Cane, M.A.; Seager, R.; Kushnir, Y. Climate change in the Fertile Crescent and implications of the recent Syrian drought. *Proc. Natl. Acad. Sci. USA* **2015**, *112*, 3241–3246. [CrossRef]
5. Zhang, J.; Mu, Q.; Huang, J. Assessing the remotely sensed Drought Severity Index for agricultural drought monitoring and impact analysis in North China. *Ecol. Indic.* **2016**, *63*, 296–309. [CrossRef]
6. Cong, D.; Zhao, S.; Chen, C.; Duan, Z. Characterization of droughts during 2001–2014 based on remote sensing: A case study of Northeast China. *Ecol. Inform.* **2017**, *39*, 56–67. [CrossRef]
7. Liang, L.; Sun, Q.; Luo, X.; Wang, J.H.; Zhang, L.P.; Deng, M.X.; Di, L.P.; Liu, Z.X. Long-term spatial and temporal variations of vegetative drought based on vegetation condition index in China. *Ecosphere* **2017**, *8*. [CrossRef]

8. Mirabbasi, R.; Anagnostou, E.N.; Fakheri-Fard, A.; Dinpashoh, Y.; Eslamian, S. Analysis of meteorological drought in northwest Iran using the Joint Deficit Index. *J. Hydrol.* **2013**, *492*, 35–48. [CrossRef]
9. Kogan, F.N. Droughts of the Late 1980s in the United States as Derived from NOAA Polar-Orbiting Satellite Data. *Bull. Am. Meteorol. Soc.* **1995**, *76*. [CrossRef]
10. Kogan, F.N. Application of Vegetation Index and Brightness Temperature for Drought Detection. *Adv. Space Res.* **1995**, *15*, 91–100. [CrossRef]
11. Kogan, F.N. Global Drought Watch from Space. *Bull. Am. Meteorol. Soc.* **1997**, *78*. [CrossRef]
12. Liang, L.; Qin, Z.; Zhao, S.; Di, L.; Zhang, C.; Deng, M.; Lin, H.; Zhang, L.; Wang, L.; Liu, Z. Estimating crop chlorophyll content with hyperspectral vegetation indices and the hybrid inversion method. *Int. J. Remote Sens.* **2016**, *37*, 2923–2949. [CrossRef]
13. Liang, L.; Di, L.; Zhang, L.; Deng, M.; Qin, Z.; Zhao, S.; Lin, H. Estimation of crop LAI using hyperspectral vegetation indices and a hybrid inversion method. *Remote Sens. Environ.* **2015**, *165*, 123–134. [CrossRef]
14. Jiao, W.Z.; Zhang, L.F.; Chang, Q.; Fu, D.J.; Cen, Y.; Tong, Q.X. Evaluating an Enhanced Vegetation Condition Index (VCI) Based on VIUPD for Drought Monitoring in the Continental United States. *Remote Sens.* **2016**, *8*, 21. [CrossRef]
15. Wang, K.Y.; Li, T.J.; Wei, J.H. Exploring Drought Conditions in the Three River Headwaters Region from 2002 to 2011 Using Multiple Drought Indices. *Water* **2019**, *11*, 20. [CrossRef]
16. Liu, W.; Wang, S.; Zhou, Y.; Wang, L.; Zhu, J.; Wang, F. Lightning-caused forest fire risk rating assessment based on case-based reasoning: a case study in DaXingAn Mountains of China. *Nat. Hazards* **2016**, *81*, 347–363. [CrossRef]
17. Li, X.; He, B.B.; Quan, X.W.; Liao, Z.M.; Bai, X.J. Use of the Standardized Precipitation Evapotranspiration Index (SPEI) to Characterize the Drying Trend in Southwest China from 1982–2012. *Remote Sens.* **2015**, *7*, 10917–10937. [CrossRef]
18. Kogan, F.; Salazar, L.; Roytman, L. Forecasting crop production using satellite-based vegetation health indices in Kansas, USA. *Int. J. Remote Sens.* **2012**, *33*, 2798–2814. [CrossRef]
19. Salazar, L.; Kogan, F.; Roytman, L. Using vegetation health indices and partial least squares method for estimation of corn yield. *Int. J. Remote Sens.* **2008**, *29*, 175–189. [CrossRef]
20. Bayarjargal, Y.; Karnieli, A.; Bayasgalan, M.; Khudulmur, S.; Gandush, C.; Tucker, C.J. A comparative study of NOAA-AVHRR derived drought indices using change vector analysis. *Remote Sens. Environ.* **2006**, *105*, 9–22. [CrossRef]
21. Unganai, L.S.; Kogan, F.N. Drought monitoring and corn yield estimation in southern Africa from AVHRR data. *Remote Sens. Environ.* **1998**, *63*, 219–232. [CrossRef]
22. Kogan, F.N. World droughts in the new millennium from AVHRR-based vegetation health indices. *Eos Trans. Am. Geophys. Union* **2002**, *83*, 557, 562–563. [CrossRef]
23. Bokusheva, R.; Kogan, F.; Vitkovskaya, I.; Conradt, S.; Batyrbayeva, M. Satellite-based vegetation health indices as a criteria for insuring against drought-related yield losses. *Agric. For. Meteorol.* **2016**, *220*, 200–206. [CrossRef]
24. Zhang, A.Z.; Jia, G.S. Monitoring meteorological drought in semiarid regions using multi-sensor microwave remote sensing data. *Remote Sens. Environ.* **2013**, *134*, 12–23. [CrossRef]
25. Rhee, J.; Im, J.; Carbone, G.J. Monitoring agricultural drought for arid and humid regions using multi-sensor remote sensing data. *Remote Sens. Environ.* **2010**, *114*, 2875–2887. [CrossRef]
26. Sandholt, I.; Rasmussen, K.; Andersen, J. A simple interpretation of the surface temperature/vegetation index space for assessment of surface moisture status. *Remote Sens. Environ.* **2002**, *79*, 213–224. [CrossRef]
27. Han, Y.; Wang, Y.; Zhao, Y. Estimating Soil Moisture Conditions of the Greater Changbai Mountains by Land Surface Temperature and NDVI. *IEEE Trans. Geosci. Remote Sens.* **2010**, *48*, 2509–2515. [CrossRef]
28. Baniya, B.; Tang, Q.H.; Xu, X.M.; Haile, G.G.; Chhipi-Shrestha, G. Spatial and Temporal Variation of Drought Based on Satellite Derived Vegetation Condition Index in Nepal from 1982–2015. *Sensors* **2019**, *19*. [CrossRef]
29. MOD13A3 MODIS/Terra vegetation Indices Monthly L3 Global 1km SIN Grid V006. Available online: https://lpdaac.usgs.gov/products/mod13a3v006/ (accessed on 14 July 2019).
30. MYD11C3 MODIS/Aqua Land Surface Temperature/Emissivity Monthly L3 Global 0.05Deg CMG V006. Available online: https://lpdaac.usgs.gov/products/myd11c3v006/ (accessed on 14 July 2019).
31. MCD12Q1 MODIS/Terra+Aqua Land Cover Type Yearly L3 Global 500m SIN Grid V006. Available online: https://lpdaac.usgs.gov/products/mcd12q1v006/ (accessed on 14 July 2019).

32. TRMM (TMPA/3B43) Rainfall Estimate L3 1 Month 0.25 Degree x 0.25 Degree V7. Available online: https://pmm.nasa.gov/data-access/downloads/trmm (accessed on 14 July 2019).
33. Du, L.T.; Tian, Q.J.; Yu, T.; Meng, Q.Y.; Jancso, T.; Udvardy, P.; Huang, Y. A comprehensive drought monitoring method integrating MODIS and TRMM data. *Int. J. Appl. Earth Obs. Geoinf.* **2013**, *23*, 245–253. [CrossRef]
34. Zhang, L.F.; Jiao, W.Z.; Zhang, H.M.; Huang, C.P.; Tong, Q.X. Studying drought phenomena in the Continental United States in 2011 and 2012 using various drought indices. *Remote Sens. Environ.* **2017**, *190*, 96–106. [CrossRef]
35. Huang, F.; Xu, S.L. Spatio-Temporal Variations of Rain-Use Efficiency in the West of Songliao Plain, China. *Sustainability* **2016**, *8*. [CrossRef]
36. Le Page, M.; Zribi, M. Analysis and Predictability of Drought In Northwest Africa Using Optical and Microwave Satellite Remote Sensing Products. *Sci. Rep.* **2019**, *9*, 1466. [CrossRef] [PubMed]
37. Karnieli, A.; Agam, N.; Pinker, R.T.; Anderson, M.; Imhoff, M.L.; Gutman, G.G.; Panov, N.; Goldberg, A. Use of NDVI and Land Surface Temperature for Drought Assessment: Merits and Limitations. *J. Clim.* **2010**, *23*, 618–633. [CrossRef]
38. Karnieli, A.; Bayasgalan, M.; Bayarjargal, Y.; Agam, N.; Khudulmur, S.; Tucker, C.J. Comments on the use of the Vegetation Health Index over Mongolia. *Int. J. Remote Sens.* **2006**, *27*, 2017–2024. [CrossRef]
39. Garcia, M.; Fernández, N.; Villagarcía, L.; Domingo, F.; Puigdefábregas, J.; Sandholt, I. Accuracy of the Temperature–Vegetation Dryness Index using MODIS under water-limited vs. energy-limited evapotranspiration conditions. *Remote Sens. Environ.* **2014**, *149*, 100–117. [CrossRef]
40. Zribi, M.; Dridi, G.; Amri, R.; Lili-Chabaane, Z. Analysis of the Effects of Drought on Vegetation Cover in a Mediterranean Region through the Use of SPOT-VGT and TERRA-MODIS Long Time Series. *Remote Sens.* **2016**, *8*. [CrossRef]
41. Long, E.; Cheng, W.-M.; Zhou, C.-H.; Yao, Y.-H.; Liu, H.-J. Extraction of landform information in Changbai Mountains based on Srtm-DEM and TM data. In In Proceedings of the 2007 IEEE International Geoscience and Remote Sensing Symposium, IGARSS, Barcelona, Spain, 23–28 July 2007.

© 2020 by the authors. Licensee MDPI, Basel, Switzerland. This article is an open access article distributed under the terms and conditions of the Creative Commons Attribution (CC BY) license (http://creativecommons.org/licenses/by/4.0/).

Article

Extraction of Spatial and Temporal Patterns of Concentrations of Chlorophyll-a and Total Suspended Matter in Poyang Lake Using GF-1 Satellite Data

Jian Xu [1], Chen Gao [2] and Yeqiao Wang [3],*

[1] School of Geography and Environment & Ministry of Education's Key Laboratory of Poyang Lake Wetland and Watershed Research, Jiangxi Normal University, Nanchang 330022, China; jianxu@jxnu.edu.cn
[2] College of Marine Science and Engineering, Nanjing Normal University, Nanjing 210046, China; 192601002@stu.njnu.edu.cn
[3] Department of Natural Resources Science, University of Rhode Island, Kingston, RI 02881, USA
* Correspondence: yqwang@uri.edu

Received: 28 December 2019; Accepted: 11 February 2020; Published: 13 February 2020

Abstract: Poyang Lake is the largest freshwater lake in China. Its ecosystem services and functions, such as water conservation and the sustaining of biodiversity, have significant impacts on the security and sustainability of the regional ecology. The lake and wetlands of the Poyang Lake are among protected aquatic ecosystems with global significance. The Poyang Lake region has recently experienced increased urbanization and anthropogenic disturbances, which has greatly impacted the lake environment. The concentrations of chlorophyll-a (Chl-a) and total suspended matter (TSM) are important indicators for assessing the water quality of lakes. In this study, we used data from the Gaofen-1 (GF-1) satellite, in situ measurements of the reflectance of the lake water, and the analysis of the Chl-a and TSM concentrations of lake water samples to investigate the spatial and temporal variation and distribution patterns of the concentrations of Chl-a and TSM. We analyzed the measured reflectance spectra and conducted correlation analysis to identify the spectral bands that are sensitive to the concentration of Chl-a and TSM, respectively. The study suggested that the wavelengths corresponding to bands 1, 3, and 4 of the GF-1 images were the most sensitive to changes in the concentration of Chl-a. The results showed that the correlation between the reflectance and TSM concentration was the highest for wavelengths that corresponded to band 3 of the GF-1 satellite images. Based on the analysis, bands 1, 3, and 4 of GF-1 were selected while using the APPEL (APProach by ELimination) model and were used to establish a model for the retrieval of Chl-a concentrations. A single-band model that was based on band 3 of GF-1 was established for the retrieval of TSM concentrations. The modeling results revealed the spatial and temporal variations of water quality in Poyang Lake between 2015 and 2016 and demonstrated the capacities of GF-1 in the monitoring of lake environment.

Keywords: chlorophyll-a concentration; total suspended matter concentration; retrieval methods; GF-1 satellite data; field-measured water spectra; Poyang Lake

1. Introduction

Lakes are valuable freshwater resources, and they are used for drinking water, fishing, agriculture, industry, and tourism [1]. Lakes also record regional environmental changes, and play important roles in regulating the regional climate and maintaining ecological balance [2]. However, the water quality of many lakes is being threatened by environmental problems that are caused by various natural and anthropogenic factors, such as eutrophication and organic and inorganic pollution [3]. Therefore, effective approaches are needed to monitor the water quality in lakes.

Laboratory analysis of lake water samples is among the main conventional methods that have been used to monitor water quality of lakes. However, this approach is time-consuming and expensive. When compared with conventional methods, satellite remote sensing technology has the advantages of providing multi-temporal and multi-spectral data with high spatial and temporal resolution [4]. Dynamic monitoring and analysis of aquatic environments while using satellite remote sensing technology have been applied in monitoring lake and wetland environments and provide warnings of aquatic environmental emergencies [5].

The monitoring of water quality while using remote sensing technology is of great importance for guiding lake management. The concentrations of chlorophyll-a (Chl-a) and total suspended matter (TSM) are important indicators for assessing the water quality of lakes [6,7]. The concentration of Chl-a can be used to estimate the degree of eutrophication and as a proxy of primary productivity in mesotrophic and eutrophic water environments. TSM directly affects the transmission of light in water and it is closely related to the optical properties of water transparency and turbidity. In the past 30 years, many studies have been reported to monitor concentrations of Chl-a and TSM in lakes while using satellite data [8–11]. Data from the Moderate Resolution Imaging Spectroradiometer (MODIS) were used to observe algal blooms and monitor seasonal variations of Chl-a concentration in Lake Taihu, China [12,13]. Medium Resolution Imaging Spectrometer (MERIS) data were used to obtain time-series of Chl-a concentration for the 50 largest standing-water bodies in South Africa [14], and long-term patterns in Poyang Lake, China [15]. Another study determined the spatial and temporal changes in the concentration of TSM in Poyang Lake while using MODIS data from 2000–2010 [16]. Shi et al. [17] integrated MODIS data from 2003–2013 and in situ observations from a number of boat-based surveys to estimate the concentrations of TSM in Lake Taihu. Semi-analytical and empirical algorithms provide indices that are sensitive with Chl-a and TSM [13–19]. The retrieval of Chl-a and TSM concentration are mainly achieved through regression relationships between the measured parameter and sensitive indices while using linear [20,21], quadratic polynomial [22], exponential [16,17], and power-law [23] regression approaches. In recent years, the emergence and application of new satellite sensors, such as Landsat-8 and Sentinel-2, have further promoted the development of the assessment of inland water quality while using remote sensing data [23–25].

Poyang Lake is the largest freshwater lake in China. Its ecosystem services and functions, such as water and biodiversity conservations, have significant impacts on ecological security and sustainability of regional ecology. Poyang Lake has experienced increased urbanization and anthropogenic disturbances, which has greatly impacted the aquatic environment of the lake and wetland system [26,27]. The eutrophication condition in Poyang Lake has been observed to increase over the past decades [28]. High spatial and temporal heterogeneity characterize the water quality of the lake [29]. Determining the dynamic changes of water quality in Poyang Lake requires satellite data with a high spatial resolution and frequency of monitoring in repeated cycle. Data from China's Gaofen-1 (GF-1) satellite possess high-resolution resolution capacity and they have the aforementioned characteristics. However, while GF-1 data have been used for the analysis of terrestrial land features, a lack of study has been implemented for water quality analysis.

This study aimed to assess the applicability of GF-1 satellite data in retrieve information about the concentrations of Chl-a and TSM of Poyang Lake. This study consisted of three steps: (1) characterize and analyze in situ water reflectance spectra and concentrations of Chl-a and TSM to determine the spectral bands or band combinations that are sensitive to retrieve Chl-a and TSM; (2) establish and evaluate algorithms while using data from the GF-1 satellite to estimate the concentrations of Chl-a and TSM; and, (3) determine the spatial and seasonal variation of water quality in the highest and lowest annual water levels between 2015 and 2016.

2. Materials and Methods

2.1. Study Area

Poyang Lake (115°49.7′~116°46.7′ E, 28°24′~29°46.7′ N) is located in the lower Yangtze River Basin. Tributaries of five rivers, namely the Ganjiang, Fuhe, Xinjiang, Raohe, and Xiushui rivers, feed the lake (Figure 1). Seasonal variation of precipitation leads to significant changes in the lake's surface area throughout the year [30]. The water surface area exceeds 3,000 km^2 during the wet season (April–September) and then drops below 1,000 km^2 during the dry season (October–March). Poyang Lake is also one of the world's most ecologically important wetlands, with millions of migratory birds, including about 98% of the world's population of Siberian cranes, inhabiting the lake area for wintering.

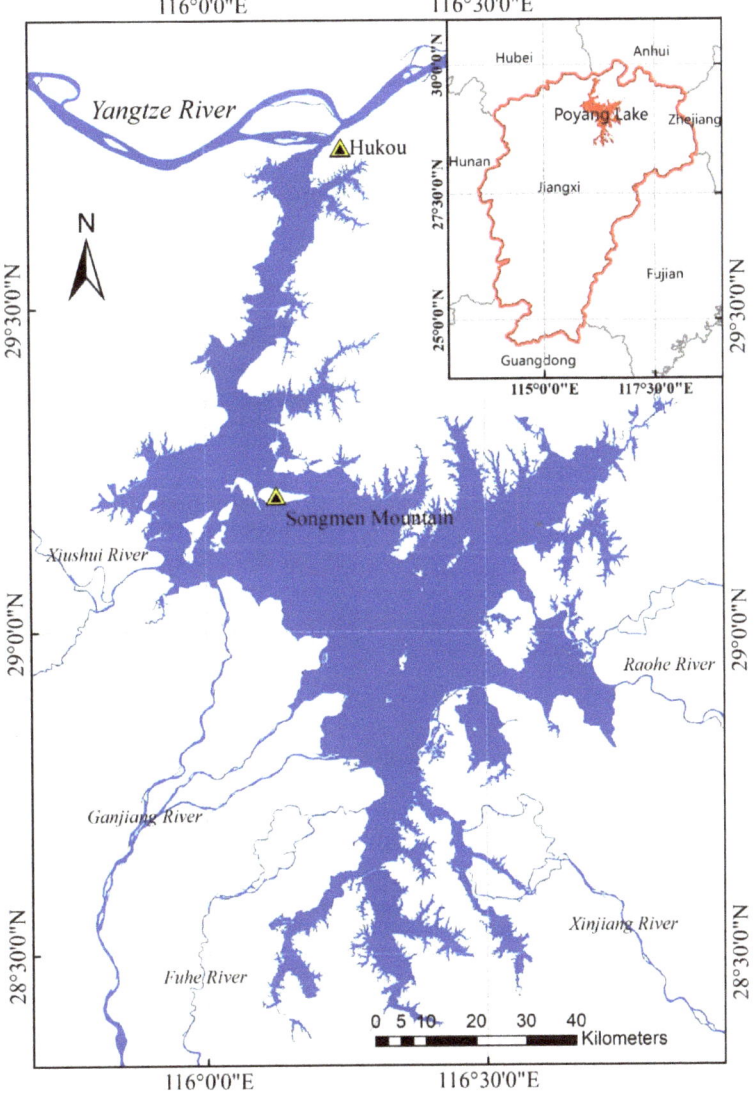

Figure 1. Map of Poyang Lake, China.

2.2. Field Sampling and Measurement

Field data were obtained from Poyang Lake in August 2015 (August 1, 3, and 5, 2015), October 2015 (October 23 and 24, 2015), and January 2016 (January 24 and 25, 2016), from 43, 33, and 26 sampling sites, respectively. Figure 2 shows the locations of the sampling sites. For each sampling exercise, water was collected from a water depth of between 0 and 50 cm. All of the samples were held on ice and stored for subsequent measurements of the concentrations of Chl-a and TSM in the laboratory. Two 500ml portions of each sample were used to filter for collecting Chl-a and TSM, respectively. Chl-a was collected while using Whatman GF/F filters (0.7 µm pore size) and extracted with hot ethanol [31]. The concentrations of Chl-a were determined by spectrophotometry using a UV-2600PC UV–vis spectrophotometer (Shimadzu, Inc., Koyto, Japan). Water samples for TSM measurement were filtered while using Whatman GF/C filters (1.2 µm pore size) under vacuum. Subsequently, the filters were weighed gravimetrically to determine the concentration of TSM. Water temperature and turbidity were measured at the sampling sites using a YSI6600 portable multi-parameter water meter (Yellow Springs Instruments, Inc., Yellow Springs, Ohio, USA).

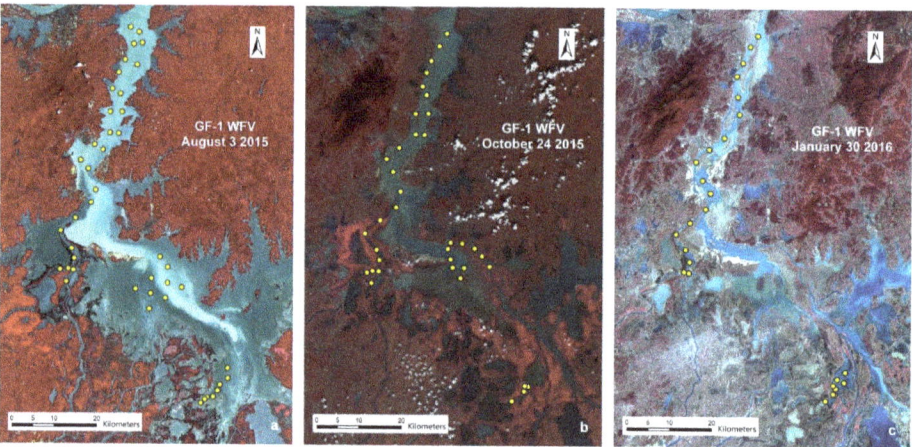

Figure 2. Pseudocolor renderings of images from the Gaofen-1 (GF-1) satellite (**a**: 770–890 nm, **b**: 630-690 nm, **c**: 520–590 nm) showing the landscape of the study area at approximately the time that the field samplings were performed.

Remote-sensing-reflectance spectra were measured above the water surface at wavelengths between 350 and 2500 nm (1 nm interval) while using a Fieldspec 4 spectroradiometer (Analytical Spectral Devices, Inc., Boulder, Colorado, USA), following standard protocols [32]. Measurements were performed between 10:00 and 14:00 on sunny windless days. A total of 66 sampling sites (35 for August 2015, 17 for October 2015, and 14 for January 2016) were measured for remote-sensing-reflectance spectra due to the limitation of measurement time. The radiances from water, sky, and a reference panel were measured at each water sampling site.

Remote sensing reflectance (R_{rs}) was determined as the ratio of water-leaving radiance (L_w) to the total downwelling irradiance [$E_d(0^+)$].

$$R_{rs} = \frac{L_w}{E_d(0^+)} = \frac{L_{SW} - \delta L_{sky}}{L_p * \pi / \rho_p} \quad (1)$$

where L_{SW} is the total upwelling radiance from water, L_{sky} is the skylight radiance, δ is a proportionality coefficient that relates L_{sky} to the reflected sky radiance determined when the detector viewed the water surface [33], L_p is the radiance from the reference panel, and ρ_p is the irradiance reflectance of the reference panel.

2.3. Model Development and Assessment

In this study, semi-analytical and empirical approaches were used for estimating the concentrations of Chl-a and TSM by sensitive indices. The retrieval algorithms were established through regression processes while using linear, quadratic polynomial, exponential, logarithmic, and power-law regression approaches. The goodness of fit was judged by the value of the coefficient of determination (R^2). The water samples were measured for remote-sensing-reflectance spectra and concentrations of Chl-a and TSM and then numbered from 1 to n, with 1 representing the samples with the highest concentration of Chl-a/TSM and n representing the samples with the lowest concentration. Subsequently, one sample was selected every three sample numbers—i.e., samples 1, 4, 7, 10, ... , were selected; the selected samples were then used for model validation, while the remaining samples were used for model calibration. The coefficient of determination (R^2), root-mean-square error (RMSE), and mean relative percentage error (MRPE) between the measured and predicted values of Chl-a or TSM concentration were calculated to assess the fitting and validation accuracy. The RMSE and MRPE were determined while using equations (2) and (3), respectively:

$$\text{RMSE} = \sqrt{\frac{1}{n} * \sum_{i=1}^{n} \left[x_{imea} - x_{ipre}\right]^2} \quad (2)$$

$$\text{MRPE} = \frac{\sum_{i=1}^{n} \left|\frac{x_{imea} - x_{ipre}}{x_{imea}}\right|}{n} * 100\% \quad (3)$$

where n is the number of samples.

2.4. Image Data and Preprocessing

The GF-1 satellite images were downloaded through the Remote Sensing Market Service Platform of the Chinese Academy of Sciences (http://www.rscloudmart.com). The GF-1 satellite was launched on 26 April 2013. The satellite has a sun-synchronous orbit with an altitude of 645 km, crossing the equator at 10:30 local time in a descending mode. The satellite carries panchromatic (2 m resolution) and multi-spectral (8 m resolution) sensor systems for high-resolution observation and four wide-field-of-view (WFV) sensors for large-scale observation. The four WFV sensors acquire multi-spectral data with a spatial resolution of 16 m, a revisit cycle of four days, and wide coverage (4 × 200 km). Table 1 shows the spectral bands of GF-1. The predicted service life of this satellite is five to eight years. The images that were employed in this study were acquired August 3 and October 24, 2015 and January 30, 2016, respectively. The images captured the lake wetland transition between the highest and lowest water levels in summer 2015 and the follow up winter. The water level change reflected the variation of aquatic environment, in particular for concentrations of Chl-a and TSM.

Table 1. Spectral bands of GF-1.

Tag	Band Order	Wavelength (nm)	Description
Panchromatic system	Pan	450–900	Panchromatic
Multi-spectral system	Band 1	450–520	Blue
Multi-spectral system	Band 2	520–590	Green
Multi-spectral system	Band 3	630–690	Red
Multi-spectral system	Band 4	770–890	Near infrared

The preprocessing of GF-1 images includes geometric correction, radiometric calibration, atmospheric correction, and water body range extraction. In this paper, image data were processed while using the ENVI 5.3 software (Exelis Visual Information Solutions, Inc., Broomfield, Colorado, USA). Geometric correction was processed using the RPC (Rational Polynomial Coefficients) Orthorectification module in ENVI 5.3. The Landsat8 OLI panchromatic image covering the Poyang Lake area was used as a reference. The GF-1 images were then radiometrically calibrated to covert DN value to radiance. Atmospheric correction was performed while using the FLAASH (Fast Line-of-Sight Atmospheric Analysis of Spectral Hypercubes) module in ENVI 5.3. FLAASH integrates MODTRAN 5 radiative transfer model with all MODTRAN atmosphere and aerosol styles to provide a unique solution for each image [34]. In this study, the mid-latitude atmosphere and rural aerosol were selected in FLAASH to correct the GF-1 images. The results of atmospheric correction were the remote sensing reflectance above the water surface. Figure 3a–c show the comparison of GF-1 bands before and after atmospheric correction, along with in situ reflectance resampled according to GF-1 band configuration. Figure 3d presents the validation result between the Band-3 reflectance of GF-1 imagery and in situ measured reflectance from August 2015. The results showed that the atmospheric interference to sensor had been effectively removed through FLAASH implementation.

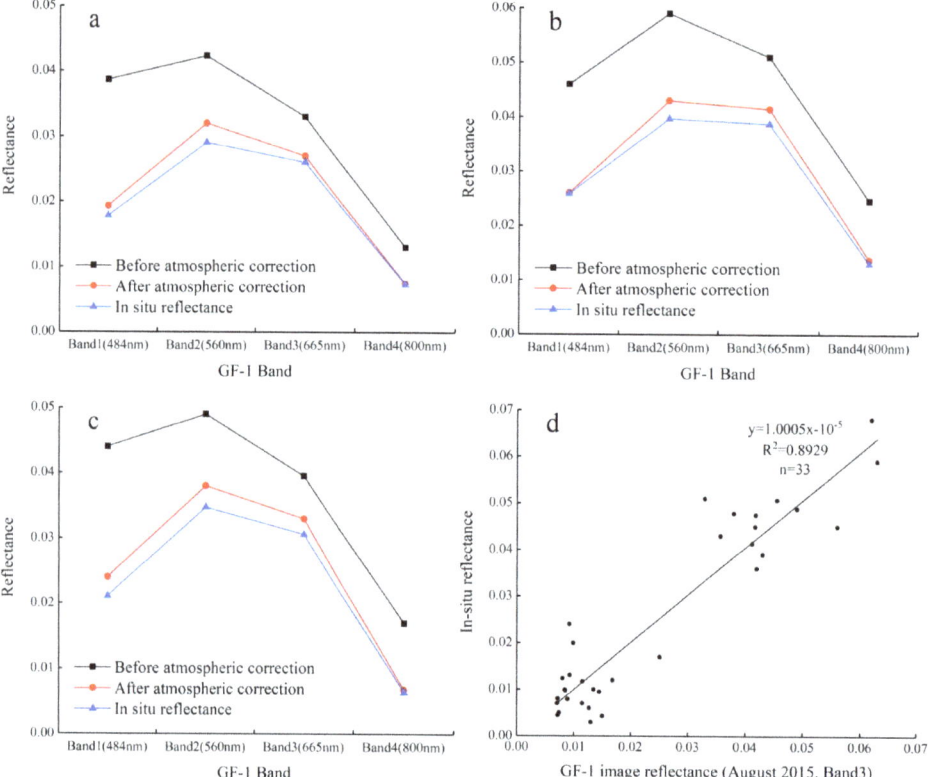

Figure 3. Comparison of GF-1 reflectance before and after atmospheric correction with in situ measured spectra, (**a**): mean value of data from August 2015; (**b**): mean value of data from October 2015; and, (**c**): mean value of data from January 2016. (**d**) Validation between the Band-3 reflectance of GF-1 imagery and in situ measured reflectance (August 2015).

3. Results

3.1. In-Situ Data

Table 2 presents descriptive statistics for the measured concentrations of Chl-a and TSM in the water samples obtained from the three samplings on Poyang Lake.

Table 2. Descriptive statistics for the measured concentrations of chlorophyll-a (Chl-a) and total suspended matter (TSM) in Poyang Lake.

Sampling Date	Number of Samples	Statistics	Chl-a Concentration (mg/m^3)	TSM Concentration (mg/L)
August 2015 (Summer)	43	Max	20.54	98.40
		Min	0.48	5.20
		Mean	5.16	28.23
		SD	3.88	22.49
		C.V.	75.17%	79.67%
October 2015 (Autumn)	33	Max	11.43	114.00
		Min	0.37	1.60
		Mean	3.10	52.20
		SD	2.43	35.79
		C.V.	78.64%	68.56%
January 2016 (Winter)	26	Max	5.02	66.25
		Min	0.80	2.40
		Mean	2.32	29.94
		SD	1.12	14.03
		C.V.	48.33%	46.86%

Note: SD is the standard deviation; C.V. is the coefficient of variation.

The water reflectance spectra were obtained by calculating the radiance that was collected at the Poyang Lake sampling sites. After removing two abnormal spectral data (with extremely low reflectance) in August 2015 sampling, a total of 33 valid spectra were obtained for the August 2015 sampling, 17 valid spectra for the October 2015 sampling, and 14 valid spectra for the January 2016 sampling (Figure 4). The reflectance spectra show a reflectance peak between 550 and 600 nm, which is related to the weak absorption of chlorophyll and carotene. The absorption of cyanophycin leads to an absorption valley between 600 and 650 nm; however, this valley was not observed in the reflectance spectra in this study due to the low concentration of chlorophyll in Poyang Lake. However, we observed an obvious absorption between 660 and 680 nm, which is caused by chlorophyll in the red-light band. The absorption becomes more obvious as the concentration of chlorophyll increases. We also observed a reflection peak near 700 nm. This reflection peak is an important feature of algae-containing water, and its position and amplitude indicate the concentration of Chl-a, with the peak moving towards longer wavelengths as the concentration of Chl-a increases. The reflectance of the spectra is low after about 730 nm, which is due to the strong absorption of Chl-a in the near-infrared band. A small reflection peak appears near 820 nm, which may be due to the scattering of suspended matter. For some sampling sites, the peaks and valleys in the reflectance spectra are not strongly pronounced, which is mainly due to the low concentration of Chl-a and the high concentration of TSM. The measured reflectance spectra vary for different seasons due to the change in the lake's water area throughout the year and the consequent change in the concentrations of Chl-a and TSM.

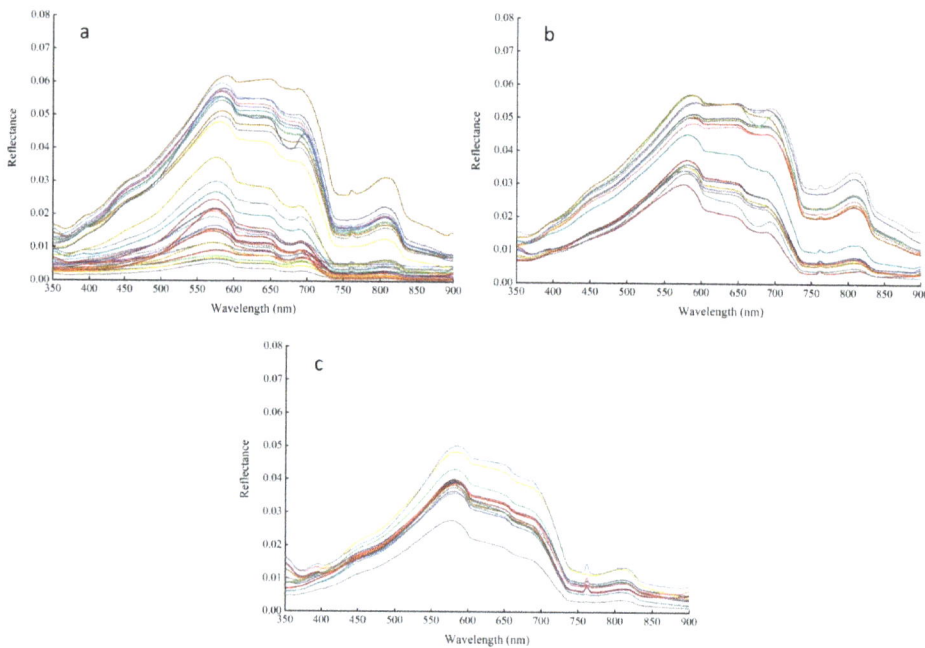

Figure 4. Measured water reflectance spectra for (**a**) August 2015, (**b**) October 2015, and (**c**) January 2016.

Of all the sampling sites that were used in this study, two sets of sampling sites were selected to determine the characteristic reflectance bands of Chl-a and TSM. Figure 5a shows reflectance spectra from three sampling sites with approximately the same Chl-a concentration (~4 mg/m^3) and different TSM concentrations (18.6 mg/L, 55.8 mg/L, and 98.4 mg/L). Meanwhile, Figure 5b shows reflectance spectra from three sampling sites with similar TSM concentrations (~6 mg/L) and different Chl-a concentrations (3.91 mg/m^3, 6.36 mg/m^3, and 9.70 mg/m^3).

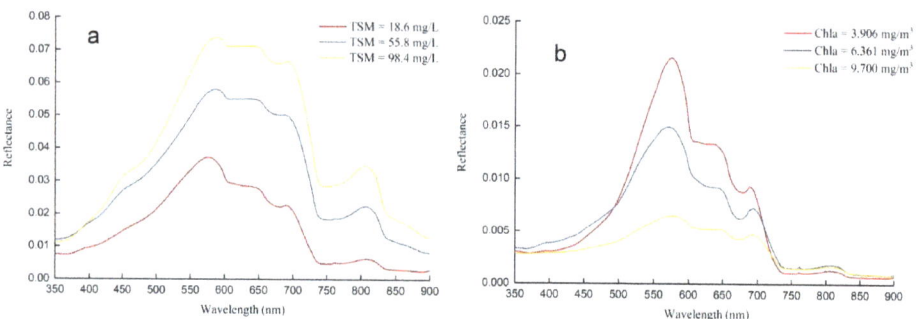

Figure 5. (**a**) The measured remote-sensing-reflectance spectra of the surface of Poyang Lake for sites with similar concentrations of Chl-a and different concentrations of TSM. (**b**) The measured remote-sensing-reflectance spectra of the surface of Poyang Lake for sites with similar concentrations of TSM and different concentrations of Chl-a.

From the spectra shown in Figure 5b, it can be concluded that the reflectance between 500 and 700 nm is inversely proportional to the concentration of Chl-a, and the reflectance beyond 700 nm is basically insensitive to the concentration of Chl-a. From the spectra that are shown in Figure 5a, it can be concluded that the reflectance is proportional to the TSM concentration. The wavelength positions of the absorption valley at the wavelength of 660 ~ 680nm and the reflection peak near 700nm are very stable, and they do not change with the change of TSM concentration. Therefore, the absorption valley at 660–680 nm and the reflection peak around 700 nm can be used as characteristic bands for the inversion of Chl-a concentration.

From the reflectance spectra that are shown in Figure 5a, it can be concluded that almost the whole reflectance of the spectra are significantly and positively correlated with TSM concentration between 350 and 900 nm. From the reflectance spectra that are shown in Figure 5b, it can be concluded that the reflectances below 700 nm are highly sensitive to Chl-a concentration. The reflectances beyond 700 nm are almost completely insensitive to Chl-a concentration, and the reflectance after 830 nm is weak and noisy. Therefore, it can be concluded that the reflectance between 700 and 830 nm can be used for the inversion of TSM concentration.

3.2. Inversion Model for Chlorophyll-a Concentration and Its Results

Figure 6 shows the Pearson correlation between normalized remote sensing reflectance and Chl-a concentration. The correlation varied significantly for different wavelengths.

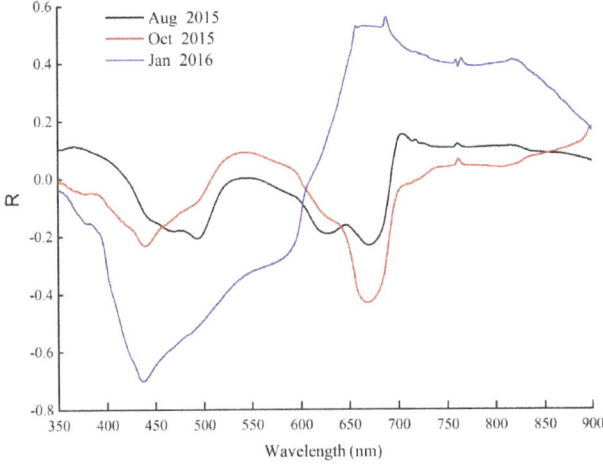

Figure 6. Pearson correlation coefficient (R) between normalized reflectance and Chl-a concentration.

Subsequently, we constructed isopotential maps of the linear correlation coefficients of determination between the spectral indices and Chl-a concentration in the spectral interval 350–900 nm based on the original reflectance spectrum by establishing the spectral indices of reflectance difference and reflectance ratio and using the least squares method to iteratively regress the spectral indices and Chl-a concentrations of Poyang Lake.

3.2.1. Analytical Results for Spectral Data from August 2015

The coefficients of determination between the ratio index and Chl-a concentration in Poyang Lake are generally low, as shown in Figure 7a. The coefficients of determination for combination of wavelengths of 700–715 nm and 690–700 nm are relatively high, but they are still at a low level (highest $R^2=0.35$). The coefficients of determination of the difference index for Chl-a concentration in Poyang Lake are higher than those of ratio index, as shown in Figure 7b. The combinations of spectral wavelengths that are sensitive to Chl-a concentration consist of the band near 650 nm and the band between 700 and 710 nm.

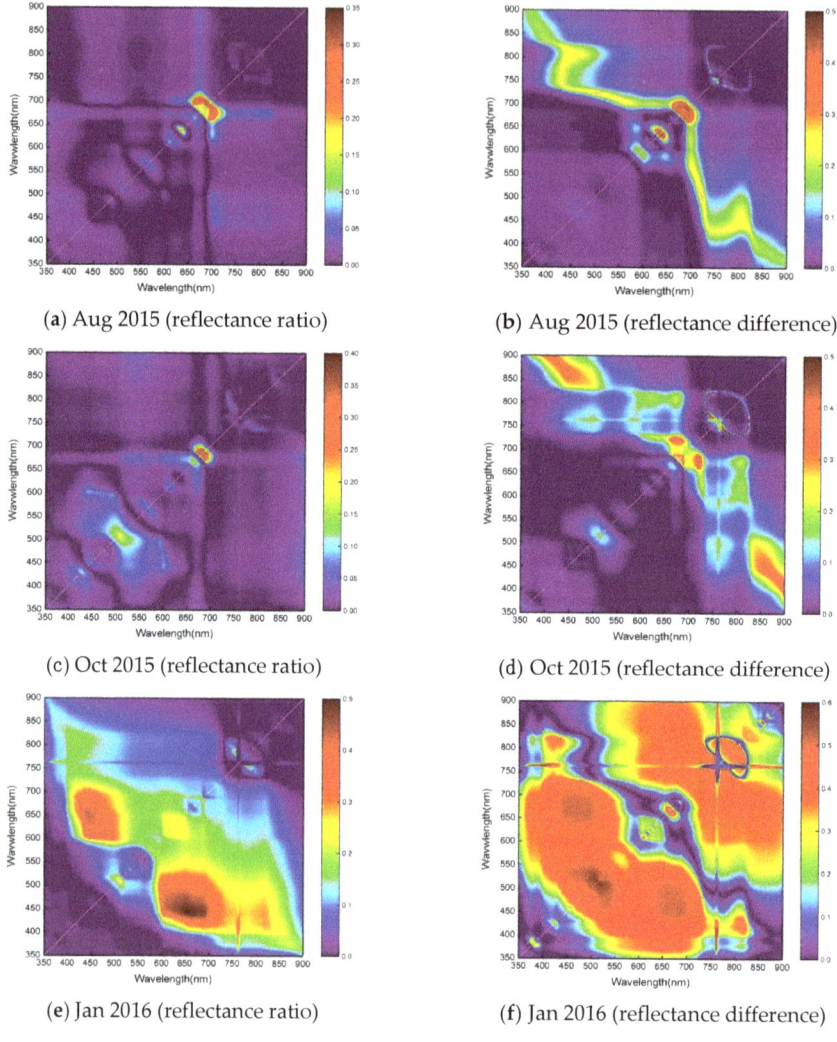

Figure 7. Linear correlation coefficients between the spectral index and Chl-a concentration. (**a**) August 2015 data (reflectance ratio). (**b**) August 2015 data (reflectance difference). (**c**) October 2015 data (reflectance ratio). (**d**) October 2015 data (reflectance difference). (**e**) January 2016 data (reflectance ratio). (**f**) January 2016 data (reflectance difference).

3.2.2. Analytical Results for Spectral Data from October 2015

As shown in the isopotential map that was based on the original spectral data from October 2015 shown in Figure 7c, the coefficients of determination between the Chl-a concentration and ratio index in Poyang Lake are low. High coefficients of determination are observed for spectral combination of wavelengths of 680–690 and 690–700 nm, however the maximum value ($R^2=0.4$) is still at a low level. The coefficients of determination between difference index and Chl-a concentration in Poyang Lake are higher than those between ratio index and Chl-a concentration, as shown in Figure 7d. The combinations of spectral wavelengths those are sensitive to Chl-a are 660–680 vs 720–730nm and 390–420 vs 870–900 nm.

3.2.3. Analytical Results for Spectral Data from January 2016

The coefficients of determination for the original spectral data from January 2016, which reached a maximum value of 0.5, were higher than the coefficients of determination for the original spectral data from August and October 2015, as shown in Figure 7e. The spectral combinations with the highest coefficients of determination are 440–470 nm vs 630–700 nm. The coefficients of determination between difference index and Chl-a concentration in Poyang Lake are higher than those of the ratio index, as shown in Figure 7f. The combinations of spectral wavelengths that are sensitive to Chl-a concentration are (1) 450–500 vs 500–540 nm, (2) 450–500 vs 650–700 nm, and (3) 360–420 nm and 760 nm. Table 3 summarizes the combinations of spectral wavelengths with the highest spectral index fit in Figure 7.

Table 3. Spectral response characteristics of Chl-a in Poyang Lake.

Data Collection Date	Spectral Index	w1 (nm)	w2 (nm)	R^2
August 2015	Reflectance Ratio (R_{w1}/R_{w2})	691	683	0.35
	Reflectance Difference ($R_{w1}-R_{w2}$)	693	695	0.50
October 2015	Reflectance Ratio (R_{w1}/R_{w2})	689	671	0.39
	Reflectance Difference ($R_{w1}-R_{w2}$)	770	737	0.49
January 2016	Reflectance Ratio (R_{w1}/R_{w2})	651	430	0.48
	Reflectance Difference ($R_{w1}-R_{w2}$)	762	403	0.58

Note: R^2 is coefficient of determination between spectral index and Chl-a concentration.

Overall, the results of the analysis show that the coefficients of determination of the linear correlation between the ratio index and the Chl-a concentration were lower than the coefficient of determination of the linear correlation between the difference index and the Chl-a concentration. The difference index corresponds to a wider range of sensitive bands than the ratio index. The spectral wavelengths that were sensitive to Chl-a concentration mainly corresponded to bands 1, 3, and 4 of the GF-1 image, and the bands of the GF-1 images have good overlap with the MODIS image. El-Alem et al. [35] have found that an APPEL (APProach by ELimination) model while using the combination of MODIS bands 1, 2, and 3 can be used to determine the Chl-a concentration in water. The maximum reflectance of Chl-a is in the near-infrared region. Furthermore, colored dissolved organic matter (CDOM), TSM, and backscattering also affect reflectance in the near-infrared band. CDOM has the maximal reflection in the blue band, so the influence of CDOM can be eliminated in the blue band. TSM is highly sensitive in the red band. Therefore, the red band can eliminate the influence of TSM on the reflectance spectrum of Chl-a. Pure water has strong absorption characteristics in the red and near-infrared bands, so the influence of backscattering can be eliminated.

The APPEL model was established based on the spectral characteristics of Chl-a, pure water, TSM, and CDOM, as follows [35]:

$$APPEL = R(b_{NIR}) - [(R(b_{BLUE}) - R(b_{NIR}))*R(b_{NIR}) + (R(b_{RED}) - R(b_{NIR}))] \quad (4)$$

The combination of GF-1 bands 1, 3, and 4 was used to establish the APPEL model. Table 4 shows the details of the models for the inversion of Chl-a concentration in different seasons.

Table 4. Details of the inversion models for the concentration of Chl-a.

Date	Model Expression	Goodness of Fit (Coefficient of Determination, R^2)
August 2015 (Summer)	$Y = 3171.2X^2 - 105.01X + 4.217$	0.6936
October 2015 (Autumn)	$Y = 2150.8X^2 - 38.504X + 2.0464$	0.6954
January 2016 (Winter)	$Y = 18594X^2 - 88.453X + 1.84$	0.6413

Note: X is calculated by the APPEL (APProach by ELimination) model: X = R(B4) - [(R(B1) - R(B4)) * R(B4) + (R(B3) - R(B4))], R(Bn) represents the reflectance of band n of the GF-1 image.

A summer inversion model for Chl-a concentration was established while using 33 sets of data that were measured in August 2015 (Figure 8a). Of the various models that were assessed, the inversion model that used quadratic polynomials had the highest fitting degree (R^2=0.6936); this model was subsequently validated while using 10 sets of measured data (Figure 8b). The results showed that the RMSE of the model was 1.158 mg/m^3 and the MRPE was 3.99%. Additionally, an autumn inversion model for Chl-a concentration (Figure 8c) was established based on 25 sets of data measured in October 2015. Again, the model that used quadratic polynomials had the highest fitting degree (R^2=0.6954). This model was validated with eight sets of measured data (Figure 8d). The results showed that the RMSE of the model was 0.90 mg/m^3 and the MRPE was 2.72%. Finally, a winter inversion model for Chl-a concentration was established while using 20 sets of data measured in January 2016 (Figure 8e). The model that used quadratic polynomials had the highest fitting degree (R^2=0.6413). This model was validated using six sets of measured data (Figure 8f). The results showed that the RMSE was 0.44 mg/m^3 and the MRPE was 9.44%.

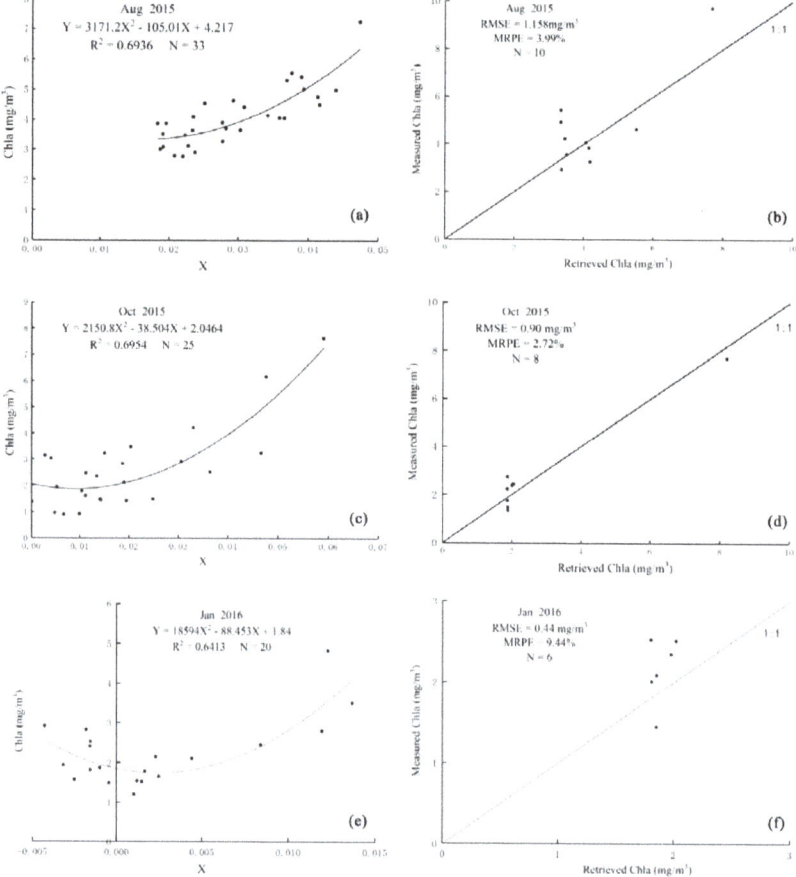

Figure 8. Test results for inversion models for Chl-a concentration for different periods: (**a,b**) calibration and validation results for August 2015; (**c,d**) calibration and validation results for October 2015; (**e,f**) calibration and validation results for January 2016. RMSE: root-mean-square error. MRPE: mean relative percentage error. R^2: coefficient of determination.

The Chl-a concentration in the entire coverage of Poyang Lake was obtained while using the ENVI 5.3 software based on the results of the summer, autumn, and winter inversion models for Chl-a concentration (Table 3). Figure 9 shows the corresponding estimates of the spatial distribution of the Chl-a concentration in Poyang Lake for each of these three seasons.

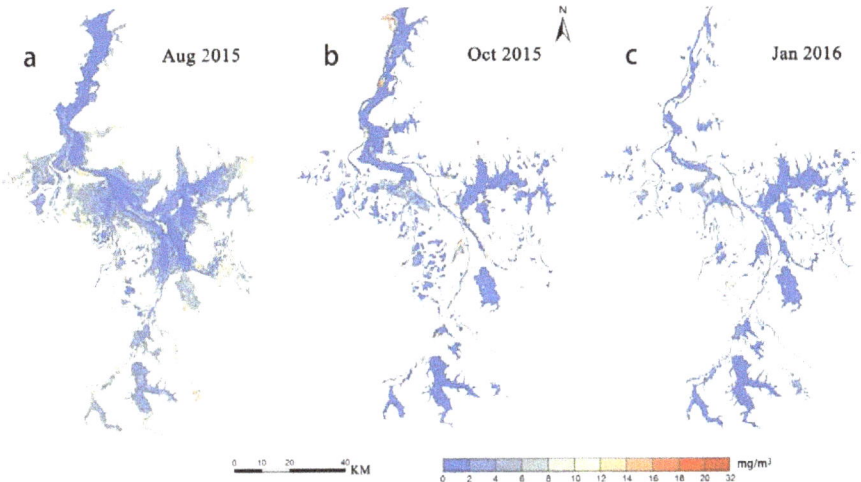

Figure 9. The estimated Chl-a concentration in Poyang Lake during (**a**) August 2015, (**b**) October 2015, and (**c**) January 2016, obtained from GF-1 satellite images using polynomial inversion models.

Figure 9a shows the results of the Chl-a concentration inversion while using the GF-1 satellite image from August 2015. In the study area, August is a summer month and is also the flooding (wet) season with the highest water level. During this month, the water temperature of Poyang Lake rises, the water velocity is the slowest, and algal growth is rapid, which causes the Chl-a concentrations to be the highest of the year, i.e., 5–30 mg/m^3. The highest concentrations of Chl-a are distributed in the waters near the shore of Poyang Lake and in the Nanji wetland national nature reserve in the south central of the lake.

Figure 9b shows the results of the Chl-a concentration inversion while using the GF-1 satellite image from October 2015. In October, which is an autumn month in the study area, the water level of Poyang Lake begins to decline and the water temperature to decrease. At this time, the estimated concentration of Chl-a in the lake decreased to between 2 and 15 mg/m^3. The highest concentrations of Chl-a are distributed around the channel in the north of Poyang Lake mouth, which connects it to the Yangtze River, and in the main channel near the center of the lake.

Figure 9c shows the results of the Chl-a concentration inversion while using the GF-1 satellite image from January 2016. In the study area, January is a winter month and is also in dry season. During this month, algae grow slowly in Poyang Lake. The estimated concentration of Chl-a was generally low, ranging from 0–11 mg/m^3, and the estimated distribution was more uniform than for August or October. The highest concentrations of Chl-a are distributed near the channel in the northern part of Poyang Lake, which connects it to the Yangtze River, and in the places where the Ganjiang, Fuhe, Xinjiang, Raohe, and Xiushui rivers flow into the lake.

3.3. Retrieval Models and Results for Total Suspended Matter Concentration

Figure 10 illustrates the Pearson correlations between normalized remote sensing reflectance and TSM concentration. The correlation coefficient shows obvious variation for different wavelengths.

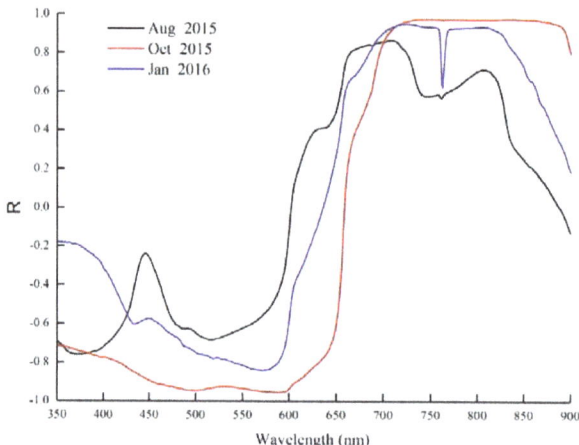

Figure 10. The Pearson correlation (R) between normalized remote sensing reflectance and TSM concentration.

Two peaks and valleys were observed in the correlation curve for August 2015 (summer), as shown in Figure 10. The two peaks are located at wavelengths of 660~730 nm and 760~820 nm, respectively, and the maximum values of the correlation coefficients for the peaks are 0.85 and 0.75, respectively. The two valleys are located at wavelengths of 350~400 nm and 500~550 nm, respectively, and the maximum values of the correlation coefficients for the valleys are −0.76 and −0.72, respectively. For the August 2015 correlation curve, the maximum correlation coefficient (0.85) appears at a wavelength of 710 nm, and the minimum correlation coefficient (−0.76) appears at a wavelength of 371 nm. The correlation curve for October 2015 (autumn) shows different trends at wavelengths below and above 650 nm, respectively. Below 650 nm, the correlation coefficient is negative, reaching its minimum value of −0.96 at a wavelength of 590 nm; above 650 nm, the correlation coefficient is positive, reaching a maximum value of 0.98 at a wavelength of 756 nm. In January 2016, the water reflectance spectra that was most sensitive to TSM concentration, was observed at a wavelength of 723 nm, with the maximum correlation coefficient of 0.95. These sensitive wavelengths most corresponded to band 3 of the GF-1 image. Therefore, band 3 of the GF-1 image was used in the model for the retrieval of TSM concentration. We assessed the performance of five mathematical models for retrieval, which used linear, quadratic polynomial, exponential, logarithmic, and power-law equations, respectively, and selected the best-fitting model (i.e., the one with the highest coefficient of determination, R^2) as the retrieval model. Table 5 describes the selected models for different seasons.

Table 5. Selected models for the retrieval of TSM concentration.

Date	Model Expression	Goodness of Fit (Coefficient of Determination, R^2)
August 2015 (summer)	$Y = 9972.1X^2 + 469.6X + 3.358$	0.9003
October 2015 (autumn)	$Y = 367347X^{2.7656}$	0.8614
January 2016 (winter)	$Y = 566102X^{2.5327}$	0.6504

Note: Y represents the TSM concentration; X represents the reflectance of band 3 of the GF-1 image.

The field data that were measured in August 2015 were used to establish the summer retrieval model (Figure 11a). The retrieval model using a quadratic polynomial was found to have the best fit (R^2=0.9003). This model was subsequently validated on 18 sets of measured data (Figure 11b). The results showed that the RMSE of the model was 6.96 mg/L and the MRPE was 27.52%. From the test results, it can be seen that the quadratic model in the single-band model predicted the summer TSM concentration of Poyang Lake well and the model had good stability.

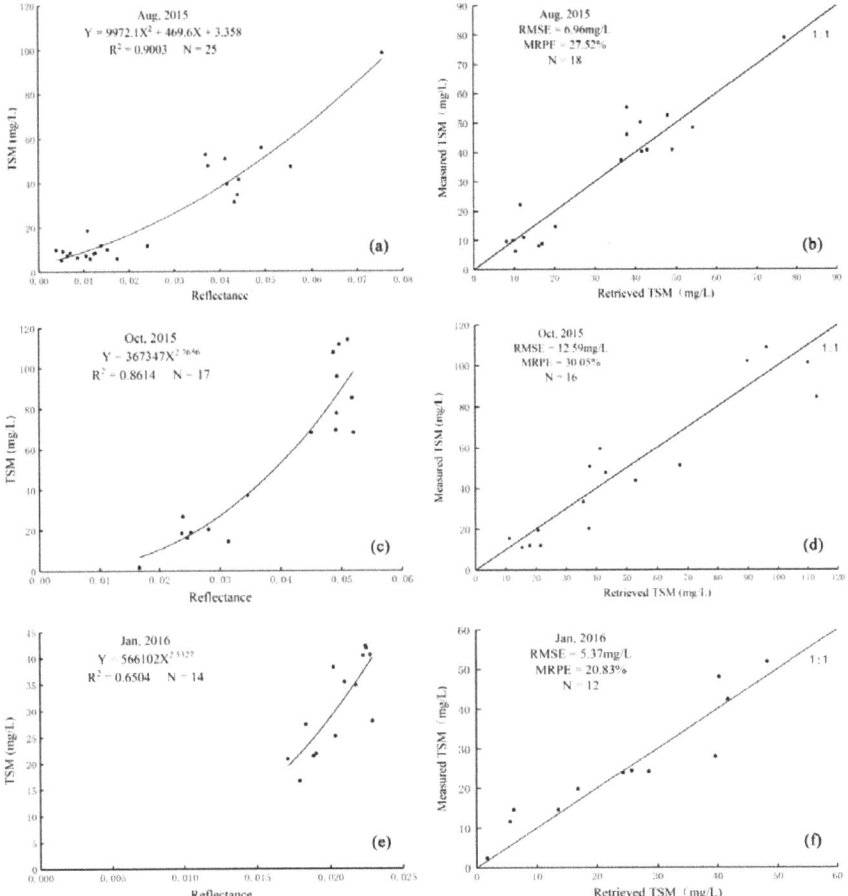

Figure 11. Test results for the models for the retrieval of TSM concentration for three seasons: (**a,b**) calibration and validation results for August 2015; (**c,d**) calibration and validation results for October 2015; (**e,f**) calibration and validation results for January 2016.

The 17 sets of data that were measured in October 2015 were used to establish the autumn retrieval model (Figure 11c). The retrieval model using a power-law equation was found to have the best fit (R^2=0.8614). The retrieval model was validated while using 16 sets of measured data (Figure 11d). The results showed that the RMSE of the model was 12.59 mg/L and the MRPE was 30.05%. It can be seen from the test results that the power-law model in the single-band model predicted the autumn TSM concentration of Poyang Lake well and the model had good stability.

The 14 sets of data that were measured in January 2016 were used to establish the winter retrieval model (Figure 11e). The retrieval model using the power-law equation was found to have the best fit (R^2=0.6504). This retrieval model was validated while using 12 sets of measured data (Figure 11f). The results showed that the RMSE of the model was 5.37 mg/L and the MRPE was 20.83%. It can be seen from the test results that the power-law model in the single-band model predicted the winter TSM concentration of Poyang Lake well and the model had good stability.

The TSM concentrations for the whole of the Poyang Lake area were calculated while using the ENVI 5.3 software based on the results of the retrieval models for TSM concentration in summer (August 2015), autumn (October 2015), and winter (January 2016) (Table 4). Figure 12 shows a map showing the spatial distribution of TSM concentration in Poyang Lake.

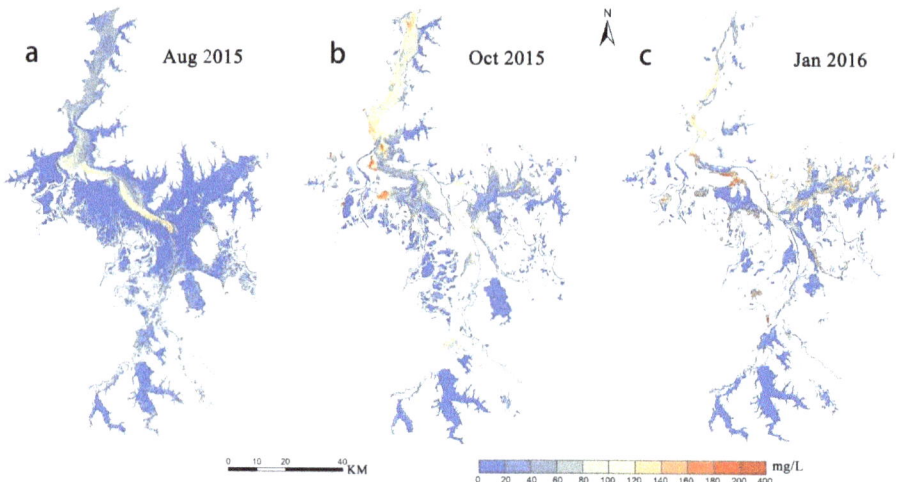

Figure 12. The estimated spatial distribution of TSM concentration in Poyang Lake for three periods: (**a**) August 2015; (**b**) October 2015; and, (**c**) January 2016.

Figure 12a shows the result of the retrieval of TSM concentration that is based on GF-1 image from August 2015. The overall level of TSM concentration in Poyang Lake was relatively low in August 2015, and the lowest TSM concentrations occurred in the eastern, western, and southern parts of Poyang Lake. The concentration of TSM in the eastern part of the lake was generally below 100 mg/L, the concentration at the junction of the Xiu River and the Ganjiang River ranged from 0~68 mg/L, and the concentration in Junshan Lake (which lies to the south of Poyang Lake) ranged from 0~46 mg/L. The concentration of TSM was relatively high in the channel, which connects the north of the lake to the Yangtze River, and in the main channel in the center of the lake, due to the influence of sand mining in Poyang Lake [36]. In the northern part of the lake, the TSM concentration ranged from about 59–80 mg/L. The highest TSM concentration that was observed in the central channel was 103 mg/L.

Figure 12b illustrated that the overall TSM concentration in Poyang Lake in October was significantly higher than that in August. The highest concentration of TSM (254.43 mg/L) was observed in the channel connecting the northern part of the lake to the Yangtze River. This can be attributed to the fact that, in August and September, increased rainfall in the Yangtze River causes the water level of the Yangtze River near Poyang Lake to increase, which suppresses water outflow from Poyang Lake and causes the water from the Yangtze River to flow back into the lake. This flow causes the TSM concentration of Poyang Lake to reach its highest levels in the channel connecting it with the Yangtze River due to the high concentration of TSM in the Yangtze River. The increases in TSM concentration that were observed in other parts of Poyang Lake can be attributed to the continuous sand mining activity in the lake area.

Figure 12c illustrated that the TSM concentration of Poyang Lake varied between 0 and 201 mg/L in January, i.e., the maximum TSM concentration higher than that in August 2015. The highest TSM concentrations were observed in the channel that connects the north of the lake with the Yangtze River, and in the main channel in the center of the lake.

4. Discussion

4.1. Spatial and Seasonal Variation of Chl-a

Figure 13 illustrates the water temperature data from the sampling sites in order to investigate the seasonal variation of the concentration of Chl-a in detail; the results of the inversion of the concentration of Chl-a; and the water level measured at the Xingzi hydrological station.

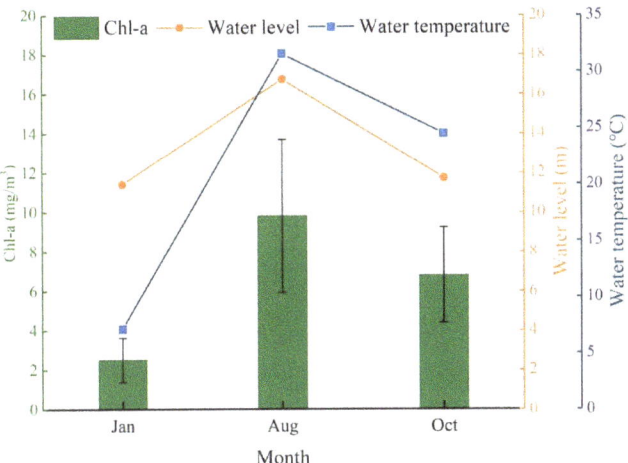

Figure 13. Relationship between Chl-a concentration, water temperature, and water level in Poyang Lake.

The comparison in Figure 13 suggested that, regarding the spatial distribution of Chl-a, the highest concentrations of Chl-a in Poyang Lake are mainly distributed near the channel in the north of the lake, which connects it to the Yangtze River, in the places where the Ganjiang, Fuhe, Xinjiang, Raohe, and Xiushui rivers flow into the lake, and in the waters near to the shore of the lake. These observations can be attributed to the transport of various pollutants into the lake by the Ganjiang, Fuhe, Xinjiang, Raohe, and Xiushui rivers, and to human activities in waters near to the lake shore [37]. Sand mining occurs during the whole year in the northern channel that connects the lake with the Yangtze River [16]. Consequently, a large amount of sewage is discharged from sand dredgers, and sand mining activities disturb the lake bottom, releasing large amounts of nutrients; both of these result in a high level of nutrients in the lake water, which in turn leads to an increase in the concentration of Chl-a [38]. The relatively high Chl-a concentration observed in the waters near the inlets of the five aforementioned rivers is mainly due to the large amount of pollutants carried by these rivers [39]. Large amounts of domestic and industrial wastewater are discharged into the lake due to the large numbers of people who live near the shores of Poyang Lake and the acceleration of industrialization and urbanization that has taken place in recent years; this discharge is also an important factor behind the increase of Chl-a concentration in the lake.

The temporal variation of Chl-a concentration in Poyang Lake is related to the lake's unique hydrological characteristics. From Figure 13, the Chl-a concentration of Poyang Lake is positively correlated with water temperature and water level. This finding is similar to the analysis of Zheng et al [37]. In the study area, summer is part of the wet season and, accordingly, the water level reached its yearly maximum in August 2015. In the wet season, the increase in water level causes Poyang Lake to enter a relatively stable state, with the water flow speed reducing and the water temperature increasing. These two factors are highly conducive to the growth of algae in the water body, which causes the concentration of Chl-a in the lake to increase. This can explain our observation that summer is the season with the highest concentration of Chl-a. On the other hand, the water level of Poyang Lake begins to decline in autumn, the water flow speed increases, and the Yangtze River begins to flow into the lake, due to the fact that the level of the river is higher than that of Poyang Lake [40]; consequently, the water temperature begins to decline in autumn, which, in turn, causes the Chl-a concentration to reduce to levels that are significantly lower than those in summer. Winter is the season with the lowest water level and the lowest water temperature in Poyang Lake. The combination of these two factors inhibits the growth of algae in the lake, which in turn causes the concentration of Chl-a to reach its lowest yearly value during this season.

4.2. Spatial and Seasonal Variation of TSM

We compared the concentration of TSM calculated while using the retrieval model, the water temperature measured at the sampling sites, and the water level measured at the Xingzi hydrological station in order to investigate the seasonal variation of the TSM concentration in Poyang Lake (Figure 14).

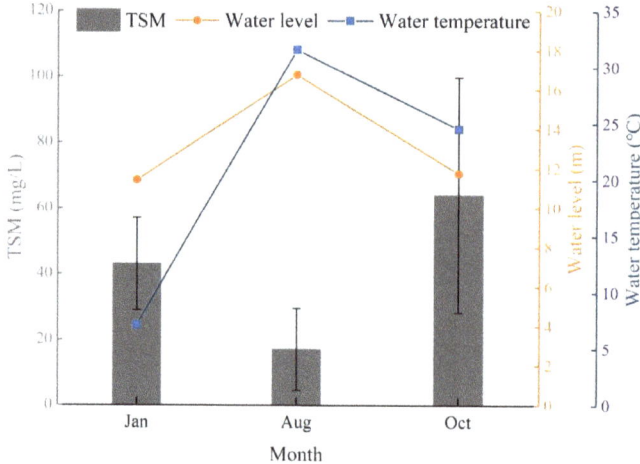

Figure 14. Plot showing the TSM concentration, water temperature, and water level of Poyang Lake.

Figure 12 concluded that the concentration of TSM was high in the channel that connects the north of Poyang Lake to the Yangtze River, and in the main central channel of the lake. This can be attributed to human activities, such as shipping and sand mining [16]. The TSM concentration in the Junshan Lake area (the south part of Poyang Lake) was at a low level throughout the year and changed little throughout the year. The concentration of TSM near the inlets of the Ganjiang, Fuhe, Xinjiang, Raohe, and Xiushui rivers changed greatly throughout the year. This can be attributed to the difference in the flow speed of these five rivers throughout the year [41].

Regarding the temporal variation of the concentration of TSM, the mean concentration was the lowest in summer (August 2015) [42]. At that time, water temperature and water level of Poyang Lake both reached their highest yearly levels. At this time, the flow speed of the lake was relatively low [41]. Although the concentration of TSM was high in the lake's main central channel, which can be attributed to the impact of sand mining activities, the TSM concentration in most of the other areas of the lake was low and relatively homogenous. The highest TSM concentration that was observed in Poyang Lake in this study was 254.43 mg/L, and it was observed in autumn (October 2015). At this time, the water level of Poyang Lake began to retreat, and the water level of the Yangtze River was higher than that of Poyang Lake. The backflow of water from the Yangtze River into Poyang Lake and the influence of sand mining activities increased the concentration of TSM in Poyang Lake [27]. The highest TSM concentrations were mainly observed in the channel that connects the north of the lake to the Yangtze River, and in the central part of the lake. In winter month (January 2016), the water temperature and the water level reached their lowest yearly levels. At this time, the TSM concentration ranged from 0 to 201 mg/L. This can be attributed to human activities, such as shipping and sand mining, which disturb the sediment at the lake bottom and thereby lead to an increase in TSM concentration [27]. In January 2016, the highest concentrations of TSM were observed in the lake's main central channel and in the channel that connects the north of the lake to the Yangtze River.

5. Conclusions

Images from the GF-1 satellite helped to establish retrieval models for concentrations of Chl-a and TSM in Poyang Lake in different seasons. The retrieval model that obtained the best fit for each season, respectively, was used to analyze the spatial and temporal variations of the concentrations of Chl-a and TSM for that season.

For Chl-a, the results showed that the wavelengths corresponding to bands 1, 3, and 4 of the GF-1 images were the most sensitive to changes in the concentration of Chl-a. Moreover, the APPEL model was used to establish the band combination, thus obtaining the retrieval models of Chl-a concentration for different seasons. The highest concentrations of Chl-a in Poyang Lake were mainly observed near the channel that connects the north of the lake to the Yangtze River, the places where the Ganjiang, Fuhe, Xinjiang, Raohe, and Xiushui rivers enter the lake, and near to the lake shore. In the central area of the lake, the concentration of Chl-a was relatively low and uniform. Regarding the temporal variation of Chl-a in Poyang Lake, the concentration was the highest in summer (August 2015), second-highest in autumn (October 2015), and lowest in winter (January 2016).

For TSM, the results showed that the correlation between the reflectance and TSM concentration was the highest for wavelengths corresponding to band 3 of the GF-1 satellite images. The highest TSM concentrations in Poyang Lake were mainly observed in the channel that connects the north of the lake to the Yangtze River, and in the lake's main central channel. The TSM concentration in Junshan Lake was relatively low and it changed little between seasons. The TSM concentrations near the inlets of the Ganjiang, Fuhe, Xinjiang, Raohe, and Xiushui rivers to Poyang Lake were lower than that in the channel that connects the north of the lake to the Yangtze River. Regarding the temporal variation of TSM concentration in Poyang Lake, the concentration was highest in autumn (October 2015), second highest in winter (January 2016), and lowest in summer (August 2015).

Author Contributions: Conceptualization, J.X. and Y.W.; Methodology, J.X.; Software, C.G.; Investigation, J.X. and C.G.; Data curation, J.X. and C.G.; Writing—original draft preparation, J.X. and C.G.; Writing—review and editing, J.X. and Y.W.; Supervision, Y.W.; Funding acquisition, J.X. All authors have read and agreed to the published version of the manuscript.

Funding: This study was funded by the National Natural Science Foundation of China (grant no. 41471298), the Key Research and Development Project of Jiangxi Province (grant no. 20192ACB70014), the Opening Fund of Key Laboratory of Poyang Lake Wetland and Watershed Research (Jiangxi Normal University), Ministry of Education (grant no. PK2018002), the Science and Technology Project of Education Department of Jiangxi Province, and the Young Talents Project of Jiangxi Normal University.

Acknowledgments: We appreciate the insightful and constructive comments and suggestions from the anonymous reviewers and the Editor that helped improve the quality of the manuscript.

Conflicts of Interest: The authors declare no conflict of interest. The funders had no role in the design of the study; in the collection, analyses, or interpretation of data; in the writing of the manuscript, or in the decision to publish the results.

References

1. Giardino, C.; Pepe, M.; Brivio, P.A.; Ghezzi, P.; Zilioli, E. Detecting chlorophyll, Secchi disk depth and surface temperature in a sub-alpine lake using Landsat imagery. *Sci. Total Environ.* **2001**, *268*, 19–29. [CrossRef]
2. Ayenew, T. Environmental implications of changes in the levels of lakes in the Ethiopian Rift since 1970. *Reg. Environ. Chang.* **2004**, *4*, 192–204. [CrossRef]

3. Wang, Y.; Qi, S.; Xu, J. Multitemporal remote sensing for inland water bodies and wetland monitoring. In *Multitemporal Remote Sensing*; Yifang Ban, Ed.; Springer: Cham, Switzerland, 2016; pp. 357–371.
4. Liu, Y.; Islam, M.A.; Gao, J. Quantification of shallow water quality parameters by means of remote sensing. *Prog. Phys. Geogr.* **2003**, *27*, 24–43. [CrossRef]
5. Wang, Y.; Yésou, H. Remote sensing of floodpath lakes and wetlands: A challenging frontier in the monitoring of changing environments. *Remote Sens.* **2018**, *10*, 1955. [CrossRef]
6. Duan, H.; Zhang, Y.; Zhang, B.; Song, K.; Wang, Z. Assessment of chlorophyll-a concentration and trophic state for Lake Chagan using Landsat TM and field spectral data. *Environ. Monit. Assess.* **2007**, *129*, 295–308. [CrossRef] [PubMed]
7. Miller, R.L.; Liu, C.-C.; Buonassissi, C.J.; Wu, A.-M. A multi-sensor approach to examining the distribution of total suspended matter (TSM) in the Albemarle-Pamlico estuarine system, NC, USA. *Remote Sens.* **2011**, *3*, 962–974. [CrossRef]
8. Torbick, N.; Hu, F.; Zhang, J.; Qi, J.; Zhang, H.; Becker, B. Mapping chlorophyll-a concentrations in west lake, China using Landsat 7 ETM. *J. Great Lakes Res.* **2008**, *34*, 559–565. [CrossRef]
9. Mishra, D.R.; Ogashawara, I.; Gitelson, A.A. *Bio-Optical Modeling and Remote Sensing of Inland Waters*; Elsevier: Amsterdam, The Netherlands, 2017.
10. Zheng, Z.; Li, Y.; Guo, Y.; Xu, Y.; Liu, G.; Du, C. Landsat-based long-term monitoring of total suspended matter concentration pattern change in the wet season for Dongting Lake, China. *Remote Sens.* **2015**, *7*, 13975–13999. [CrossRef]
11. Molkov, A.A.; Fedorov, S.V.; Pelevin, V.V.; Korchemkina, E.N. Regional models for high-resolution retrieval of chlorophyll a and TSM concentrations in the Gorky Reservoir by Sentinel-2 imagery. *Remote Sens.* **2019**, *11*, 1215. [CrossRef]
12. Wang, M.; Shi, W. Satellite-observed algae blooms in China's Lake Taihu. *EosTransactions Am. Geophys. Union* **2008**, *89*, 201–202. [CrossRef]
13. Wang, M.; Shi, W.; Tang, J. Water property monitoring and assessment for China's inland Lake Taihu from MODIS-Aqua measurements. *Remote Sens. Environ.* **2011**, *115*, 841–854. [CrossRef]
14. Matthews, M.W. Eutrophication and cyanobacterial blooms in South African inland waters: 10years of MERIS observations. *Remote Sens. Environ.* **2014**, *155*, 161–177. [CrossRef]
15. Feng, L.; Hu, C.; Han, X.; Chen, X.; Qi, L. Long-term distribution patterns of chlorophyll-a concentration in China's largest freshwater lake: MERIS full-resolution observations with a practical approach. *Remote Sens.* **2014**, *7*, 275–299. [CrossRef]
16. Feng, L.; Hu, C.; Chen, X.; Tian, L.; Chen, L. Human induced turbidity changes in Poyang Lake between 2000 and 2010: Observations from MODIS. *J. Geophys. Res. Ocean.* **2012**, *117*. [CrossRef]
17. Shi, K.; Zhang, Y.; Zhu, G.; Liu, X.; Zhou, Y.; Xu, H.; Qin, B.; Liu, G.; Li, Y. Long-term remote monitoring of total suspended matter concentration in Lake Taihu using 250m MODIS-Aqua data. *Remote Sens. Environ.* **2015**, *164*, 43–56. [CrossRef]
18. Salem, S.I.; Higa, H.; Kim, H.; Kazuhiro, K.; Kobayashi, H.; Oki, K.; Oki, T. Multi-algorithm indices and look-up table for chlorophyll-a retrieval in highly turbid water bodies using multispectral data. *Remote Sens.* **2017**, *9*, 556. [CrossRef]
19. Van Nguyen, M.; Lin, C.-H.; Chu, H.-J.; Muhamad Jaelani, L.; Aldila Syariz, M. Spectral feature selection optimization for water quality estimation. *Int. J. Environ. Res. Public Health* **2020**, *17*, 272. [CrossRef] [PubMed]
20. Zhou, L.; Roberts, D.A.; Ma, W.; Zhang, H.; Tang, L. Estimation of higher chlorophylla concentrations using field spectral measurement and HJ-1A hyperspectral satellite data in Dianshan Lake, China. *ISPRS J. Photogramm. Remote Sens.* **2014**, *88*, 41–47. [CrossRef]
21. Matthews, M.W.; Bernard, S.; Robertson, L. An algorithm for detecting trophic status (chlorophyll-a), cyanobacterial-dominance, surface scums and floating vegetation in inland and coastal waters. *Remote Sens. Environ.* **2012**, *124*, 637–652. [CrossRef]
22. Shen, F.; Zhou, Y.-X.; Li, D.-J.; Zhu, W.-J.; Suhyb Salama, M. Medium resolution imaging spectrometer (MERIS) estimation of chlorophyll-a concentration in the turbid sediment-laden waters of the Changjiang (Yangtze) Estuary. *Int. J. Remote Sens.* **2010**, *31*, 4635–4650. [CrossRef]
23. Liu, H.; Li, Q.; Shi, T.; Hu, S.; Wu, G.; Zhou, Q. Application of Sentinel 2 MSI images to retrieve suspended particulate matter concentrations in Poyang Lake. *Remote Sens.* **2017**, *9*, 761. [CrossRef]

24. Bresciani, M.; Cazzaniga, I.; Austoni, M.; Sforzi, T.; Buzzi, F.; Morabito, G.; Giardino, C. Mapping phytoplankton blooms in deep subalpine lakes from Sentinel-2A and Landsat-8. *Hydrobiologia* **2018**, *824*, 197–214. [CrossRef]
25. Xu, J.; Fang, C.; Gao, D.; Zhang, H.; Gao, C.; Xu, Z.; Wang, Y. Optical models for remote sensing of chromophoric dissolved organic matter (CDOM) absorption in Poyang Lake. *ISPRS J. Photogramm. Remote Sens.* **2018**, *142*, 124–136. [CrossRef]
26. de Leeuw, J.; Shankman, D.; Wu, G.; De Boer, W.F.; Burnham, J.; He, Q.; Yesou, H.; Xiao, J. Strategic assessment of the magnitude and impacts of sand mining in Poyang Lake, China. *Reg. Environ. Chang.* **2010**, *10*, 95–102. [CrossRef]
27. Gao, J.H.; Jia, J.; Kettner, A.J.; Xing, F.; Wang, Y.P.; Xu, X.N.; Yang, Y.; Zou, X.Q.; Gao, S.; Qi, S. Changes in water and sediment exchange between the Changjiang River and Poyang Lake under natural and anthropogenic conditions, China. *Sci. Total Environ.* **2014**, *481*, 542–553. [CrossRef] [PubMed]
28. Deng, X.; Zhao, Y.; Wu, F.; Lin, Y.; Lu, Q.; Dai, J. Analysis of the trade-off between economic growth and the reduction of nitrogen and phosphorus emissions in the Poyang Lake watershed, China. *Ecol. Model.* **2011**, *222*, 330–336. [CrossRef]
29. Li, B.; Yang, G.; Wan, R.; Hörmann, G.; Huang, J.; Fohrer, N.; Zhang, L. Combining multivariate statistical techniques and random forests model to assess and diagnose the trophic status of Poyang Lake in China. *Ecol. Indic.* **2017**, *83*, 74–83. [CrossRef]
30. Xu, J.; Wang, Y.; Gao, D.; Yan, Z.; Gao, C.; Wang, L. Optical properties and spatial distribution of chromophoric dissolved organic matter (CDOM) in Poyang Lake, China. *J. Great Lakes Res.* **2017**, *43*, 700–709. [CrossRef]
31. Lorenzen, C.J. Determination of chlorophyll and pheo-pigments: Spectrophotometric equations. *Limnol. Oceanogr.* **1967**, *12*, 343–346. [CrossRef]
32. Mueller, J.; Fargion, G. *Ocean Optics Protocols for Satellite Ocean Color Sensor Validation, Revision 3*; Goddard Space Flight Center: Greenbelt, Maryland, 2002.
33. Mobley, C.D. Estimation of the remote-sensing reflectance from above-surface measurements. *Appl. Opt.* **1999**, *38*, 7442–7455. [CrossRef]
34. Song, K.; Li, L.; Wang, Z.; Liu, D.; Zhang, B.; Xu, J.; Du, J.; Li, L.; Li, S.; Wang, Y. Retrieval of total suspended matter (TSM) and chlorophyll-a (Chl-a) concentration from remote-sensing data for drinking water resources. *Environ. Monit. Assess.* **2012**, *184*, 1449–1470. [CrossRef] [PubMed]
35. El-Alem, A.; Chokmani, K.; Laurion, I.; El-Adlouni, S.E. Comparative analysis of four models to estimate chlorophyll-a concentration in case-2 waters using moderate resolution imaging spectroradiometer (MODIS) imagery. *Remote Sens.* **2012**, *4*, 2373–2400. [CrossRef]
36. Yao, J.; Zhang, D.; Li, Y.; Zhang, Q.; Gao, J. Quantifying the hydrodynamic impacts of cumulative sand mining on a large river-connected floodplain lake: Poyang Lake. *J. Hydrol.* **2019**, *579*, 124156. [CrossRef]
37. Zheng, L.; Wang, H.; Huang, M.; Liu, Y. Relationships between temporal and spatial variations of water quality and water level changes in Poyang Lake based on 5 consecutive years' monitoring. *Appl. Ecol. Environ. Res.* **2019**, *17*, 11687–11699. [CrossRef]
38. Wu, Z.; He, H.; Cai, Y.; Zhang, L.; Chen, Y. Spatial distribution of chlorophyll a and its relationship with the environment during summer in Lake Poyang: A Yangtze-connected lake. *Hydrobiologia* **2014**, *732*, 61–70. [CrossRef]
39. Wu, Z.; Cai, Y.; Liu, X.; Xu, C.P.; Chen, Y.; Zhang, L. Temporal and spatial variability of phytoplankton in Lake Poyang: The largest freshwater lake in China. *J. Great Lakes Res.* **2013**, *39*, 476–483. [CrossRef]
40. Wu, Z.; Zhang, D.; Cai, Y.; Wang, X.; Zhang, L.; Chen, Y. Water quality assessment based on the water quality index method in Lake Poyang: The largest freshwater lake in China. *Sci. Rep.* **2017**, *7*, 1–10. [CrossRef]
41. Gu, C.; Mu, X.; Gao, P.; Zhao, G.; Sun, W.; Li, P. Effects of climate change and human activities on runoff and sediment inputs of the largest freshwater lake in China, Poyang Lake. *Hydrol. Sci. J.* **2017**, *62*, 2313–2330. [CrossRef]
42. Wang, J.; Chen, E.; Li, G.; Zhang, L.; Cao, X.; Zhang, Y.; Wang, Y. Spatial and temporal variations of suspended solid concentrations from 2000 to 2013 in Poyang Lake, China. *Environ. Earth Sci.* **2018**, *77*, 590. [CrossRef]

 © 2020 by the authors. Licensee MDPI, Basel, Switzerland. This article is an open access article distributed under the terms and conditions of the Creative Commons Attribution (CC BY) license (http://creativecommons.org/licenses/by/4.0/).

Article

A Multi-Temporal Object-Based Image Analysis to Detect Long-Lived Shrub Cover Changes in Drylands

Emilio Guirado [1,2,*], **Javier Blanco-Sacristán** [1,3,4], **Juan Pedro Rigol-Sánchez** [5], **Domingo Alcaraz-Segura** [1,3,6] **and Javier Cabello** [1,5]

[1] Andalusian Center for Assessment and Monitoring of Global Change (CAESCG), University of Almería, 04120 Almería, Spain; javier.blanco@unimib.it (J.B.-S.); dalcaraz@ugr.es (D.A.-S.); jcabello@ual.es (J.C.)
[2] Multidisciplinary Institute for Environment Studies "Ramon Margalef", University of Alicante, San Vicente del Raspeig, 03690 Alicante, Spain
[3] Department of Botany, Faculty of Science, University of Granada Campus Fuentenueva, 18071 Granada, Spain
[4] Remote Sensing of Environmental Dynamics Lab, University of Milano – Bicocca, Piazza della Scienza 1, 20126 Milan, Italy
[5] Department of Biology and Geology, University of Almería, 04120 Almería, Spain; jprigol@ual.es
[6] Inter-university Institute for Earth System Research (IISTA), University of Granada, Av. Mediterráneo, s/n. 18006 Granada, Spain
* Correspondence: e.guirado@ual.es

Received: 10 October 2019; Accepted: 10 November 2019; Published: 13 November 2019

Abstract: Climate change and human actions condition the spatial distribution and structure of vegetation, especially in drylands. In this context, object-based image analysis (OBIA) has been used to monitor changes in vegetation, but only a few studies have related them to anthropic pressure. In this study, we assessed changes in cover, number, and shape of *Ziziphus lotus* shrub individuals in a coastal groundwater-dependent ecosystem in SE Spain over a period of 60 years and related them to human actions in the area. In particular, we evaluated how sand mining, groundwater extraction, and the protection of the area affect shrubs. To do this, we developed an object-based methodology that allowed us to create accurate maps (overall accuracy up to 98%) of the vegetation patches and compare the cover changes in the individuals identified in them. These changes in shrub size and shape were related to soil loss, seawater intrusion, and legal protection of the area measured by average minimum distance (AMD) and average random distance (ARD) analysis. It was found that both sand mining and seawater intrusion had a negative effect on individuals; on the contrary, the protection of the area had a positive effect on the size of the individuals' coverage. Our findings support the use of OBIA as a successful methodology for monitoring scattered vegetation patches in drylands, key to any monitoring program aimed at vegetation preservation.

Keywords: arid zones; drylands; object-based; seawater intrusion; soil loss; time series classification; very high-resolution images; *Ziziphus lotus*; Cabo de Gata-Níjar Natural Park; Southeast Spain

1. Introduction

A fundamental fact of ecological observation is that most living organisms do not show random distributions. In fact, environmental controls and anthropogenic impacts are determinants of the spatial patterns of these organisms. This implies that it is possible to know the performance of ecosystems through the study of the spatial distribution patterns of the organisms that live in them [1,2]. This is particularly important in drylands where, as a result of water scarcity and edaphic limitations, vegetation appears to form isolated patches of one or more plant individuals [3,4]. In these ecosystems, it has been observed that the spatial pattern of these patches determines key aspects of ecosystem functioning such as primary production [5], water and nutrient cycles [6], and biotic interactions [7,8]. Tools to produce accurate vegetation maps at the appropriate spatial scale over time could be very useful for gaining knowledge about the health and dynamics of dryland ecosystems.

Remote sensing has proven to be the most useful tool for monitoring changes in vegetation, as it is cost-effective, allows repeated mapping, and produces information on a large scale [9–11]. Within this technique, pixel-based methods are the most commonly used. However, these methods show several limitations for producing accurate maps of vegetation patches or plant individuals in drylands. First, pixel-based methods do not consider the spatial context in which the pixels are framed, making it difficult to identify isolated image elements. Second, they often result in a final overlap of such elements from automatic classifications, when the analysis is based on high spatial resolution images [12]. In drylands, the land surface is characterized by scattered vegetation in a matrix of bare soil and scattered shrubs, so contextual information is very useful for image classification [13]. Both characteristics limit the possibility of identifying and classifying patches of vegetation and individual plant elements.

Several methods have emerged as an alternative to pixel-based methods for mapping individuals or vegetation patches. For example, in the case of forests, light detection and ranging (LiDAR) and very high frequency (VHF) synthetic aperture radar (SAR) images allow the characterization of various attributes of individual trees from their three-dimensional structure (e.g., [14–16]). However, this method is difficult to use when vegetation shows reduced aerial volume such as in drylands. In these cases, object-based image analysis (OBIA) can be a good solution for mapping patches of vegetation and individual plants [17], particularly because there is currently a wide variety of freely available high spatial resolution orthoimages. OBIA can provide a more accurate and realistic identification of scattered vegetation in drylands because of the combined spectral information of each pixel with the spatial context [18,19]. This method has yielded good results in the monitoring of spatial patterns, functioning, and structure of vegetation in these environments [20,21].

OBIA may be particularly useful for assessing the dynamics of populations of long-lived plants of conservation concern. In this case, it is difficult and costly to assess the environmental controls of population dynamics due to their high persistence and sometimes low rate of regeneration, which requires very long-term studies [22,23]. It has been proposed that the maintenance of long-lived plant populations is the result of a balance between regeneration (replacement of individuals by recruiting new recruits) and persistence (maintenance of individuals in space, physically and temporarily), or a combination of both strategies [24,25], depending on abiotic stress and biotic competition [26]. Monitoring populations of persistent individuals over time is complicated, as there are continuous disturbances in the environment that can alter their performance [24]. However, the availability of the analysis of historical aerial orthophotographs and high spatial resolution satellite images with OBIA provides a good opportunity to reconstruct the interannual dynamics of long-lived plant populations over long periods of time, thus enabling the evaluation of changes experienced by these shrub populations.

Ziziphus lotus (L.) Lam, a long-lived shrub from Mediterranean drylands [27], shows characteristics for a multi-temporal analysis of the spatial distribution in its populations with OBIA. This shrub species depends on groundwater [28], forms fertility islands, and is considered an engineering species [29] of an ecosystem of interest for conservation at the European level (Directive 92/43/CEE). The main European population of the shrub species is located in a flat coastal area surrounded by greenhouses in the Cabo de Gata-Níjar Natural Park, Spain. This population has been affected by several threats for many decades, including sand mining [30,31], reducing the amount of sand available to develop the *Z. lotus* fertility islands; urban pressure [32], which has reduced the potential distribution of *Z. lotus*; and the expansion of intensive agriculture [33,34], responsible for the decline in the level of the aquifer's water table, which may have caused the seawater intrusion [35,36]. Since 1944, several studies have evaluated this community of *Z. lotus*. Shrub patterns to identify groundwater dependence [28], the formations of shrubs in dunes [37], shrub spatial aggregation and consequences for reproductive success [29], and mutual positive effects between shrubs [38] have been researched. Yet, the monitoring of the shrub population dynamics has never been studied.

Despite most of the shrub population being located within the protected area, its temporal dynamics could be affected by several human-induced disturbances. However, due to the slow growth of shrubs and the inertia in the extinction of individuals, it is difficult to assess such dynamics without considering the population structure of the shrub species over the last several decades. This work proposes the use of remote sensing methods to map the spatial distributions of shrubs and to analyze their size and shape as a means of identifying anthropic disturbances. Our guiding hypothesis was that *Z. lotus*, phreatophytic shrubs, were affected by soil loss and seawater intrusion that decreased their cover area. On the contrary, after the legal protection of the area in 1987, the shrubs increased their cover area. Within this framework, the objectives of this work were as follows: (i) to make precision maps of scattered shrubs from historical remotely sensed images using OBIA and (ii) to extract information on changes in the shape, size, and spatial distribution of shrubs, and thus infer their relationships with human disturbances over a period of 60 years (1956–2016).

2. Materials and Methods

2.1. Study Case

We used a reliable and reproducible methodology to monitor structural changes in scattered vegetation of a dryland coastal zone using very high spatial resolution images and OBIA. The temporal dynamics of the *Ziziphus lotus* (L.) Lam population in a semi-arid coastal zone was evaluated to infer the effects of human disturbances on the shape and size of individuals over a period of 60 years. Two human disturbances were evaluated: (i) the extraction of coastal sands in the 1970s [30], which eliminated the aeolian sands found in the upper layer of the soil using heavy equipment and created roads and dirt tracks in the area, and (ii) the seawater intrusion in the mid-1980s caused by groundwater withdrawals for greenhouse irrigation. The withdrawals resulted in the water table of the main aquifer dropping by 30 m [35]. In addition, we evaluated the impacts of the protection of the area in 1987 in the shrub species.

The study area is located on a coastal aeolian plain in the Cabo de Gata-Níjar Natural Park, Spain (36°49′46.3″N, 2°17′37.1″W; Figure 1). This area is one of the driest in Europe, with a mean annual precipitation, temperature, and potential evapotranspiration (PET) of 200 mm, 19 °C, and 1390 mm, respectively [39]. The area is a hydrogeological complex located between two wadis with numerous fractures [40,41] and shows the typical landscape of arid areas with bare soil and patches of shrubs dominated by *Z. lotus* shrubs [29,37]. This vegetation is supported by a shallow, unconfined coastal aquifer composed of gravel and sand deposits located in the discharge zone at the end of the two wadis. A major fault hydrologically separates this aquifer from the main regional one (Hornillo-Cabo de Gata, [32]). Consequently, inflows to the aquifer come mainly from scarce local rainfall.

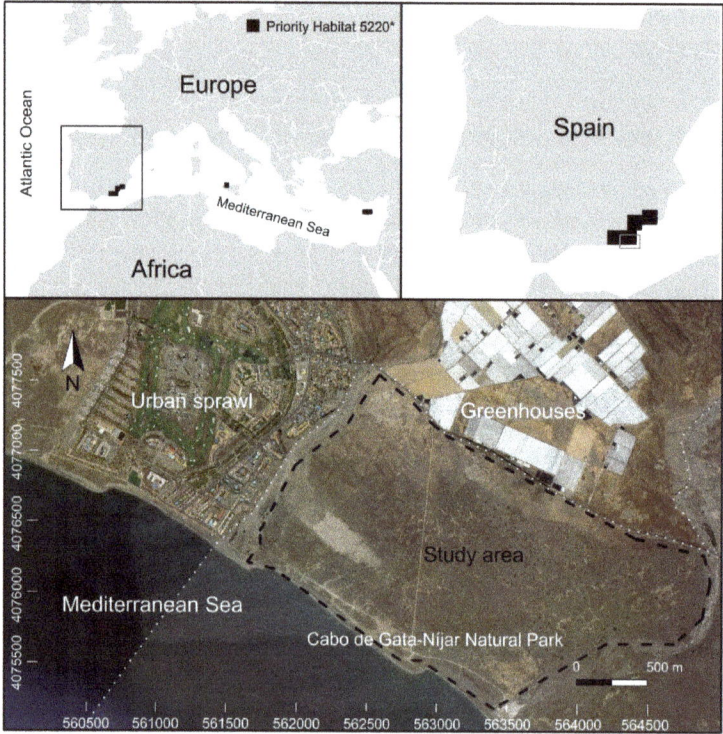

Figure 1. Upper images: the distribution of *Ziziphus lotus* priority habitat 5220* in the Mediterranean area. Lower image: Dashed line shows the study area under urban and intensive agricultural pressure in 2016. UTM projection Zone 30N; WGS 1984 Datum. Map data: Google, Maxar.

The scattered shrubs of *Ziziphus lotus* in SE Spain form the largest population of this shrub species in Europe. This population is protected by the Habitat Directive (5220* habitat, 92/43/CEE) and the Water Framework Directive (WFD) in Europe [31]. In this area, *Z. lotus* forms intricate structures of 1–3 m tall, accumulating sand under its cover called nebkas. This forms favorable microclimatic conditions under its cover with respect to the outside and increases the water availability due to hydraulic lift [38], carbon exchanges, and energy cycles [29], creating islands of fertility [42], which increases the diversity of animal and plant species. For this reason, *Z. lotus* is considered an ecosystem engineering species in this environment [31,33].

2.2. Datasets and Ground Truth

Eight orthoimages from two sources were used, namely, six orthoimages from the Andalusian Environmental Information Network (REDIAM) with a spatial resolution of 1 m/pixel from 1956 and 1977 (panchromatic images) and 0.5 m/pixel from 1984, 1997, 2004, and 2008 (multispectral images), and two Google Earth orthoimages from 2013 and 2016, with a resolution of 0.5 m/pixel (multispectral images). To work with the same spatial and spectral resolution, we homogenized the images to the lowest spatial resolution (i.e., 1 m/pixel) and transformed them into panchromatic images (1 band) from the three spectral band images with QGIS software v. 3.8 (manufacturer, city, state abbreviation, country). For sand mining mapping, we used airborne LiDAR data with 1 m point spacing obtained in 2011. A summary of the dataset is shown in Table 1.

Table 1. Data sources, spatial resolution, band numbers and year of the data used in the object-based shrub mapping and sand extraction estimation from airborne laser scanning (light detection and ranging (LiDAR)).

Data Source	Spatial Resolution	Band Number	Year
Andalusian Environmental Information Network (REDIAM)	1 m/pixel	1	1956
	1 m/pixel	1	1977
	0.5 m/pixel	3	1984
	0.5 m/pixel	3	1997
	0.5 m/pixel	3	2004
	0.5 m/pixel	3	2008
Google Earth[TM]	0.5 m/pixel	3	2013
	0.5 m/pixel	3	2016
Airborne laser scanning	1 m point spacing	-	2011

Two hundred perimeters of *Z. lotus* and 200 points of bare soil with scarce vegetation were randomly taken as the ground truth. A submeter precision GPS (Leica GS20 Professional Data Mapper; Leica, Wetzlar, Germany) was used. To do this, 12 longitudinal transects along the coast with a separation of 150 m between them were followed. The perimeter was taken with a distance of 1 m between nodes and the bare soil points were taken with a separation of at least 2 m from the nearest shrub. In addition, 200 shrub perimeters were digitized in each historical image with a distance of 1 m between nodes coinciding with the pixel size of the orthoimages in QGIS software v. 3.8.

2.3. Object-Based Image Analysis

OBIA consists of two phases, namely, the segmentation of the image into almost homogeneous objects and its subsequent classification based on similarities of shape, spectral information, and contextual information [17]. In the segmentation phase, it is necessary to establish an appropriate scale level depending on the size of the object studied in the image [43]; for example, low values for small vegetation and high values for large constructions [44,45]. During the classification, the segmented objects are classified to obtain cartographies of the classes of interest using algorithms such as nearest neighbor [46]. The success of the classification depends on the accuracy of the previous segmentation [47].

2.3.1. Image Segmentation

To segment the images, we used the multiresolution segmentation algorithm implemented in eCognition v. 8.9 software (Definiens, Munich, Germany). This algorithm depends on three parameters: (i) Scale, which controls the amount of spatial variation within objects and therefore their output size; (ii) Shape, which considers the form and color of objects; if it is set to high values, the form will be considered and if it is close to 0, the color will be considered instead; (iii) Compactness, a weighting to represent the smoothness of objects formed during the segmentation; if it is set to high values, the compactness will be considered complex and if it is set to values close to 0, the smoothness will be considered as simple [48]. To obtain the optimal value for each segmentation parameter, we used a ruleset in eCognition v8.9 that segmented the image by systematically increasing the Scale parameter in steps of 5 and the Shape and Compactness parameters in steps of 0.1 [49]. The Scale ranged from 5 to 50, and the Shape and the Compactness ranged from 0.1 to 0.9. A total of 6480 shapefiles were generated with possible segmentations of *Z. lotus* shrubs in a computer with a Core i7-4790K, 4 GHz and 32G of RAM memory (Intel, Santa Clara, CA, USA).

To evaluate the accuracy of all segmentations, we developed an R script to calculate the Euclidean Distance v.2 (ED2; [50]; Equation (1)), measuring the arithmetic and geometric discrepancies between the 200 reference polygons of *Z. lotus* and the corresponding segmented objects:

$$ED2 = \sqrt{(PSE)^2 + (NSR)^2}. \tag{1}$$

ED2 optimizes geometric and the arithmetic discrepancies with the "Potential Segmentation Error" (PSE; Equation (2)) and the "Number-of-Segments Ratio" (NSR; Equation (3)), respectively. According to [50], values of ED2 close to 0 indicate good arithmetic and geometric coincidence, whereas high values indicate a mismatch between them:

$$\text{PSE} = \frac{\sum |s_i - r_k|}{|r_k|}, \qquad (2)$$

where r_k is the area of the reference polygon and s_i is the overestimated area of the segment obtained during the segmentation;

$$\text{NSR} = \frac{abs(m - v)}{m}, \qquad (3)$$

where NSR is the arithmetic discrepancy between the polygons of the resulting segmentation and the reference polygons and *abs* is the absolute value of the difference of the number of reference polygons, *m*, and the number of segments obtained, *v*.

2.3.2. Classification and Validation of Segments

We used the *k*-nearest neighbors algorithm to classify the best segmentations (lowest ED2 values) in two classes, that is, (i) *Ziziphus lotus* shrub (Z) and (ii) Bare soil with sparse vegetation patches (S). In order to train the classification algorithm, 70% of the ground-truth samples (140 Z and 140 S) and the features of greatest separability (J) between them, obtained using the separability and threshold (SEaTH) algorithm, were used [51,52]. The remaining 30% of the ground-truth samples (60 Z and 60 S) were used as the validation set [53,54], and the accuracy of the classifications was evaluated using error and confusion matrices, extracting Cohen's kappa index of accuracy (KIA) [55] and the overall accuracy (OA) of them. Finally, errors in shrub segmentation were evaluated by estimation of the root-mean-square Error (RMSE) and the mean bias error (MBE) between reference polygons and segments classified as *Z. lotus* shrubs.

2.4. Sand Extraction Curvature Analysis

The evaluation of areas affected by sand extraction within the study area in the 1970s was performed using a geomorphometric analysis of the land surface [56]. The analyses were based on a LiDAR-derived digital elevation model (DEM) dataset, generated using an ArcGIS toolbox for multiscale DEM geomorphometric analysis. This toolbox allows the generation of a number of curvature-related land surface variables [57], including plane, profile, mean, minimum profile, maximum profile, tangential, non-sphericity, and total Gaussian curvature; positive and negative openness; and signed average relief. Several maps were derived for each curvature variable at different spatial scales. The sizes of the analysis window ranged from 3 × 3 m to 101 × 101 m with a 14 m interval. Univariate and bivariate statistics were calculated for variables related to curvature [56,58].

A window size of 61 m was selected for the geomorphometric analysis of the curvature of the surface, which provided a good compromise between the size of land surface depressions resulting from sand mining operations and spatial generalization. The sand extraction areas were located and digitized on a final map and validated with a field survey. In addition, an estimation of the volume of soil loss resulting from sand extractions was performed. To calculate the volume of soil loss, a new digital surface model was generated without the soil loss zones extracted with the previous curvature analysis. Then, the difference was applied to the initial surface model with the areas identified as soil loss and to the digital surface model without soil loss, obtaining the volume of the previously identified soil loss areas.

2.5. Shrub Area and Shape Dynamics

Variations in the size and number of shrubs were determined by calculating the number of shrubs lost and differences in shrub cover area between consecutive image pairs. To calculate losses and gains in the coverage of the individuals, we assumed that a resulting negative area meant a loss of surface coverage, whereas a positive area meant a gain of coverage. To determine the edge effect and the health indicator on shrubs [59], the round shape index was calculated as the ratio between the cover area and the perimeter of each shrub in different years [60]. In order to evaluate whether the shrub cover reduction that occurred between 1956 and 1977 was related to sand extraction, the average minimum distance (AMD) and average random distance (ARD) between shrubs and sand extraction areas were calculated using PASSaGE v.2 software (The Biodesign Institute, Arizona State University, Tempe, AZ, USA) [61] with 999 permutations. We assumed that the shrubs in the 1956 image rather than the 1977 image were removed during sand mining, and those that reduced their cover were affected during this process. To evaluate whether the shrub population was affected by seawater intrusion between 1977 and 1984, the AMD and the ARD between the shrubs and the coastline were calculated as previously. Shrubs affected by sand extractions (those appearing in the 1956 image but not in the 1977 one) were not included in this analysis. When calculating the AMD, the shrubs that showed a reduction in cover over the corresponding period were used, whereas for the calculation of the ARD simulated shrubs were used. In order to evaluate the effects of protecting the shrubs within the Natural Park in 1987, reduction in shrub cover and number of shrubs in the 1984–2016 period was determined.

3. Results

3.1. Segmentation Accuracy

The average values of the Scale and the ED2 were 25 and 0.45, respectively (Table 2). The most precise segmentations were from 1977 and 2016, with a Scale of 20 and an ED2 of 0.35 in both. The least accurate segmentations were the ones of 1956 and 2004, with ED2 values of 0.59 and 0.51, respectively. The lowest RMSE was obtained in the image of 2016, with a value of 46.38 m^2 and an MBE of -6.36 m^2, overestimating the cover area of the shrubs. The highest RMSE was up to 120.64 m^2, and in 5 of the 8 years (1956, 1977, 1997, 2004, and 2008) the area of the shrubs was underestimated as indicated by the MBE. With a computation time of 20 s per segmentation, we spent 36 h for a total of 6480 segmentations.

Table 2. Parameters used for the segmentation and their accuracies. RMSE, root-mean-square error; MBE, mean bias error; ED2, Euclidean Distance v.2. Lower (better) values of RMSE and ED2 are highlighted in bold type.

Year	1956	1977	1984	1997	2004	2008	2013	2016
Scale parameter	30	20	35	30	15	20	20	20
RMSE (m^2)	112.91	69.96	110.04	120.64	80.57	84.26	97.02	**46.38**
MBE (m^2)	14.07	15.07	−6.59	0.78	32.78	15.74	−9.96	−6.37
ED2	0.59	**0.35**	0.43	0.49	0.51	0.47	0.42	**0.35**

3.2. Classification and Characteristics of Ziziphus lotus Shrubs

The analysis of class separability and threshold with the SEaTH algorithm showed that the best features for discriminating between classes (i.e., those with the highest separability) were mainly related to texture (i.e., the family of features related to the Gray-Level Co-Occurrence Matrix (GLCM)) and brightness of objects (Table 3).

Table 3. Features used in the classifications and separability between them using the separability and threshold (SEaTH) algorithm. In bold, the two features for each year with the highest separability used to classify the images.

Year	1956	1977	1984	1997	2004	2008	2013	2016
Separability (J)								
Brightness	**0.92**	1.24	0.88	0.97	1.04	1.13	0.53	**0.86**
GLCM Homogeneity	0.03	0.87	0.47	0.29	0.34	0.56	0.45	0.8
GLCM Contrast	0.25	0.77	0.14	0.29	0.94	0.26	1.01	0.79
GLCM Entropy	0.06	0.38	0.38	0.24	0	0.41	**1.24**	0.56
GLCM Mean	**0.91**	**1.32**	**0.89**	**1**	**1.03**	1.14	0.57	**0.86**
GLCM SD	0.06	0.35	0.2	0.3	0.06	0.45	0.84	0.26
GLCM Correlation	0.01	0.06	0.06	0.2	0.35	**1.93**	0.94	0.59
Area	0.55	0.15	**0.92**	0.04	0.38	0.7	1.08	0.19
Border Length	0.7	0.54	0.27	0.35	0.06	1.37	0.79	0.07
Border Index	0.79	0.69	0.51	0.5	0.17	1.7	0.28	0.32
Compactness	0.73	0.63	0.46	0.51	0.5	1.38	0.29	0.46
Density	0.32	**1.11**	0.52	0.82	0.86	0.64	**1.42**	0.85
Roundness	0.63	0.35	0.25	0.36	0.22	1.44	0.29	0.2
Shape Index	0.82	0.7	0.52	0.57	0.25	**1.74**	0.22	0.4

All the classifications were highly accurate, with values of OA and KIA close to 1 (Table 4). The most accurately segmented image (OA = 0.98; KIA = 0.97) was the image of 2004, whereas the worst one was the image of 1956 (OA = 0.89; KIA = 0.79).

Table 4. Error matrix of all the classified images in the study. Z, *Ziziphus lotus*; S, Bare soil with sparse vegetation patches; Uncl., Unclassified; Prod., Producer's accuracy; User, User's accuracy; Held, Helden; KIA-c, KIA per class; AO, Overall accuracy; KIA, Kappa index of agreement. Highest (best) KIA values for Z (*Z. lotus*) and S (Bare soil) classes are highlighted in bold type.

Year	1956			1977			1984			1997		
Class	Z	S	sum	Z	S	sum	Z	S	sum	Z	S	sum
Z	52	5	57	47	0	47	59	2	61	57	1	58
S	7	55	62	13	60	73	0	58	58	3	59	62
Uncl.	1	0	1	0	0	0	1	0	1	0	0	0
Sum	60	60		60	60		60	60		60	60	
Prod.	0.86	0.92		0.78	1		0.98	0.97		0.95	0.98	
User	0.92	0.88		1	0.82		0.97	1		0.98	0.95	
						Total						
KIA-c	0.74	0.84		0.64	1		0.95	0.94		0.9	0.96	
OA	0.89	0.89		0.89	0.89		0.97	0.97		0.97	0.97	
KIA	0.79	0.79		0.78	0.78		0.95	0.95		0.93	0.93	
Year	2004			2008			2013			2016		
Class	Z	S	sum	Z	S	sum	Z	S	sum	Z	S	sum
Z	59	1	60	55	0	55	57	1	58	58	1	59
S	1	59	60	5	60	65	3	59	62	2	59	61
Uncl.	0	0	0	0	0	0	0	0	0	0	0	0
Sum	60	60		60	60		60	60		60	60	

Table 4. Cont.

Prod.	0.98	0.99		0.92	1		0.95	0.99	0.96	0.99
User	0.99	0.98		0.99	0.92		0.99	0.95	0.99	0.96
					Total					
KIA-c	0.95	0.98		0.84	0.99		0.9	0.98	0.92	0.99
OA	0.98	0.98		0.96	0.96		0.97	0.97	0.98	0.96
KIA	**0.97**	0.97		0.91	0.91		0.94	0.94	0.96	**0.99**

3.3. Shrub Number, Area, and Shape Dynamics

During the 60-year period evaluated, the number of shrubs decreased by 742. The moment of highest shrub population was 1977, with 2625 shrubs. Conversely, the lowest number of shrubs was detected in 2016, with 1883 shrubs (Table 5). However, the total shrub area between 1956 and 2016 increased by 3692 m². In addition, we observed an increase in the maximum cover area value of shrubs after 1997. Finally, the most circular shrubs appeared in 1956 (i.e., the lowest values of the round shape index) and the high values of the round shape index increased over the years (Table 5).

Table 5. The number of shrubs detected each year and their cover-related average statistics. The highest number of bushes, the maximum area, the total cover area, and the lowest (best) round shape index are highlighted in bold type.

Year	1956	1977	1984	1997	2004	2008	2013	2016
Number of shrubs	2055	**2625**	2434	2345	2071	2078	1999	1883
Average area (m²)	82.62	78.32	87.98	76.51	93.78	100.57	99.5	111.31
SD area (m²)	67.65	67.23	79.45	66.91	76.77	82.28	77.84	83.22
Minimum area (m²)	8	5	7	4	6	7	8	6
Maximum area (m²)	525	570	643	701	586	**742**	678	658
Total cover area (m²)	152,932	208,702	209,763	177,706	**223,322**	208,889	198,708	212,394
Round shape index	**1.32**	1.52	1.47	1.53	1.71	1.76	2.04	1.94
SD round shape index	0.16	0.26	0.26	0.27	0.36	0.38	0.47	0.46

In general, the cover area of shrubs between pairs of years showed an increase, with a trend of smaller individuals to lose more cover area than larger shrubs. In the period 1984–1997, 1423 shrubs reduced their cover area. In the period 1977–1984 (Table 6), 1650 shrubs increased their cover area.

Table 6. Change of cover and frequency of the difference in *Ziziphus lotus* area in the studied years (1956–2016). The negative and positive areas are the result of the subtraction between the year and its predecessor. The highest lost area, the largest positive area, the balance between greater areas, the positive frequency and the negative frequency of shrubs are highlighted in bold type.

	Difference of Area Between 1956–1977	Difference of area Between 1977–1984	Difference of Area Between 1984–1997	Difference of area Between 1997–2004	Difference of area Between 2004–2008	Difference of area Between 2008–2013	Difference of area Between 2013–2016
Negative area (m²)	−27,797	−18,917	**−56,838**	−25,968	−32,426	−34,902	−29,684
Positive area (m²)	42,753	69,207	31,594	**84,898**	60,870	44,147	45,101
Balance of areas (m²)	14,956	50,290	−25,244	**58,930**	28,444	9245	15,417
Negative frequency (n)	752	903	**1423**	824	726	861	893
Positive frequency (n)	1158	**1650**	871	1405	1319	1150	978

3.4. Sand Extraction Mapping and Curvature Analysis

The results of the analyses indicated that more than 187 m³ of sand were extracted in the study area (4.2 km²). According to [30], a visual analysis of resulting maps also suggested that sand extractions were distributed spatially following a connected network and following existing roads in the area (Figure 2).

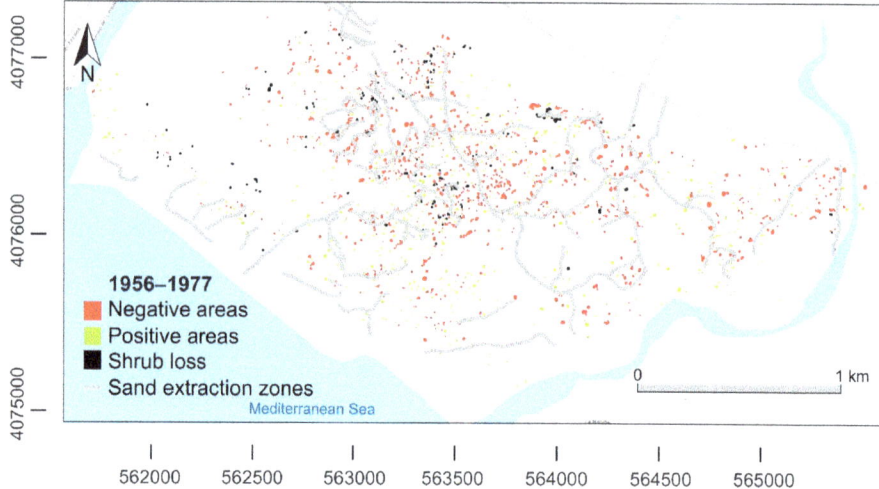

Figure 2. Sand extraction areas and differences of cover areas of *Ziziphus lotus* shrubs in the 1956–1977 period, when massive sand extractions took place in the study area. In red are shown negative areas, in green, positive areas, and in black, shrub loss. Spatial coordinate system, WGS84/UTM Zone 30 N.

3.5. Spatial Relationships of Shrubs with Sand Extractions, Coastline (Seawater Intrusion), and Protected Area

Between 1956–1977, 752 shrubs reduced their cover area in the sand mining event. The AMD between the shrubs and the zones of sand extractions presented an average minimum distance of 25.57 ± 37.49 m. The ARD analysis showed an average minimum distance of 127.48 m ± 23.68 m between the random simulated shrubs and the zones of the sand extractions (Figure 2).

Seawater intrusion (1977–1984) reduced the cover area by 903 shrubs. The AMD analysis showed an average minimum distance of 681.32 m ± 50.15 m to the coastline. The ARD analysis showed an average minimum distance of 882.67 m ± 57.66 m between the random simulated shrubs and the coastline (Figure 3).

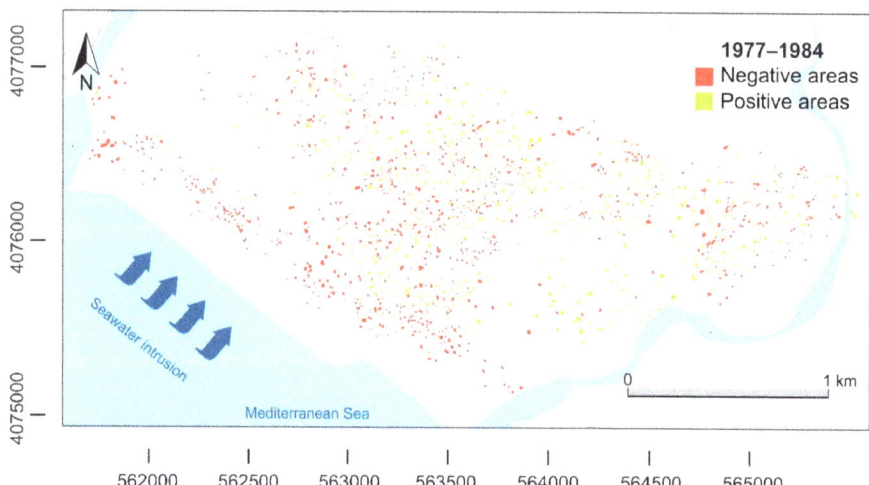

Figure 3. Differences of cover areas of *Ziziphus lotus* shrubs in the 1977–1984 period, when massive groundwater withdrawals took place in the study area. In red are shown negative areas and in green, positive areas. Spatial coordinate system, WGS84/UTM Zone 30 N.

In the period 1984–2016, 551 individuals were lost (Figure 4), but in the area there was a total gain of more than 23 m² (Table 5), coinciding with the protection of the study zone under the Cabo de Gata-Níjar Natural Park.

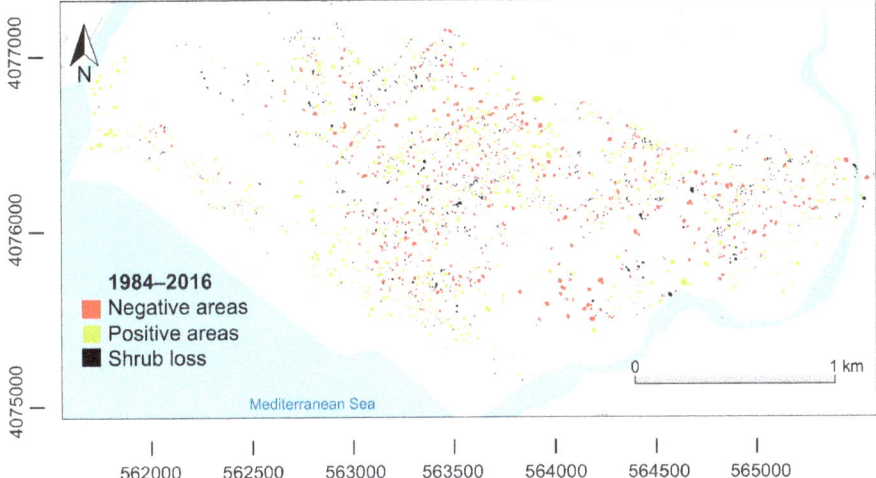

Figure 4. Differences of cover areas of *Ziziphus lotus* shrubs in the 1984–2016 period, when the area was protected under the Cabo de Gata-Níjar Natural Park. In red are shown negative areas, in green, positive areas, and in black, shrub loss. Spatial coordinate system, WGS84/UTM Zone 30 N.

4. Discussion

The first step in inferring human disturbances in vegetation was to generate an accurate object-based map of the study area. The high accuracies obtained in the segmentations, evaluated with the ED2 index, facilitated the classifications of the shrubs, obtaining similar accuracies to previous studies in which other species of shrubs with OBIA were detected [21,62]. In the classification step, the segments of *Z. lotus* and bare soil with sparse vegetation showed high separability, using features related to their brightness and texture as a consequence of clear spectral differences between vegetation and bare soil [63]. According to [64] and [52], the best spectral-related features for discriminating between vegetation and bare soil with sparse vegetation were Brightness, and the GLCM family of features, which present high separability values in well-differentiated classes, such as vegetation and soil [52,65,66]. The worst features to discriminate (i.e., lowest separability) were the geometry-related ones (e.g., Roundness, Area). This is contrary to previous studies, in which Roundness has been suggested as a potential feature to discriminate between rounded shrubs and bare soil [21,66,67]. This discrepancy with previous studies could be explained by the high heterogeneity in the vegetation form that can present after disturbance [5]. However, care must be taken interpreting these results, since shadows generated by large individuals may result in an overestimation of shrub cover area [68,69], and individuals growing together may be underestimated as a result of appearing in the images as one [62].

The spatial distribution of sand extraction areas was unrelated to the topography of the area but was related to the location of older roads and tracks. This suggests that sand extractions were preferentially located to previous sand extractions in an effort to minimize labor costs. This observation is consistent with a previous study on sand extraction in the region (e.g., [30]). According to [70], the negative effect on shrubs by sand mining was shown in the low values of AMD calculated in the period 1956–1977. This reduction in population cover area could be related to sand mining during the 1960s and 1970s [30], which might confirm the positive effect that sands have on the health of *Z. lotus* shrubs in the study area [31] and in other areas of North Africa [71,72].

The lower value of the AMD between shrubs with reduced cover in the period 1977–1984 to the coastline suggested by the ARD that this reduction could be related to a decrease in the freshwater table and the intrusion of seawater into the aquifer [35,73]. The smallest shrubs were the most affected, which can be related to difficulties of access to groundwater due to their smaller roots compared to larger and more developed individuals [28,74]. These results agree with previous studies evaluating the negative effect that seawater intrusion has on vegetated communities and groundwater-dependent ecosystems (e.g., [75,76]).

In addition, the results of this work could be affected by other natural conditions or affections, for example, shrubs could be affected by herbivory [5,77], climate change [78], or uncontrolled use of pesticides [79]. We argue that in order to better understand the results obtained in this work, it is necessary to complement remote sensing techniques with in situ works. For example, complementing the results obtained with the presence–absence of isotopes and relating them with the seawater intrusion [80,81] could provide a better understanding of how this phreatophytic community responds to anthropic perturbations over time.

Although 742 *Z. lotus* individuals were lost during the study period, their average size and the round shape of the shrubs were higher and bigger at the end than at the beginning of the period. However, the variability of these characteristics also increased over time, which means that a greater variety of shapes and sizes was observed in the population. This could be explained by the 1987 declaration regarding the Cabo de Gata-Níjar Natural Park, where the study area is located. This protection, in addition to a slow recovery of the aquifer after undergoing seawater intrusion between 1977 and 1984, could have contributed to a slow but continued development in time by adults, which might have better access to fresh water from the aquifer due to a more developed root system [82] up to a length of 60 m [83]. Furthermore, the fact that the largest shrubs were the most developed in time supports the longevity character of this species through longevity [27], which is an important strategy for its survival in the Mediterranean region [26]. This, together with the anthropic pressure on the system, may explain the development of adult individuals, but the lack of recruitment of juveniles, as observed by [27] not only in this area but also in other regions in SE Spain.

5. Conclusions

The combination of very high-resolution historical images and OBIA is a powerful tool for identifying and monitoring communities of sparse vegetation in drylands [62]. Our results suggest that monitoring changes in the number and the cover of a shrub community in a semi-arid ecosystem could help to infer anthropogenic disturbances that affect its health. The vegetation conditions showed that the loss of sandy substrate affected *Z. lotus* negatively, either by reducing its cover or by eliminating individuals by direct sand extraction processes. In addition, seawater intrusion into the aquifer influenced the cover and structure of the shrubs close to the coastline in a period of massive groundwater extraction [35], negatively affecting the smallest shrubs for the most part. However, the legal protection of the area had a positive effect on the health of the remaining individuals, which increased their coverage. The implementation of semi-automatic methods to infer the effects of human activities on shrub populations, such as the one evaluated in this study, could help improve the monitoring programs of existing protected areas. This could reduce the cost of these activities, not only in economic terms but also from a human perspective, which is key to the long-term preservation of any protected area.

Author Contributions: E.G. and J.B.-S. designed the experiments, collected the data, implemented the codes, ran the models, analyzed the results, and wrote the first draft. E.G., J.B.-S., D.A.-S., J.P.R.-S., and J.C. provided guidance, and wrote the final manuscript.

Funding: E.G. and D.A-S. received support from the European LIFE Project ADAPTAMED LIFE14 CCA/ES/000612, and from ERDF and the Andalusian Government under the project GLOCHARID (Global Change in Arid Zones - 852/2009/M/00). E.G. is supported by the European Research Council grant agreement nº 647038 (BIODESERT). J.B.-S. received funding from the European Union's Horizon 2020 research and innovation 514 programme under the Marie Sklodowska-Curie grant agreement No. 721995. D.A.-S. received support from the NASA

Work Programme on Group on Earth Observations - Biodiversity Observation Network (GEOBON) under grant 80NSSC18K0446, from project ECOPOTENTIAL, funded by the European Union's Horizon 2020 Research and Innovation Programme under grant agreement No. 641762, and from the Spanish Ministry of Science under project CGL2014-61610-EXP and grant JC2015-00316.

Acknowledgments: We would like to express our appreciation to the four reviewers for their comments and suggestions which served to substantially improve this research paper.

Conflicts of Interest: The authors declare no conflict of interest.

References

1. Tilman, D.; Kareiva, P. *Spatial Ecology: The Role of Space in Population Dynamics and Interspecific Interactions (MPB-30)*; Princeton University Press: Princeton, NJ, USA, 2018.
2. Maestre, F.T.; Escudero, A.; Martinez, I.; Guerrero, C.; Rubio, A. Does spatial pattern matter to ecosystem functioning? Insights from biological soil crusts. *Funct. Ecol.* **2005**, *19*, 566–573. [CrossRef]
3. Ludwig, J.A.; Wilcox, B.P.; Breshears, D.D.; Tongway, D.J.; Imeson, A.C. Vegetation Patches and Runoff–Erosion as Interacting Ecohydrological Processes in Semiarid Landscapes. *Ecology* **2005**, *86*, 288–297. [CrossRef]
4. Thompson, S.E.; Harman, C.J.; Troch, P.A.; Brooks, P.D.; Sivapalan, M. Spatial scale dependence of ecohydrologically mediated water balance partitioning: A synthesis framework for catchment ecohydrology. *Water Resour. Res.* **2011**, *47*. [CrossRef]
5. Aguiar, M.R.; Sala, O.E. Patch structure, dynamics and implications for the functioning of arid ecosystems. *Trends Ecol. Evol.* **1999**, *14*, 273–277. [CrossRef]
6. Puigdefábregas, J. The role of vegetation patterns in structuring runoff and sediment fluxes in drylands. *Earth Surf. Process. Landf.* **2005**, *30*, 133–147. [CrossRef]
7. Reynolds, J.F.; Virginia, R.A.; Kemp, P.R.; De Soyza, A.G.; Tremmel, D.C. Impact of drought on desert shrubs: Effects of seasonality and degree of resource island development. *Ecol. Monogr.* **1999**, *69*, 69–106. [CrossRef]
8. Berdugo, M.; Maestre, F.T.; Kéfi, S.; Gross, N.; Le Bagousse-Pinguet, Y.; Soliveres, S. Aridity preferences alter the relative importance of abiotic and biotic drivers on plant species abundance in global drylands. *J. Ecol.* **2019**, *107*, 190–202. [CrossRef]
9. Zhang, X.; Friedl, M.A.; Schaaf, C.B.; Strahler, A.H.; Hodges, J.C.; Gao, F.; Reed, B.C.; Huete, A. Monitoring vegetation phenology using MODIS. *Remote Sens. Environ.* **2003**, *84*, 471–475. [CrossRef]
10. Tsai, Y.; Stow, D.; Chen, H.; Lewison, R.; An, L.; Shi, L. Mapping vegetation and land use types in fanjingshan national nature reserve using Google Earth Engine. *Remote Sens.* **2018**, *10*, 927. [CrossRef]
11. Minasny, B.; Setiawan, B.I.; Saptomo, S.K.; McBratney, A.B. Open digital mapping as a cost-effective method for mapping peat thickness and assessing the carbon stock of tropical peatlands. *Geoderma* **2018**, *313*, 25–40.
12. Schiewe, J.; Tufte, L.; Ehlers, M. Potential and problems of multi-scale segmentation methods in remote sensing. *GeoBIT/GIS* **2001**, *6*, 34–39.
13. Zheng, X.; Wu, B.; Weston, M.; Zhang, J.; Gan, M.; Zhu, J.; Deng, J.; Wang, K.; Teng, L. Rural settlement subdivision by using landscape metrics as spatial contextual information. *Remote Sens.* **2017**, *9*, 486. [CrossRef]
14. Hallberg, B.; Smith-Jonforsen, G.; Ulander, L.M. Measurements on individual trees using multiple VHF SAR images. *IEEE Trans. Geosci. Remote Sens.* **2005**, *43*, 2261–2269. [CrossRef]
15. Maksymiuk, O.; Schmitt, M.; Auer, S.; Stilla, U. Single tree detection in millimeterwave SAR data by morphological attribute filters. *Proc. Jahrestag. DGPF* **2014**, *34*.
16. Hamraz, H.; Contreras, M.A.; Zhang, J. Vertical stratification of forest canopy for segmentation of understory trees within small-footprint airborne LiDAR point clouds. *ISPRS J. Photogramm. Remote Sens.* **2017**, *130*, 385–392. [CrossRef]
17. Blaschke, T. Object based image analysis for remote sensing. *ISPRS J. Photogramm. Remote Sens.* **2010**, *65*, 2–16. [CrossRef]
18. Deblauwe, V.; Barbier, N.; Couteron, P.; Lejeune, O.; Bogaert, J. The global biogeography of semi-arid periodic vegetation patterns. *Glob. Ecol. Biogeogr.* **2008**, *17*, 715–723. [CrossRef]

19. Kéfi, S.; Rietkerk, M.; Alados, C.L.; Pueyo, Y.; Papanastasis, V.P.; ElAich, A.; de Ruiter, P.C. Spatial vegetation patterns and imminent desertification in Mediterranean arid ecosystems. *Nature* **2007**, *449*, 213–217. [CrossRef]
20. Burnett, C.; Blaschke, T. A multi-scale segmentation/object relationship modelling methodology for landscape analysis. *Ecol. Model.* **2003**, *168*, 233–249. [CrossRef]
21. Hellesen, T.; Matikainen, L. An Object-Based Approach for Mapping Shrub and Tree Cover on Grassland Habitats by Use of LiDAR and CIR Orthoimages. *Remote Sens.* **2013**, *5*, 558–583. [CrossRef]
22. Kallio, T. Protection of spruce stumps against Fomes annosus (Fr.) Cooke by some wood-inhabiting fungi. *Acta For. Fenn.* **1971**, *117*, 1–20. [CrossRef]
23. Eriksson, O. Regional dynamics of plants: A review of evidence for remnant, source-sink and metapopulations. *Oikos* **1996**, *77*, 248–258. [CrossRef]
24. Bellingham, P.J.; Sparrow, A.D. Resprouting as a life history strategy in woody plant communities. *Oikos* **2000**, *89*, 409–416. [CrossRef]
25. Bond, W.J.; Midgley, J.J. Ecology of sprouting in woody plants: The persistence niche. *Trends Ecol. Evol.* **2001**, *16*, 45–51. [CrossRef]
26. García, D.; Zamora, R. Persistence, multiple demographic strategies and conservation in long-lived Mediterranean plants. *J. Veg. Sci.* **2003**, *14*, 921–926. [CrossRef]
27. Rey, P.J.; Cancio, I.; Manzaneda, A.J.; González-Robles, A.; Valera, F.; Salido, T.; Alcántara, J.M. Regeneration of a keystone semiarid shrub over its range in Spain: Habitat degradation overrides the positive effects of plant–animal mutualisms. *Plant Biol.* **2018**, *20*, 1083–1092. [CrossRef] [PubMed]
28. Guirado, E.; Alcaraz-Segura, D.; Rigol-Sánchez, J.P.; Gisbert, J.; Martínez-Moreno, F.J.; Galindo-Zaldívar, J.; González-Castillo, L.; Cabello, J. Remote-sensing-derived fractures and shrub patterns to identify groundwater dependence. *Ecohydrology* **2018**, *11*, 1933. [CrossRef]
29. Tirado, R.; Pugnaire, F.I. Shrub spatial aggregation and consequences for reproductive success. *Oecologia* **2003**, *136*, 296–301. [CrossRef]
30. Martínez-Lage, A.V. Las extracciones de áridos en el litoral de almería para su utilización en la agricultura intensiva (1956-1997). In Proceedings of the Actas de Las Jornadas Sobre el Litoral de Almería: Caracterización, Ordenación y Gestión de un Espacio Geográfico Celebradas, Andalusia, Spain, 20–24 May 1997; Instituto de Estudios Almerienses: Andalusia, Spain, 1999; pp. 83–110.
31. Tirado, R. *5220 Matorrales arborescentes con Ziziphus*. VV AA Bases Ecológicas Prelim; Para Conserv. Los Tipos Hábitat Interés Comunitario En España; Dir. Gral. de Medio Natural; Ministerio de Medio Ambiente, y Medio Rural y Marino: Madrid, Spain, 2009.
32. Daniele, L.; Sola, F.; Izquierdo, A.V.; Bosch, A.P. Coastal aquifers and desalination plants: Some interpretations to new situations. In Proceedings of the Conference on Water Observation and Information System for Decision Support, Balwois, Ohrid, Republic of Macedonia, 25–29 May 2010.
33. Cancio, I.; González-Robles, A.; Bastida, J.M.; Manzaneda, A.J.; Salido, T.; Rey, P.J. Habitat loss exacerbates regional extinction risk of the keystone semiarid shrub Ziziphus lotus through collapsing the seed dispersal service by foxes (Vulpes vulpes). *Biodivers. Conserv.* **2016**, *25*, 693–709. [CrossRef]
34. Martín-Rosales, W.; Gisbert, J.; Pulido-Bosch, A.; Vallejos, A.; Fernández-Cortés, A. Estimating groundwater recharge induced by engineering systems in a semiarid area (southeastern Spain). *Environ. Geol.* **2007**, *52*, 985–995. [CrossRef]
35. García García, J.P.; Sánchez Caparós, A.; Castillo, E.; Marín, I.; Padilla, A.; Rosso, J.I. Hidrogeoquímica de las aguas subterráneas en la zona de Cabo de Gata. In *Tecnología de la Intrusión de Agua de Mar en Acuíferos Costeros: Países Mediterráneos*; IGME: Granada, Spain, 2003.
36. Mendoza-Fernández, A.J.; MartíNez-Hernández, F.; Pérez-García, F.J.; Garrido-Becerra, J.A.; Benito, B.M.; Salmerón-Sánchez, E.; Guirado, J.; Merlo, M.E.; Mota, J.F. Extreme habitat loss in a Mediterranean habitat: Maytenus senegalensis subsp. europaea. *Plant Biosyst.-Int. J. Deal. Asp. Plant Biol.* **2015**, *149*, 503–511. [CrossRef]
37. Rivas Goday, S.; Bellot, F. Las formaciones de Ziziphus lotus (L.) Lamk. en las dunas del Cabo de Gata. *An. Inst. Esp. Edafol. Ecol. Fisiol. Veg.* **1944**, *3*, 109–126.
38. Pugnaire, F.I.; Armas, C.; Maestre, F.T. Positive plant interactions in the Iberian Southeast: Mechanisms, environmental gradients, and ecosystem function. *J. Arid Environ.* **2011**, *75*, 1310–1320. [CrossRef]

39. Oyonarte, C.; Rey, A.; Raimundo, J.; Miralles, I.; Escribano, P. The use of soil respiration as an ecological indicator in arid ecosystems of the SE of Spain: Spatial variability and controlling factors. *Ecol. Indic.* **2012**, *14*, 40–49. [CrossRef]
40. Goy, J.L.; Zazo, C. Synthesis of the quaternary in the almeria littoral neotectonic activity and its morphologic features, western betics, Spain. *Tectonophysics* **1986**, *130*, 259–270. [CrossRef]
41. Sola, F.; Daniele, L.; Sánchez Martos, F.; Vallejos, A.; Urízar, R.; Pulido Bosch, A. Influencia de la desaladora de Rambla Morales (Almería) sobre las características hidrogeológicas del acuífero del que se abastece. *Los Acuíferos Costeros Retos Soluc.* **2007**, *1*, 997–1004.
42. Tirado, R.; Bråthen, K.A.; Pugnaire, F.I. Mutual positive effects between shrubs in an arid ecosystem. *Sci. Rep.* **2015**, *5*, 14710. [CrossRef]
43. Aksoy, S.; Tilton, J.C.; Tarabalka, Y. Image segmentation algorithms for land categorization. In *Remote Sensing Handbook V.1 Remotely Sensed Data Characterization, Classification, and Accuracies*; Taylor & Francis: Abingdon, UK, 2015; pp. 317–342. ISBN 978-1-4822-1786-5.
44. Benz, U.C.; Hofmann, P.; Willhauck, G.; Lingenfelder, I.; Heynen, M. Multi-resolution, object-oriented fuzzy analysis of remote sensing data for GIS-ready information. *ISPRS J. Photogramm. Remote Sens.* **2004**, *58*, 239–258. [CrossRef]
45. Drăguţ, L.; Csillik, O.; Eisank, C.; Tiede, D. Automated parameterisation for multi-scale image segmentation on multiple layers. *ISPRS J. Photogramm. Remote Sens.* **2014**, *88*, 119–127. [CrossRef]
46. Kavzoglu, T.; Tonbul, H. A comparative study of segmentation quality for multi-resolution segmentation and watershed transform. In Proceedings of the 2017 8th International Conference on Recent Advances in Space Technologies (RAST), Istanbul, Turkey, 19–22 June 2017; pp. 113–117.
47. Zhan, Q.; Molenaar, M.; Tempfli, K.; Shi, W. Quality assessment for geo-spatial objects derived from remotely sensed data. *Int. J. Remote Sens.* **2005**, *26*, 2953–2974. [CrossRef]
48. Tian, J.; Chen, D.-M. Optimization in multi-scale segmentation of high-resolution satellite images for artificial feature recognition. *Int. J. Remote Sens.* **2007**, *28*, 4625–4644. [CrossRef]
49. Kavzoglu, T.; Yildiz, M. Parameter-based performance analysis of object-based image analysis using aerial and Quikbird-2 images. *ISPRS Ann. Photogramm. Remote Sens. Spat. Inf. Sci.* **2014**, *2*, 31. [CrossRef]
50. Liu, Y.; Bian, L.; Meng, Y.; Wang, H.; Zhang, S.; Yang, Y.; Shao, X.; Wang, B. Discrepancy measures for selecting optimal combination of parameter values in object-based image analysis. *ISPRS J. Photogramm. Remote Sens.* **2012**, *68*, 144–156. [CrossRef]
51. Nussbaum, S.; Menz, G. SEaTH–A New Tool for Feature Analysis. In *Object-Based Image Analysis and Treaty Verification: New Approaches in Remote Sensing–Applied to Nuclear Facilities in Iran*; Nussbaum, S., Menz, G., Eds.; Springer: Dordrecht, The Netherlands, 2008; pp. 51–62. ISBN 978-1-4020-6961-1.
52. Gao, Y.; Mas, J.F.; Kerle, N.; Pacheco, J.A.N. Optimal region growing segmentation and its effect on classification accuracy. *Int. J. Remote Sens.* **2011**, *32*, 3747–3763. [CrossRef]
53. Dobrowski, S.Z.; Safford, H.D.; Cheng, Y.B.; Ustin, S.L. Mapping mountain vegetation using species distribution modeling, image-based texture analysis, and object-based classification. *Appl. Veg. Sci.* **2008**, *11*, 499–508. [CrossRef]
54. Wang, X.Z.; Ashfaq, R.A.R.; Fu, A.M. Fuzziness based sample categorization for classifier performance improvement. *J. Intell. Fuzzy Syst.* **2015**, *29*, 1185–1196. [CrossRef]
55. Cohen, J. Weighted kappa: Nominal scale agreement provision for scaled disagreement or partial credit. *Psychol. Bull.* **1968**, *70*, 213–220. [CrossRef]
56. Jordan, G. Morphometric analysis and tectonic interpretation of digital terrain data: A case study. *Earth Surf. Process. Landf.* **2003**, *28*, 807–822. [CrossRef]
57. Rigol-Sanchez, J.P.; Stuart, N.; Pulido-Bosch, A. ArcGeomorphometry: A toolbox for geomorphometric characterisation of DEMs in the ArcGIS environment. *Comput. Geosci.* **2015**, *85*, 155–163. [CrossRef]
58. Evans, I.S. General geomorphology, derivatives of altitude and descriptive statistics. In *Spatial Analysis in Geomorphology*; Chorley, R.J., Ed.; Harper and Row: Manhattan, NY, USA, 1972; pp. 17–90.
59. Collinge, S.K.; Palmer, T.M. The influences of patch shape and boundary contrast on insect response to fragmentation in California grasslands. *Landsc. Ecol.* **2002**, *17*, 647–656. [CrossRef]
60. Schumaker, N.H. Using Landscape Indices to Predict Habitat Connectivity. *Ecology* **1996**, *77*, 1210–1225. [CrossRef]

61. Rosenberg, M.S.; Anderson, C.D. PASSaGE: Pattern Analysis, Spatial Statistics and Geographic Exegesis. Version 2. *Methods Ecol. Evol.* **2011**, *2*, 229–232. [CrossRef]
62. Laliberte, A.S.; Rango, A.; Havstad, K.M.; Paris, J.F.; Beck, R.F.; McNeely, R.; Gonzalez, A.L. Object-oriented image analysis for mapping shrub encroachment from 1937 to 2003 in southern New Mexico. *Remote Sens. Environ.* **2004**, *93*, 198–210. [CrossRef]
63. Fernández-Buces, N.; Siebe, C.; Cram, S.; Palacio, J.L. Mapping soil salinity using a combined spectral response index for bare soil and vegetation: A case study in the former lake Texcoco, Mexico. *J. Arid Environ.* **2006**, *65*, 644–667. [CrossRef]
64. Yu, Q.; Gong, P.; Clinton, N.; Biging, G.; Kelly, M.; Schirokauer, D. Object-based Detailed Vegetation Classification with Airborne High Spatial Resolution Remote Sensing Imagery. *Photogramm. Eng. Remote Sens.* **2006**, *7*, 799–811. [CrossRef]
65. Murray, H.; Lucieer, A.; Williams, R. Texture-based classification of sub-Antarctic vegetation communities on Heard Island. *Int. J. Appl. Earth Obs. Geoinf.* **2010**, *12*, 138–149. [CrossRef]
66. Laliberte, A.S.; Browning, D.M.; Rango, A. A comparison of three feature selection methods for object-based classification of sub-decimeter resolution UltraCam-L imagery. *Int. J. Appl. Earth Obs. Geoinf.* **2012**, *15*, 70–78. [CrossRef]
67. Laliberte, A.S.; Rango, A. Texture and Scale in Object-Based Analysis of Subdecimeter Resolution Unmanned Aerial Vehicle (UAV) Imagery. *IEEE Trans. Geosci. Remote Sens.* **2009**, *47*, 761–770. [CrossRef]
68. Asner, G.P.; Knapp, D.E.; Kennedy-Bowdoin, T.; Jones, M.O.; Martin, R.E.; Boardman, J.; Hughes, R.F. Invasive species detection in Hawaiian rainforests using airborne imaging spectroscopy and LiDAR. *Remote Sens. Environ.* **2008**, *112*, 1942–1955. [CrossRef]
69. Walsh, S.J.; McCleary, A.L.; Mena, C.F.; Shao, Y.; Tuttle, J.P.; González, A.; Atkinson, R. QuickBird and Hyperion data analysis of an invasive plant species in the Galapagos Islands of Ecuador: Implications for control and land use management. *Remote Sens. Environ.* **2008**, *112*, 1927–1941. [CrossRef]
70. Partridge, T.R. Vegetation recovery following sand mining on coastal dunes at Kaitorete Spit, Canterbury, New Zealand. *Biol. Conserv.* **1992**, *61*, 59–71. [CrossRef]
71. Tengberg, A.; Chen, D. A comparative analysis of nebkhas in central Tunisia and northern Burkina Faso. *Geomorphology* **1998**, *22*, 181–192. [CrossRef]
72. Wang, X.; Yang, F.; Yang, D.; Chen, X. Relationship between the growth of Tamarix ramosissima and morphology of nebkhas in oasis-desert ecotones. In *Global Climate Change and Its Impact on Food & Energy Security in the Drylands, Proceedings of the Eleventh International Dryland Development Conference, Beijing, China, 18–21 March 2013*; International Dryland Development Commission (IDDC): Beijing, China, 2014; pp. 616–628.
73. Sánchez, F.J.T. El uso del agua en Nijar: Implicaciones ambientales del modelo actual de gestión. *Rev. Estud. Reg.* **2008**, *83*, 145–176.
74. Houérou, H.N.L. Agroforestry and sylvopastoralism: The role of trees and shrubs (Trubs) in range rehabilitation and development. *Sci. Chang. Planétaires Sécher.* **2006**, *17*, 343–348.
75. Howard, J.; Merrifield, M. Mapping Groundwater Dependent Ecosystems in California. *PLoS ONE* **2010**, *5*, e11249. [CrossRef] [PubMed]
76. Ponce, V.M. Effect of groundwater pumping on the health of arid vegetative ecosystems. *Online Report, December*. 2014. Available online: http://ponce.sdsu.edu/effect_of_groundwater_pumping.html (accessed on 10 November 2019).
77. Roques, K.G.; O'connor, T.G.; Watkinson, A.R. Dynamics of shrub encroachment in an African savanna: Relative influences of fire, herbivory, rainfall and density dependence. *J. Appl. Ecol.* **2001**, *38*, 268–280. [CrossRef]
78. Sturm, M.; Racine, C.; Tape, K. Increasing shrub abundance in the Arctic. *Nature* **2001**, *411*, 546–547. [CrossRef]
79. Leonard, J.A.; Yeary, R.A. Exposure of Workers Using Hand-Held Equipment During Urban Application of Pesticides to Trees and Ornamental Shrubs. *Am. Ind. Hyg. Assoc. J.* **1990**, *51*, 605–609. [CrossRef]
80. Bear, J.; Cheng, A.H.-D.; Sorek, S.; Ouazar, D.; Herrera, I. *Seawater Intrusion in Coastal Aquifers: Concepts, Methods and Practices*; Springer Science & Business Media: Berlin, Germany, 1999; Volume 14.

81. Mahlknecht, J.; Merchán, D.; Rosner, M.; Meixner, A.; Ledesma-Ruiz, R. Assessing seawater intrusion in an arid coastal aquifer under high anthropogenic influence using major constituents, Sr and B isotopes in groundwater. *Sci. Total Environ.* **2017**, *587*, 282–295. [CrossRef]
82. Jackson, R.B.; Canadell, J.; Ehleringer, J.R.; Mooney, H.A.; Sala, O.E.; Schulze, E.D. A global analysis of root distributions for terrestrial biomes. *Oecologia* **1996**, *108*, 389–411. [CrossRef]
83. Nègre, R. *Recherches Phytogéographiques Sur L'étage de Végétation Méditerranéen Aride (Sous-Étage Chaud) au Maroc Occidental*; Société des Sciences Naturelles et Physiques du Maroc: Rabat, Morocco, 1959.

© 2019 by the authors. Licensee MDPI, Basel, Switzerland. This article is an open access article distributed under the terms and conditions of the Creative Commons Attribution (CC BY) license (http://creativecommons.org/licenses/by/4.0/).

Article

High Spatial Resolution Remote Sensing for Salt Marsh Mapping and Change Analysis at Fire Island National Seashore

Anthony Campbell [1,2] and Yeqiao Wang [1,*]

1. Department of Natural Resources Science, University of Rhode Island, Kingston, RI 02881, USA; anthony.d.campbell@yale.edu
2. Currently at the Yale School of Forestry & Environmental Studies, New Haven, CT 06511, USA
* Correspondence: yqwang@uri.edu

Received: 3 April 2019; Accepted: 6 May 2019; Published: 9 May 2019

Abstract: Salt marshes are changing due to natural and anthropogenic stressors such as sea level rise, nutrient enrichment, herbivory, storm surge, and coastal development. This study analyzes salt marsh change at Fire Island National Seashore (FIIS), a nationally protected area, using object-based image analysis (OBIA) to classify a combination of data from Worldview-2 and Worldview-3 satellites, topobathymetric Light Detection and Ranging (LiDAR), and National Agricultural Imagery Program (NAIP) aerial imageries acquired from 1994 to 2017. The salt marsh classification was trained and tested with vegetation plot data. In October 2012, Hurricane Sandy caused extensive overwash and breached a section of the island. This study quantified the continuing effects of the breach on the surrounding salt marsh. The tidal inundation at the time of image acquisition was analyzed using a topobathymetric LiDAR-derived Digital Elevation Model (DEM) to create a bathtub model at the target tidal stage. The study revealed geospatial distribution and rates of change within the salt marsh interior and the salt marsh edge. The Worldview-2/Worldview-3 imagery classification was able to classify the salt marsh environments accurately and achieved an overall accuracy of 92.75%. Following the breach caused by Hurricane Sandy, bayside salt marsh edge was found to be eroding more rapidly ($F_{1, 1597}$ = 206.06, $p < 0.001$). However, the interior panne/pool expansion rates were not affected by the breach. The salt marsh pannes and pools were more likely to revegetate if they had a hydrological connection to a mosquito ditch ($\chi 2$ = 28.049, $p < 0.001$). The study confirmed that the NAIP data were adequate for determining rates of salt marsh change with high accuracy. The cost and revisit time of NAIP imagery creates an ideal open data source for high spatial resolution monitoring and change analysis of salt marsh environments.

Keywords: Salt marsh; change analysis; Worldview-2; Worldview-3; NAIP aerial data; Topobathymetric LiDAR; Fire Island National Seashore

1. Introduction

Salt marshes are defined by daily tidal inundation and dominated by halophytic vegetation. These ecosystems are the boundary between terrestrial and nearshore aquatic environments their unique location on the landscape and vegetation composition provides ecosystem services such as denitrification, filtration of pollutants, nursey habitat, coastal resilience, and carbon storage and sequestration [1,2]. Historically, salt marshes have displayed high rates of loss due to land reclamation and disturbances such as mosquito ditching [3,4]. Currently, salt marshes along the mid-Atlantic coastal region of the United States are at risk of loss due to sea level rise (SLR), eutrophication, nutrient enrichment, sediment availability, tidal range, and herbivory and human disturbances [5–12]. Recent studies have demonstrated regional and site-specific salt marsh changes including degradation in the Mid-Atlantic [13], proliferation of salt marsh pools in Maryland [14], loss coupled with increased Phragmites on Long Island [15], and loss driven by SLR in New England [16]. However, salt marsh change is a complex combination of persistence, migration, and loss. In the Chesapeake Bay, conversion of uplands to wetlands has mitigated past salt marsh losses [17]. Future salt marsh change is uncertain with some models predicting that salt marsh migration in response to SLR will result in increased salt marsh area [18]. Salt marsh monitoring is necessary for improved understanding of how these ecosystems are changing which in turn can inform their management.

Salt marshes are changing in a variety of ways necessitating a shared nomenclature to discuss these changes. In New England, four types of salt marsh losses have been identified channel widening, interior die-off, shoreline erosion, and loss in the bay head region [16]. These distinctions are dependent on the location of the change. In remote sensing and this study, two major categories were evident change along the edge and change in the interior of the salt marsh. Two types of interior salt marsh loss have been identified: sudden vegetation dieback and drowning. Sudden vegetation dieback in salt marshes is a rapid onset event that persists for a brief period ($\approx$2 years) [19]. These die-offs are predominately located in the mid-marsh and have been documented across the eastern Atlantic coast [20]. In contrast, interior drowning driven by sea level rise is outside the scope of these rapid die-off events and represents a fundamental shift in the ecosystem [20]. Monitoring salt marsh change is further complicated by drowning appearing similar to ponding in microtidal salt marshes [21] and pools changing shape dynamically, draining, and revegetating [22]. Monitoring and differentiating between sudden vegetation dieback, drowning, and ponding necessitates high spatial resolution monitoring to assess expansion and recovery dynamics of interior salt marsh areas.

In this study, we focused on estuarine persistent emergent vegetation, i.e. intertidal areas with perennial salt marsh vegetation [23]. We are interested in identifying pools and pannes and how they changed. Pannes are recessed areas of the salt marsh which drain at mean lowest low water (MLLW) and can be vegetated or nonvegetated, in this study we are referring to nonvegetated pannes unless otherwise stated. In comparison pools are those areas of persistent water. Ponding, pools, and pannes are natural elements of the salt marsh landscape, however long-term and widespread loss of vegetation is not. An in situ study of Plum Island estuary, Massachusetts, found pools were in equilibrium with vegetation regrowth occurring after a few years or at most a decade [24]. However, vegetation regrowth is not guaranteed with site-specific characteristics, such as low sediment input, small tidal range, and high regional SLR, contributing to lack of vegetation growth in pannes and slow filling of pools [25]. Identifying these changes in situ is time consuming and non-tenable for large geographic areas. This study presents a method for using satellite and aerial imagery to differentiate between drowning and ponding by monitoring for decades, a temporal period beyond the expected recovery time.

This study is focused on Fire Island, NY, a barrier island, salt marshes on barrier islands have limited land for salt marsh migration. As a result, salt marsh persistence in place is of particular concern, and is a key component of understanding the long-term stability of barrier island systems. Storm events are critical for shaping the geomorphology of barrier islands, e.g., Hurricane Sandy caused overwash on 41 percent of Fire Island depositing an estimated 508,354 m^3 of sediment [26] and breached the barrier island. Both overwash and inlet creation are essential sources of sediment accretion in bayside salt marsh environments [27,28]. Even thick overwash deposits (>50 cm) result in quick recovery of the salt marsh vegetation [29]. Mapping salt marsh change following a storm event and breach is critical for our understanding of salt marsh persistence on barrier islands. A previous breach on Fire Island was documented with radiometric dating of salt marsh cores, finding a connection between back-barrier salt marsh formation and inlet creation and the resulting processes [30]. Recent research on anthropogenic alterations of an inlet, i.e., jetty creation, demonstrated changes in local mean sea level (LMSL) and tidal range, which resulted in stabilization of salt marsh directly surrounding the inlet [31]. Similar changes to LMSL and tidal range could be affecting the salt marshes of Fire Island.

The decision to allow the Fire Island breach to evolve naturally facilitates monitoring to understand the effect of this process on the surrounding salt marshes.

Remote sensing monitoring can quantify essential attributes of the salt marsh landscape. Remote sensing has been used to understand differences in salt marsh pond density and the total surface area between ditched and unditched marshes [32]. Imagery and Light Detection and Ranging (LiDAR) have been used to quantify pools and pannes and their landscape location [33]. Salt marsh change analysis has been used to identify complex patterns of spatial and temporal variation [14], dramatic conversions from high to low marsh [34], impacts of salt marsh restoration and Hurricane Sandy [35], and identify the effect of SLR on salt marsh communities [36]. The combination of very high resolution (VHR) satellite imagery and aerial imagery provided the necessary spatial and temporal resolution to understand the dynamic coastal environment.

This study mapped salt marsh habitat on Fire Island National Seashore (FIIS) utilizing object-based image analysis (OBIA) with VHR satellite imagery. Multitemporal OBIA was utilized to analyze the development and geospatial dynamics of pool and pannes with high spatial resolution aerial imagery from 1994, 2011, 2013, 2015, and 2017. Change to pannes and pools and edge erosion were analyzed to determine if they were significant components of salt marsh loss at FIIS. The use of remote sensing to determine how protected areas are changing with high accuracy is vital to better manage these areas. The objectives of this study are to (1) classify FIIS salt marsh with OBIA and VHR remote sensing imagery data; (2) determine change rates for interior salt marsh pannes and pools and edge erosion from 1994-2017; (3) determine the relationship between hydrological connectivity and panne expansion; and (4) determine if edge erosion or panne/pool expansion increased surrounding the Hurricane Sandy breach between 2011–2017.

2. Materials and Methods

2.1. Study Area

Fire Island is a barrier island along the southern coast of Long Island, New York, of which 7924 hectares are managed and protected by the National Park Service (NPS) (Figure 1). Salt marsh is the dominant land cover on the island comprising 26% of the protected area [37]. The bayside environment is polyhaline [38]. The island's geomorphology has been altered by urbanization, beach replenishment, and inlet stabilization [39]. More frequent medium intensity storms have been linked with salt marsh edge erosion [40]. Hurricane Sandy was a 1-in-500-year storm surge event that impacted the northeast Atlantic coastal region on 24 Oct 2012 [41]. In 2003, Quickbird-2 satellite images were used to map the study area with a focus on terrestrial and submerged aquatic vegetation [42]. Due to the limited bayside tidal exchange, Fire Island's salt marshes have a small tidal range of approximately 45.5 cm between MLLW and MHHW [43]. SLR is outpacing accretion on the salt marshes of Fire Island as determined by Surface Elevation Tables (SET) [5].

Recent estimates of salt marsh change using aerial imagery from 1974 to 2005/2008 found a 14.1% loss of salt marsh vegetation in a region including Fire Island [15]. The protected areas of FIIS include the William Floyd Estate on the mainland and large bayside islands of Sexton, West Fire and East Fire Island (Figure 1).

Figure 1. (**a**) A mosaic of Worldview-2 and Worldview 3 imagery of Fire Island and Sentinel-2 satellite image of the surrounding area (NIR, G, B displayed as R, G, B). (**b**) A post-Hurricane Sandy oblique aerial view acquired by U.S. Geological Survey (USGS) showing the bayside salt marshes surrounding the breach [44]. (**c**) A field photo of the Fire Island National Seashore (FIIS) salt marsh landscape.

2.2. Data

Satellite imageries were collected on 4/15/2015 and 5/25/2015 with Worldview-3 and Worldview-2 sensors, respectively. Four-band National Agriculture Imagery Program (NAIP) data acquired between 2011 and 2017 and the true color National Aerial Photography Program (NAPP) data acquired in 1994 were used to classify the salt marshes (Table 1). The aerial imageries were collected at a range of tidal stages (Table 1). Change analysis was also conducted with a 1997 classification of FIIS performed with true color aerial photos for the island and verified with in situ field assessment with a highest achieved overall accuracy of 87.5% [45,46].

Table 1. The description of data used by acquisition date, spectral resolution (band wavelength when available), sensor or program, and spatial resolution. R = red, G = green, B = blue, NIR = near infrared, CB = coastal blue, Y = yellow, RE = Red edge.

Date	Spectral Resolution (nm)	Type	Spatial Resolution (m)	Tidal Stage (NAVD 1988) *
8 April 1994	R, G, B	NAPP	1	NA
7 May 2011	R, G, B, NIR	LEICA ADS-52 (NAIP)	1	0.02 m
21 June 2013–22 June 2013	R, G, B, NIR	ADS40-SH51 (NAIP)	1	−0.21 m
7 May 2015	R (619–651), G (525–585), B (435–495), NIR (808–882)	Leica ADS-100 (NAIP)	0.5	−0.29 m
9 August 2017 & 27 August 2017	R (619–651), G (525–585), B (435–495), NIR (808–882)	Leica ADS-100 (NAIP)	1	0.09 m
25 May 2015 & 15 April 2015	CB (400–450), B (450–510), G (510–580), Y (585–625), R (630–690), RE (705–745), NIR1 (770–895), NIR2 (860–1040)	Worldview-2/3	0.5	−0.29 m
8 January 2014–22 May 2014	Green and NIR	Leica ALS 50-II & Riegl sensor 9999609	0.5 Digital Elevation Model	NA

* Average tidal stage across the acquisition as derived from USGS tidal gauge 01309225 [47].

2.3. Tidal Stage Effects

The effects of the tidal stage at the time of imagery acquisition on mapped salt marsh extent has long been recognized e.g., [48,49]. In this study, the tidal stage could impact the edge erosion calculations and the panne analysis. Therefore, the highest tidal stage out of all the imageries was analyzed using topobathymetric LiDAR, bathtub models, and the 2015 FIIS classification [50]. The method utilized the topobathymetric LIDAR-derived Digital Elevation Model (DEM) to create a bathtub model, i.e., a binary raster of inundated and non-inundated pixels, at the target tidal stage. The highest tidal stage of our images was 35.66 cm MLLW or 14.3 cm above the North American Vertical Datum 1988 (NAVD 88) occurring in 2017. The bathtub model was then used to determine areas mapped as vegetated in 2015 that were likely inundated at the 2017 image's tidal stage. The method has been applied in Jamaica Bay, New York, and was utilized in this study to understand the potential impact tidal stage had on the aerial image classifications. These data provide an understanding of the uncertainty derived from inundation that could occur in this analysis.

2.4. Object-Based Image Analysis

OBIA begins with an unsupervised classification or segmentation dividing the image into areas with similar spectral characteristics and spatial proximity [51]. This method used mean shift segmentation: a hierarchical segmentation with demonstrated success in remote sensing and other disciplines [52,53]. OBIA allows for the combination of spectral, spatial, and ancillary data and has been shown to increase classification accuracy compared to pixel methodologies when using VHR satellite imagery [54–56]. In this study, a multiscale segmentation approach was used, selecting under segmented areas and resegmenting them at a finer segmentation scale [57,58] (Figure 2). This study's final segmentation was dual scale with 80% of objects segmented at a spectral radius of 13 and minimum size of five pixels. The other 20% were segmented at a spectral radius of 8 and minimum size of five pixels. Data processing and segmentation were conducted with Python 2.7 [59] and Orfeo Toolbox 5.2 [60].

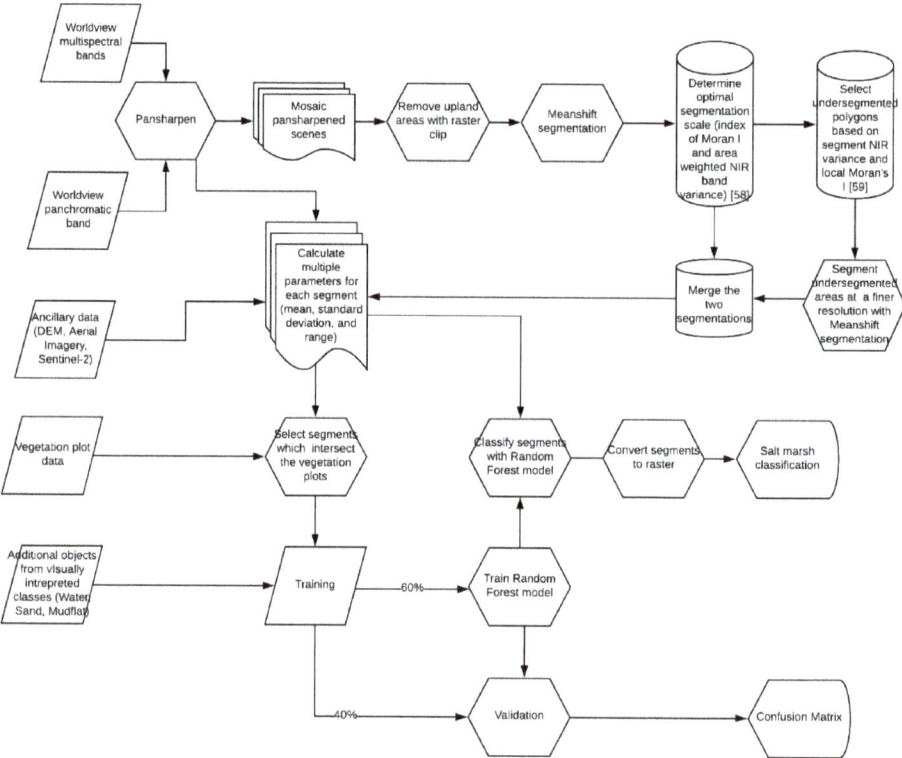

Figure 2. The data processing and classification workflow for classification of the Worldview-2 and Worldview-3 imagery.

The classification was composed of 10 categories including *S. alterniflora*, patchy *S. alterniflora*, high marsh, upland, dune vegetation, sand, mudflat, water, *Phragmites*, and wrack. A one-thousand-nine-hundred-and-thirteen 1-m^2 vegetation plot data were adapted to create training data. The training data were composed of plots with a Braun–Blanquet percent cover greater than ≥50%. Objects that intersected training points were selected, resulting in a total of 1964 training samples. The species included in the high marsh category were *S. patens*, *D. spicata*, *I. frutescens*, and *J. gerardii*. Percent cover differentiated the two *S. alterniflora* classes with the patchy *S. alterniflora* class being between 49 and 10% cover, and the *S. alterniflora* class being ≥ 50% cover. The vegetation plots were predominantly within the salt marsh environment leading to water, sand and upland classes being trained from samples gathered by visual interpretation and field knowledge. The Random Forest (RF) classifier was used to classify the 2015 Worldview-2/Worldview-3 image data for vegetation mapping and the panne and edge classifications.

Accuracy assessment was conducted for the 2015 Worldview-2/Worldview-3 image classification using a subset of the total training data. The data were randomly split 60% training and 40% for testing. The training data were used to train the RF model. An error matrix including kappa, producers, users, and overall accuracies was computed using the testing data.

The 2015 image classification was compared with a 1997 classification based on aerial imagery [45]. The 1997 classes of Reed grass marsh, high marsh, low marsh, and mosquito ditches were compared with the 2015 classes of *Phragmites*, high marsh, and *S. alterniflora* classes accordingly. Mosquito ditches in 1997 were included as vegetated area due to their small average width reported from 25.4 to 50.8 cm [61]. This width is below the minimum mapping unit of 0.25 ha for the 1997 classification [45] and the three-pixel width of the VHR classification. Change rate was calculated between 2015 and the 1997 salt marsh.

2.5. Change Analysis (1994–2017)

The panne and pool analysis, subsequently referred to as panne analysis, was conducted on imageries from 1994 to 2017. The edge erosion change analysis was conducted for 1994 to 2011 and 2011 to 2017. Each image collection was segmented at a spectral radius of 10 and shape radius of 5 with mean shift segmentation. This segmentation scale was adequate given the lower spectral resolution of the aerial images. ArcMap 10.5 [62] was used to select segments which intersected the interior mud or water areas of 2015 Worldview-2/Worldview-3 classifications, these segments comprised the 2015 pannes. The segments were then merged, creating a multitemporal segmentation. The classification parameters included mean, median, standard deviation, simple indices (i.e., red band/blue band), normalized difference vegetation index (NDVI) for those years with NIR, and the difference of each band for the year of interest and subsequent year. The panne and edge classifications were trained with objects that were either vegetated or non-vegetated. Their accuracies were verified with 522 randomly selected points, which were assessed as vegetated or non-vegetated for each time period.

2.6. Statistical Analysis

Edge erosion was calculated for two periods from 1994 to 2011 and 2011 to 2017. The salt marsh edge erosion was calculated based on starting edge length and the total area lost for each edge erosion object. Welch ANOVAs were used to compare rates of change from 1994 to 2011 and 2011 to 2017 for both pannes and edge erosion. Least-squares means was used to compare the two edge erosion rates and to see if edge erosion had increased following the hurricane breach and how this varied between barrier island, bayside islands, and the mainland. A two-way ANOVA was used to compare panne yearly change rates before and after the breach and between barrier island, bayside islands, and the mainland. Annual salt marsh change rates were compared with a Kruskal–Wallis rank sum test, a nonparametric statistical test, followed by a Wilcoxon rank sum test. Linear regression models were used to test the relationship between edge erosion and distance from the breach for the period 2011 to 2017 for 2000 m in both directions. A beta regression ANOVA-like table was produced testing the relationship between the percent change from 2015 to 2017 of interior areas with and without mosquito ditch hydrological connections. All statistical analysis was done within the R 3.5.1 statistical environment [63].

3. Results

3.1. OBIA Classification

The 2015 Worldview-2/Worldview-3 image classification achieved an overall accuracy of 92.75% (Table 2). Visual inspections revealed an appealing result with an appropriate gradient from *S. alterniflora*, high marsh, Phragmites, and upland vegetation (Figure 3). Errors were mostly between the two types of *S. alterniflora* due to their spectral similarities and early seasonal acquisition date of the imagery. There was no confusion between mudflat/water and other classes making this classification an ideal baseline for change analysis.

The 2015 Worldview-2/Worldview-3 imagery classification was compared with a 1997 classification conducted with aerial imagery. When comparing salt marsh vegetation between the two periods a reduction from 531.27 ha to 505.95 ha was observed or 1.41 ha y^{-1}.

Table 2. An error matrix for the 2015 Worldview-2/Worldview-3 classification of Fire Island National Seashore.

	Class	Dune Vegetation	High Marsh	Mudflat	Phragmites	Sand	Patchy S. alterniflora	S. alterniflora	Upland	Water	Wrack	User's Accuracy
Classified data	Dune Veg.	36	0	0	0	0	0	0	0	0	0	100
	High marsh	0	61	0	1	0	0	6	2	0	0	87.1
	Mudflat	0	0	30	0	0	0	0	0	0	0	100
	Phragmites	0	0	0	15	0	0	0	0	0	0	100
	Sand	0	0	0	0	26	0	0	0	0	0	100
	Patchy *S. alterniflora*	0	0	0	0	0	34	14	0	0	0	70.8
	S. alterniflora	0	3	0	0	0	0	79	0	0	0	96.3
	Upland	0	0	0	0	0	0	0	40	0	0	100
	Water	0	0	0	0	0	0	0	0	40	0	100
	Wrack	0	1	0	0	1	0	0	0	0	10	83.3
	Producer's Accuracy	100	93.8	100	93.8	96.3	100	79.8	95.2	100	100	OA: 92.75

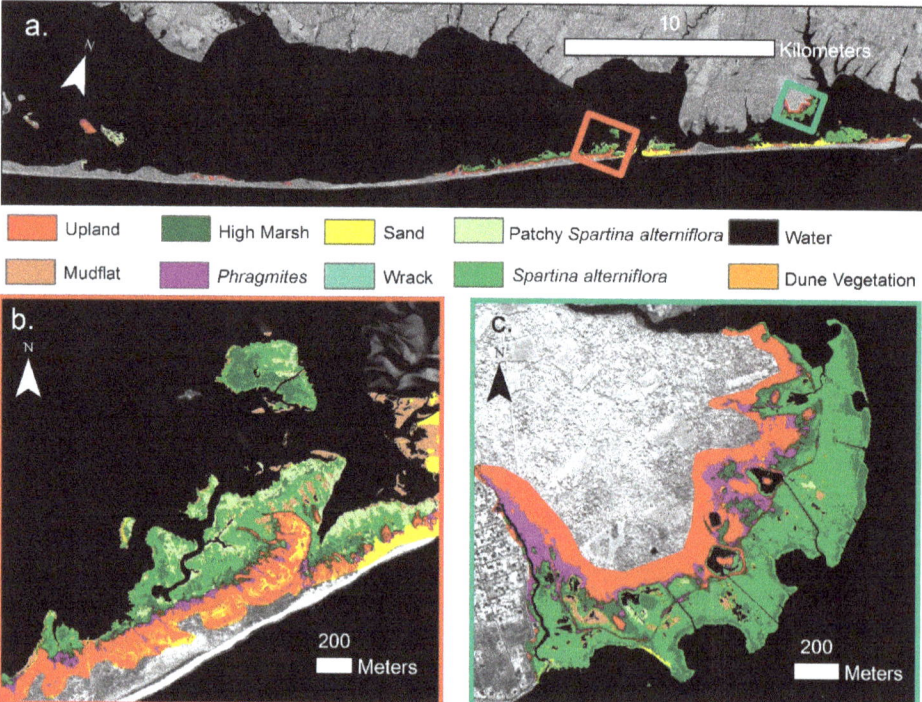

Figure 3. (**a**) The 2015 Worldview-2/Worldview-3 image classification of salt marshes in FIISs. (**b**) A section of FIIS directly to the west of the old inlet breach. (**c**) The Floyd Bennet estate, a section of FIIS on the mainland.

3.2. Tidal Stage Effect

The 2017 NAIP imageries were acquired at an approximate tidal stage of 35.66 cm above MLLW at U.S. Geological Survey (USGS) 01305575 at Watch Hill [43]. The 2014 DEM derived from topobathymetric LiDAR was used to determine how much inundation of *S. alternilflora* would be expected at this elevation. The analysis found 7.39% of the 2015 classification's *S. alterniflora* classes were inundated. The indundation could be subcanopy and have little impact on the 2017 classification.

The areas of modeled inundation were most prevalent in mosquito ditches, sandbars, and interior mudflats (Figure 4).

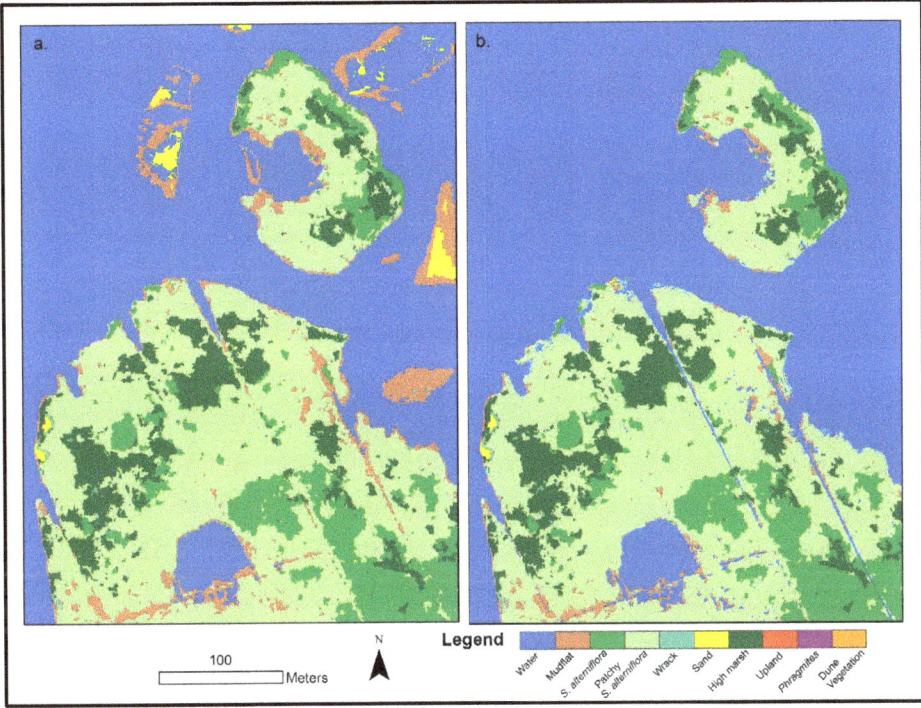

Figure 4. (**a**) The 2015 Worldview-2 and Worldview-3 Classification for a section of Fire Island east of the breach. (**b**) The modeled tidal inundation of the 2015 classification at a tidal stage of 14.3 cm above NAVD 88, which corresponded to the highest tidal stage of the imagery used in the study.

3.3. Change Analysis (1994–2017)

The panne change analysis achieved overall accuracy > 85% for all years (Table 3). However since these classifications were being used in tandem it is important to note that propogated percent error calculated by the square root of the sum of squares was 16.7, 10.2, 11.8, and 13.1 for 1994–2011, 2011–2013, 2013–2015, and 2015–2017, respectively. Models with low overall accuracies were evaluated and manually digitized when necessary. The analysis included 475 pannes mapped in 2015 that were > 10 square meters. In 2015, the mean panne size was 441.6 m compared to the 1994 mean size of 121.4 m. In 1994, 257 of the 475 pannes were vegetated. The mean yearly rates of panne change for each of the periods were 25.00, 5.34, 27.98, and 15.91 $m^2 y^{-1}$ for 2015–2017, 2013–2015, 2011–2013, and 1994–2011, respectively (Figure 5). There were statistical differences between the yearly change rates (H $_{(3)}$ = 30.097, $p < 0.001$) were compared with the Wilcoxon rank sum test (Table 4). There were significant differences between edge erosion rates between 1994 and 2011 and 2011 and 2017 ($F_{1, 1597}$ = 206.06, $p < 0.001$). There were no significant differences between panne change in 1994–2011 and 2011–2017 (F $_{1,948}$ = 0.13, $p = 0.72$). The edge erosion rates for the mainland, bayside islands, and barrier island locations were compared before and after the Hurricane Sandy breach with least-square means (Figure 6). Pannes, in general, became larger from 1994 to 2017, the temporal resolution from 2011–2017 shows fluctuation in these increases (Figure 7). In 2015–2017, pannes were more likely to revegetate if they had a hydrological connection to a mosquito ditch ($\chi2=28.049$, $p < 0.001$). Edge erosion to the east of the breach from 2011 to 2017 had a significant linear trend ($F_{1, 27} = 28.2$, $p < 0.001$) and an R^2 of 0.51.

Edge erosion to the west of the breach from 2011–2017 had no trend ($F_{1,94} = 1.5$, $p = 0.22$) and an R^2 of 0.02.

Table 3. Panne and edge overall classification accuracies, 1994–2015.

Year	Overall Accuracy (%)	Image Source	Spatial Resolution (m)
2017	92.4	NAIP	1
2015	89.3	NAIP	0.5
2013	95.0	NAIP	1
2011	91.1	NAIP	1
1994	85.88	NAPP	1

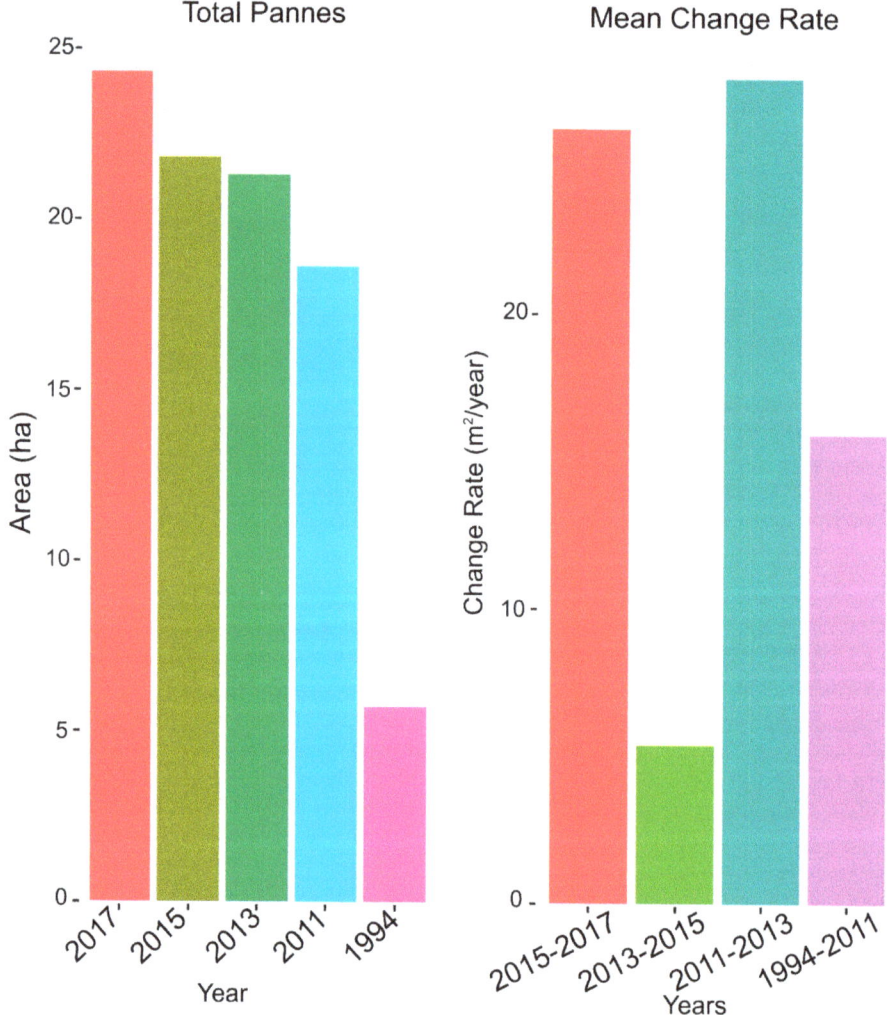

Figure 5. Interior pannes total area and change rates from (1994–2017) both the average yearly change rates of a time period and total area of pannes and pools.

Table 4. Wilcoxon rank sum test between annual panne change rates.

Years	2017–2015	2015–2013	2013–2011
2015–2013	$p < 0.05$		
2013–2011	$p = 0.72$	$p < 0.72$	
2011–1994	$p < 0.001$	$p < 0.72$	$p < 0.001$

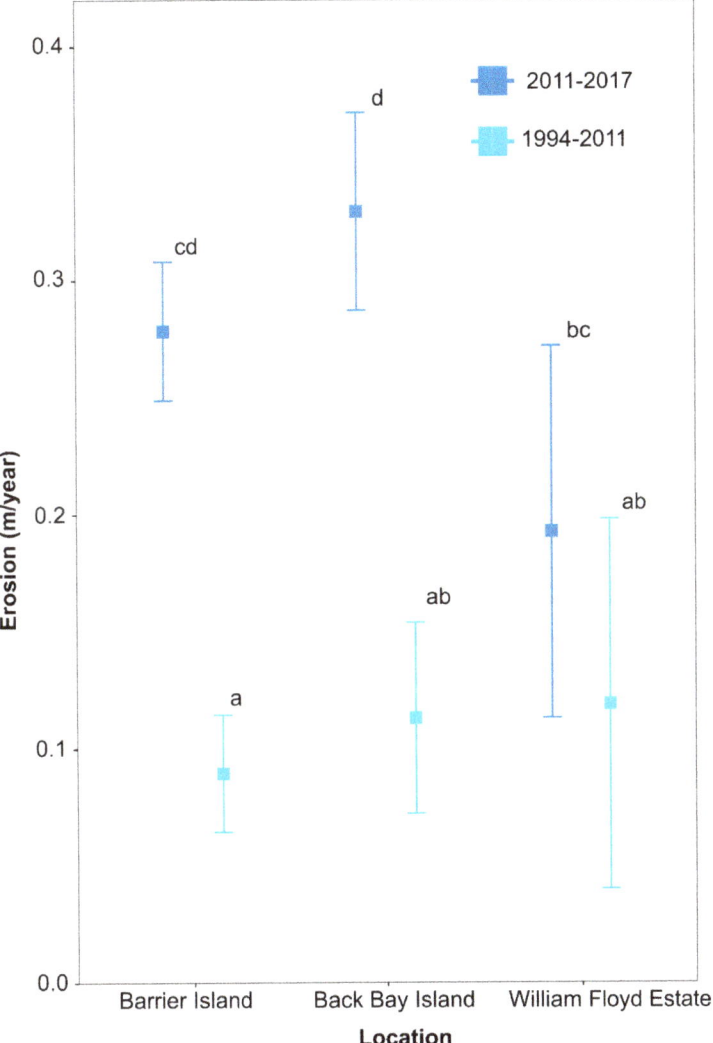

Figure 6. Edge erosion rates compared by time period (1994–2011, 2011–2017) and location (barrier island, mainland, or back bay island) with least square means with Bonferroni p-value adjustment. Location/dates that share letters did not demonstrate significant differences ($p > 0.05$).

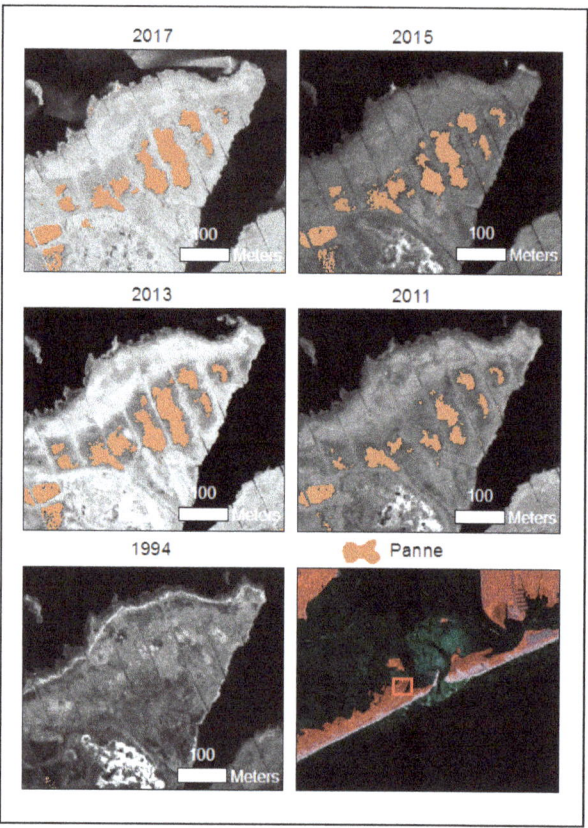

Figure 7. Panne classification from 1994 to 2017 for an area to the west of the old inlet breach. Each inset has the corresponding years NIR in panchromatic or red band in 1994. The locus map is a Sentinel-2 image from 5/21/2016.

4. Discussion

The change analysis between the 1997 classification and the 2015 classification revealed salt marsh loss (Figure 8). High marsh area fell from 199.6 ha to 109.8 ha. Previous studies mapping salt marsh change from 1974 to 2005/2008 for the entirety of Long Island, NY found similar change, including a 35.5% reduction in the high marsh for a region from Fire Island inlet to Smith Point, and a decrease in Phragmites on the south shore [15]. The conversion of upland and Phragmites to low and high marsh categories suggests salt marsh migration in response to SLR. The utility of the comparison between 1997 and 2015 was limited due to the different classification schemes.

In general, pannes/pools demonstrated several periods of statistically significant expansion. Of the 475 pannes, 46% were present in 1994. Meaning there was a doubling of pools and pannes from 1994 to 2015. Two hundred and twelve of the 475 pannes were in areas classified as high marsh in 1997. These pannes accounted for 12.51 ha out of a total of 21.81 ha, i.e., the largest area of pannes occurred in the high marsh. These non-vegetated pannes/pools are essentially tidal mudflats which provide some essential ecosystems services. However, ecosystem service valuations suggest salt marsh to be over five times more valuable than mudflats [64].

The expected evolution of an interior salt marsh pool is expansion until hydrological connectivity is established leading to drainage and possible vegetation regrowth [22]. In our analysis, pannes/pools connected to mosquito ditches in 2015 had a mean change rate of -3.52 m^2 y^{-1} compared to 30.87 m^2 y^{-1}

for non-hydrologically connected pannes/pools. This is encouraging for the possibility of vegetation regrowth. However, natural creeks are infrequent landscape features having remained relatively stable from 1930 to 2007 [61]. In contrast mosquito ditches are common across Fire Island, leading to hydrological connectivity with mosquito ditches being common. However, besides providing a hydrological connection, mosquito ditches likely drive drowning by altering marsh hydrology, and plugged mosquito ditches cause subsidence and loss of salt marsh function [65]. Additionally, the berms surrounding ditches can lead to poor drainage [66]. Highly variable accumulation of sediment in Fire Island's mosquito ditches has led to the infill of some ditches and little to no accumulation in others [61] (Figure 8). The varied rate of infilling could be influencing observed rates of panne/pool expansion. The landscape legacy of the mosquito ditches is a site-specific factor that is critical for understanding salt marsh change on Fire Island.

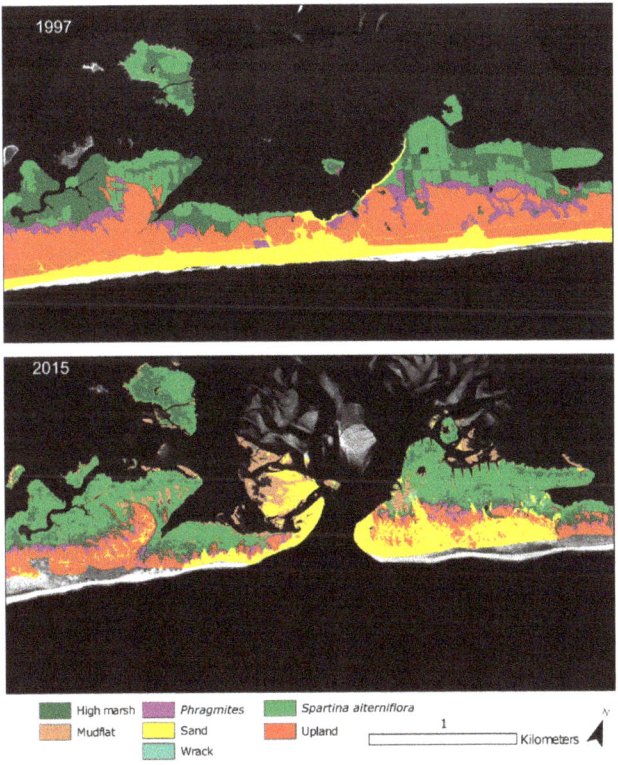

Figure 8. Land cover of the area surrounding the 2012 Fire Island breach for 1997 and 2015. Land cover change both due to the breach and overwash are evident in the 2015 classification. *Spartina alterniflora* classes are shown as a single class due to the 1997 class no differentiating between percent cover. Upland and dune vegetation classes are also shown as a single class.

Whether vegetation regrowth occurred within the pannes/pools is a critical question. Vegetation regrowth is limited by the growth range of *S. alterniflora* at the site. The lowest elevation of living *S. alterniflora* at the site was 25.7 cm below NAVD 1988 [67]. The minimum growing elevation of *S. alterniflora* for FIIS was determined using the tidal range of 45.5 cm and the methods of [10]. Finding a minimum growth elevation of 12.7 cm below NAVD 1988, which is a more conservative estimate than the observed minimum growth elevation. Six of the 475 pannes analyzed were below the vegetation range of *S. alterniflora* at the site, meaning vegetation could grow on nearly all of the observed pannes.

However, only 30 of the 475 2015 pannes/pools were entirely vegetated in 2017. Complete vegetation regrowth was rare but did occur in the pannes and pools analyzed.

Significant increases in edge erosion were observed following the breach. These areas likely experienced changes in currents, wave energy, or LMSL. For example, the William Floyd estate site saw no significant difference in edge erosion before and after the breach. This site is approximately 8 km away from the breach and approximately 5 km from the stabilized Moriches Inlet. In contrast, the area immediately surrounding the breach to the east experienced significant loss from increased edge erosion. An increase in edge erosion as you neared the breach was evident towards the east. However, there was no such trend to the west of the breach. The high variability of the bayside salt marsh erosion demonstrates the importance of geospatial monitoring to understand how these systems are changing spatially. As previous studies reported, Surface Elevation Table (SET)-derived accretion estimates at the site are below the rates of SLR [5]. FIIS' wilderness areas have little infrastructure limiting the migration of salt marsh. However, the islands width and interconnectedness of the barrier island systems means salt marsh migration alone will not maintain the barrier island.

The bayside of barrier islands have low energy and small tidal range (Watch Hill, NY on Fire Island's tidal range is 45.5 cm between MLLW and MHHW [43]), which can result in slower expansion of pools due to edge erosion [25]. The establishment of an inlet can cause an increase in tidal range; however, there were no statistical differences between panne/pool expansion rates between the examined time periods (1994–2011 and 2011–2017). Possibly due to the scarcity of pools in our analysis which would be expected to expand more rapidly with increased tidal range. The breach caused by Hurricane Sandy did not appear to accelerate or slow the interior salt marsh change. However, edge erosion significantly increased following the breach. Continued monitoring is necessary to determine if the observed trend continues.

5. Conclusions

This study evaluated panne/pool development and fluctuations with remote sensing, identifying spatial and temporal patterns of coastal marsh habitat change in a protected National Seashore. Remote sensing methods were essential for understanding how these protected salt marshes changed from 1994 to 2017. This analysis was contingent on the proliferation of remote sensing data which allowed for the synthesis of multiple data types to better understand salt marsh trends and dynamics. Change analysis demonstrated that panne/pool expansion and edge erosion accounted for the majority of salt marsh loss. The losses were partly driven by an increase in edge erosion observed following the breach. Vegetation regrowth occurred with pannes/pools demonstrating increased regrowth when hydrologically connected to a mosquito ditch or channel. The pannes/pools analyzed were not in equilibrium in the two decades analyzed instead demonstrating a long-term trend of expansion.

There is a need for increased salt marsh monitoring for determining where, when, and how salt marshes are changing. This study presents a methodology for salt marsh classification and change analysis of pannes and edge erosion. The aerial imagery classifications achieved satisfactory overall accuracies (> 85%) as suggested by Thomlison et al. [68], however, propagated error when conducting the change analyses was a concern. NAIP imagery is an ideal data source in regards to spatial, temporal and spectral resolution with several caveats. The lack of a NIR band led to a decrease in accuracy due to vegetated and non-vegetated pannes appearing spectrally similar. Additionally, aerial image acquisitions had variable quality and tidal stages at time of acquisition which limited the accuracy of particular years. Finally, the data are only available for the USA. The workflow used in this study allowed for rapid classification and change analysis of salt marsh environments. The biennial collection of NAIP imagery makes it uniquely suited for the low-cost continuation of high-resolution salt marsh monitoring into the future.

Author Contributions: Conceptualization, A.C. and Y.W.; Formal Analysis, Y.W.; Funding Acquisition, Y.W.; Investigation, A.C. and Y.W.; Methodology, A.C. and Y.W.; Project administration, Y.W.; Resources, Y.W.; Software,

Y.W.; Supervision, Y.W.; Validation, A.C.; Visualization, A.C.; Writing—Original Draft, A.C.; Writing—Review & Editing, A.C. and Y.W.

Funding: This research was funded by U.S. National Park Service, grant number: P14AC00230.

Acknowledgments: This project was funded by the Northeast Coastal and Barrier Network (NCBN) of the National Park Service (NPS)'s Inventory & Monitoring Program under the Task Agreement Number P14AC00230. We appreciate the guidance and support of Sara Stevens, Dennis Skidds, Bill Thompson, and Charles Roman of the NPS NCBN and North Atlantic Coast CESU. The authors appreciate the assistance by administrators and professionals from Fire Island National Seashore (FIIS), particularly Jordan Raphael for his expertise, insights, field guidance, and logistic support. Additionally, the authors would like to thank the three anonymous reviewers for their comments and constructive criticism.

Conflicts of Interest: The authors declare no conflict of interest.

References

1. Zedler, J.B.; Kercher, S. Wetland Resources: Status, Trends, Ecosystem Services, and Restorability. *Annu. Rev. Environ. Resour.* **2005**, *30*, 39–74. [CrossRef]
2. Barbier, E.B.; Hacker, S.D.; Kennedy, C.; Koch, E.W.; Stier, A.C.; Silliman, B.R. The Value of Estuarine and Coastal Ecosystem Services. *Ecol. Monogr.* **2011**, *81*, 169–193. [CrossRef]
3. Gedan, K.B.; Silliman, B.R.; Bertness, M.D. Centuries of Human-Driven Change in Salt Marsh Ecosystems. *Annu. Rev. Mar. Sci.* **2009**, *1*, 117–141. [CrossRef] [PubMed]
4. Crain, C.M.; Gedan, K.B.; Dionne, M. Tidal Restrictions and Mosquito Ditching in New England Marshes. In *Human Impacts on Salt Marshes: A Global Perspective*; University of California Press: Berkeley, CA, USA, 2009; pp. 149–169.
5. Crosby, S.C.; Sax, D.F.; Palmer, M.E.; Booth, H.S.; Deegan, L.A.; Bertness, M.D.; Leslie, H.M. Salt Marsh Persistence is Threatened by Predicted Sea-Level Rise. *Estuar. Coast. Shelf Sci.* **2016**, *181*, 93–99. [CrossRef]
6. Watson, E.B.; Raposa, K.B.; Carey, J.C.; Wigand, C.; Warren, R.S. Anthropocene Survival of Southern New England's Salt Marshes. *Estuaries Coasts* **2017**, *40*, 617–625. [CrossRef] [PubMed]
7. Wigand, C.; Roman, C.T.; Davey, E.; Stolt, M.; Johnson, R.; Hanson, A.; Watson, E.B.; Moran, S.B.; Cahoon, D.R.; Lynch, J.C. Below the Disappearing Marshes of an Urban Estuary: Historic Nitrogen Trends and Soil Structure. *Ecol. Appl.* **2014**, *24*, 633–649. [CrossRef] [PubMed]
8. Altieri, A.H.; Bertness, M.D.; Coverdale, T.C.; Herrmann, N.C.; Angelini, C. A Trophic Cascade Triggers Collapse of a Salt-marsh Ecosystem with Intensive Recreational Fishing. *Ecology* **2012**, *93*, 1402–1410. [CrossRef]
9. Deegan, L.A.; Johnson, D.S.; Warren, R.S.; Peterson, B.J.; Fleeger, J.W.; Fagherazzi, S.; Wollheim, W.M. Coastal Eutrophication as a Driver of Salt Marsh Loss. *Nature* **2012**, *490*, 388–392. [CrossRef]
10. Kirwan, M.L.; Guntenspergen, G.R. Influence of Tidal Range on the Stability of Coastal Marshland. *J. Geophys. Res. Earth Surf.* **2010**, *115*. [CrossRef]
11. Kirwan, M.L.; Murray, A.B.; Boyd, W.S. Temporary Vegetation Disturbance as an Explanation for Permanent Loss of Tidal Wetlands. *Geophys. Res. Lett.* **2008**, *35*. [CrossRef]
12. Holdredge, C.; Bertness, M.D.; Altieri, A.H. Role of Crab Herbivory in Die-Off of New England Salt Marshes. *Conserv. Biol.* **2009**, *23*, 672–679. [CrossRef]
13. Kearney, M.S.; Rogers, A.S.; Townshend, J.R.; Rizzo, E.; Stutzer, D.; Stevenson, J.C.; Sundborg, K. Landsat Imagery shows Decline of Coastal Marshes in Chesapeake and Delaware Bays. *EOS Trans. Am. Geophys. Union* **2002**, *83*, 173–178. [CrossRef]
14. Schepers, L.; Kirwan, M.; Guntenspergen, G.; Temmerman, S. Spatio-temporal Development of Vegetation Die-off in a Submerging Coastal Marsh. *Limnol. Oceanogr.* **2017**, *62*, 137–150. [CrossRef]
15. Cameron Engineering and Associates. *Long Island Tidal Wetlands Trends Analysis*; New England Interstate Water Pollution Control Commission: Lowell, MA, USA, 2015; 207p.
16. Watson, E.B.; Wigand, C.; Davey, E.W.; Andrews, H.M.; Bishop, J.; Raposa, K.B. Wetland Loss Patterns and Inundation-Productivity Relationships Prognosticate Widespread Salt Marsh Loss for Southern New England. *Estuaries Coasts* **2017**, *40*, 662–681. [CrossRef]
17. Schieder, N.W.; Walters, D.C.; Kirwan, M.L. Massive Upland to Wetland Conversion Compensated for Historical Marsh Loss in Chesapeake Bay, USA. *Estuaries Coasts* **2018**, *41*, 940–951. [CrossRef]

18. Schuerch, M.; Spencer, T.; Temmerman, S.; Kirwan, M.L.; Wolff, C.; Lincke, D.; McOwen, C.J.; Pickering, M.D.; Reef, R.; Vafeidis, A.T. Future Response of Global Coastal Wetlands to Sea-Level Rise. *Nature* **2018**, *561*, 231. [CrossRef]
19. Marsh, A.; Blum, L.K.; Christian, R.R.; Ramsey, E.; Rangoonwala, A. Response and Resilience of Spartina Alterniflora. *J. Coast. Conserv.* **2016**, *20*, 335–350. [CrossRef]
20. Alber, M.; Swenson, E.M.; Adamowicz, S.C.; Mendelssohn, I.A. Salt Marsh Dieback: An Overview of Recent Events in the US. *Estuar. Coast. Shelf Sci.* **2008**, *80*, 1–11. [CrossRef]
21. Kearney, M.S.; Turner, R.E. Microtidal Marshes: Can these Widespread and Fragile Marshes Survive Increasing Climate–sea Level Variability and Human Action? *J. Coast. Res.* **2016**, *32*, 686–699. [CrossRef]
22. Wilson, K.R.; Kelley, J.T.; Croitoru, A.; Dionne, M.; Belknap, D.F.; Steneck, R. Stratigraphic and Ecophysical Characterizations of Salt Pools: Dynamic Landforms of the Webhannet Salt Marsh, Wells, ME, USA. *Estuaries Coasts* **2009**, *32*, 855–870. [CrossRef]
23. Cowardin, L.M.; Carter, V.; Golet, F.C.; LaRoe, E.T. *Classification of Wetlands and Deepwater Habitats of the United States*; U.S. Department of the Interior: Washington, DC, USA, 1979; 131p.
24. Wilson, C.A.; Hughes, Z.J.; FitzGerald, D.M.; Hopkinson, C.S.; Valentine, V.; Kolker, A.S. Saltmarsh Pool and Tidal Creek Morphodynamics: Dynamic Equilibrium of Northern Latitude Saltmarshes? *Geomorphology* **2014**, *213*, 99–115. [CrossRef]
25. Mariotti, G. Revisiting Salt Marsh Resilience to Sea Level Rise: Are Ponds Responsible for Permanent Land Loss? *J. Geophys. Res. Earth Surf.* **2016**, *121*, 1391–1407. [CrossRef]
26. Hapke, C.J.; Brenner, O.; Hehre, R.; Reynolds, B.J. *Coastal Change from Hurricane Sandy and the 2012–13 Winter Storm Season—Fire Island*; Geological Survey Open-File Report; U.S. Geological Survey: Reston, VA, USA, 2013.
27. Leatherman, S.P. Geomorphic and Stratigraphic Analysis of Fire Island, New York. *Mar. Geol.* **1985**, *63*, 173–195. [CrossRef]
28. Friedrichs, C.T.; Perry, J.E. Tidal Salt Marsh Morphodynamics: A Synthesis. *J. Coast. Res.* **2001**, 7–37. Available online: https://www.jstor.org/stable/25736162 (accessed on 15 January 2019).
29. Courtemanche, R.P., Jr.; Hester, M.W.; Mendelssohn, I.A. Recovery of a Louisiana Barrier Island Marsh Plant Community Following Extensive Hurricane-Induced Overwash. *J. Coast. Res.* **1999**, *15*, 872–883.
30. Roman, C.T.; King, D.R.; Cahoon, D.R.; Lynch, J.C.; Appleby, P.G. *Evaluation of Marsh Development Processes at Fire Island National Seashore: Recent and Historic Perspectives*; National Park Service: Boston, MA, USA, 2007.
31. Silvestri, S.; D'Alpaos, A.; Nordio, G.; Carniello, L. Anthropogenic Modifications can significantly Influence the Local Mean Sea Level and Affect the Survival of Salt Marshes in Shallow Tidal Systems. *J. Geophys. Res. Earth Surf.* **2018**, *123*, 996–1012.
32. Adamowicz, S.C.; Roman, C.T. New England Salt Marsh Pools: A Quantitative Analysis of Geomorphic and Geographic Features. *Wetlands* **2005**, *25*, 279–288. [CrossRef]
33. Millette, T.L.; Argow, B.A.; Marcano, E.; Hayward, C.; Hopkinson, C.S.; Valentine, V. Salt Marsh Geomorphological Analyses Via Integration of Multitemporal Multispectral Remote Sensing with LIDAR and GIS. *J. Coast. Res.* **2010**, *26*, 809–816. [CrossRef]
34. Smith, S.M. Vegetation Change in Salt Marshes of Cape Cod National Seashore (Massachusetts, USA) between 1984 and 2013. *Wetlands* **2015**, *35*, 127–136. [CrossRef]
35. Campbell, A.; Wang, Y.; Christiano, M.; Stevens, S. Salt Marsh Monitoring in Jamaica Bay, New York from 2003 to 2013: A Decade of Change from Restoration to Hurricane Sandy. *Remote Sens.* **2017**, *9*, 131. [CrossRef]
36. Sun, C.; Fagherazzi, S.; Liu, Y. Classification Mapping of Salt Marsh Vegetation by Flexible Monthly NDVI Time-Series using Landsat Imagery. *Estuar. Coast. Shelf Sci.* **2018**, *213*, 61–80. [CrossRef]
37. McElroy, A.; Benotti, M.; Edinger, G.; Feldmann, A.; O'Connell, C.; Steward, G.; Swanson, R.L.; Waldman, J. *Assessment of Natural Resource Conditions: Fire Island National Seashore*; Natural Resource Report NPS/NRPC/NRR; National Park Service: Fort Collins, CO, USA, 2009.
38. Hair, M.E.; Buckner, S. *An Assessment of the Water Quality Characteristics of Great South Bay and Contiguous Streams*; Adelphi University Institute of Marine Science: Garden City, NY, USA, 1973.
39. Lentz, E.E.; Hapke, C.J. Geologic Framework Influences on the Geomorphology of an Anthropogenically Modified Barrier Island: Assessment of Dune/Beach Changes at Fire Island, New York. *Geomorphology* **2011**, *126*, 82–96. [CrossRef]

40. Leonardi, N.; Ganju, N.K.; Fagherazzi, S. A Linear Relationship between Wave Power and Erosion Determines Salt-Marsh Resilience to Violent Storms and Hurricanes. *Proc. Natl. Acad. Sci. USA* **2016**, *113*, 64–68. [CrossRef]
41. Aerts, J.C.; Lin, N.; Botzen, W.; Emanuel, K.; de Moel, H. Low-Probability Flood Risk Modeling for New York City. *Risk Anal.* **2013**, *33*, 772–788. [CrossRef]
42. Wang, Y.; Traber, M.; Milstead, B.; Stevens, S. Terrestrial and Submerged Aquatic Vegetation Mapping in Fire Island National Seashore using High Spatial Resolution Remote Sensing Data. *Mar. Geod.* **2007**, *30*, 77–95. [CrossRef]
43. U.S. Geological Survey (USGS). USGS Surface-Water Daily Data for New York. Watch Hill, NY, USA, 2017. Available online: https://waterdata.usgs.gov/ny/nwis/uv?site_no=01305575 (accessed on 15 January 2019).
44. Morgan, K.L.M.; Krohn, M.D. *Post-Hurricane Sandy Coastal Oblique Aerial Photographs Collected from Cape Lookout, North Carolina, to Montauk, New York, November 4–6, 2012*; U.S. Geological Survey Data Series 858; U.S. Geological Survey: Reston, VA, USA, 2014. [CrossRef]
45. Conservation Management Institute, NatureServe, and New York Natural Heritage Program. *Fire Island National Seashore Vegetation Inventory Project—Spatial Vegetation Data*; National Park Service: Denver, CO, USA, 2002. Available online: https://www.sciencebase.gov/catalog/item/541ca909e4b0e96537e0a41e (accessed on 15 January 2019).
46. Klopfer, S.D.; Olivero, A.; Sneddon, L.; Lundgren, J. *Final Report of the NPS Vegetation Mapping Project at Fire Island National Seashore*; Conservation Management Institute—Virginia Tech: Blacksburg, VA, USA, 2002.
47. U.S. Geological Survey (USGS). USGS Surface-Water Daily Data for New York. Lindenhurst, NY, USA, 2017. Available online: https://waterdata.usgs.gov/ny/nwis/uv?site_no=01309225 (accessed on 15 January 2019).
48. Jensen, J.R.; Cowen, D.J.; Althausen, J.D.; Narumalani, S.; Weatherbee, O. The Detection and Prediction of Sea Level Changes on Coastal Wetlands using Satellite Imagery and a Geographic Information System. *Geocarto Int.* **1993**, *8*, 87–98. [CrossRef]
49. Jensen, J.R.; Cowen, D.J.; Althausen, J.D.; Narumalani, S.; Weatherbee, O. An Evaluation of the CoastWatch Change Detection Protocol in South Carolina. *Photogramm. Eng. Remote Sens.* **1993**, *59*, 1039–1044.
50. Campbell, A.; Wang, Y. Examining the Influence of Tidal Stage on Salt Marsh Mapping using High-Spatial-Resolution Satellite Remote Sensing and Topobathymetric Lidar. *IEEE Trans. Geosci. Remote Sens.* **2018**, *56*, 5169–5176. [CrossRef]
51. Hay, G.J.; Castilla, G. Geographic Object-Based Image Analysis (GEOBIA): A new name for a new discipline. In *Object-Based Image Analysis*; Blaschke, T., Lang, S., Hay, G.J., Eds.; Springer: Berlin, Germany, 2008; pp. 75–89.
52. Bo, S.; Ding, L.; Li, H.; Di, F.; Zhu, C. Mean Shift-based Clustering Analysis of Multispectral Remote Sensing Imagery. *Int. J. Remote Sens.* **2009**, *30*, 817–827. [CrossRef]
53. Yang, G.; Pu, R.; Zhang, J.; Zhao, C.; Feng, H.; Wang, J. Remote Sensing of Seasonal Variability of Fractional Vegetation Cover and its Object-Based Spatial Pattern Analysis Over Mountain Areas. *ISPRS J. Photogramm. Remote Sens.* **2013**, *77*, 79–93. [CrossRef]
54. Castillejo-González, I.L.; López-Granados, F.; García-Ferrer, A.; Peña-Barragán, J.M.; Jurado-Expósito, M.; de la Orden, M.S.; González-Audicana, M. Object- and Pixel-Based Analysis for Mapping Crops and their Agro-Environmental Associated Measures using QuickBird Imagery. *Comput. Electron. Agric.* **2009**, *68*, 207–215. [CrossRef]
55. Myint, S.W.; Gober, P.; Brazel, A.; Grossman-Clarke, S.; Weng, Q. Per-Pixel vs. Object-Based Classification of Urban Land Cover Extraction using High Spatial Resolution Imagery. *Remote Sens. Environ.* **2011**, *115*, 1145–1161. [CrossRef]
56. Lantz, N.J.; Wang, J. Object-Based Classification of Worldview-2 Imagery for Mapping Invasive Common Reed, Phragmites Australis. *Can. J. Remote Sens.* **2013**, *39*, 328–340. [CrossRef]
57. Espindola, G.M.; Câmara, G.; Reis, I.A.; Bins, L.S.; Monteiro, A.M. Parameter Selection for Region-growing Image Segmentation Algorithms using Spatial Autocorrelation. *Int. J. Remote Sens.* **2006**, *27*, 3035–3040. [CrossRef]
58. Johnson, B.; Xie, Z. Unsupervised Image Segmentation Evaluation and Refinement using a Multi-Scale Approach. *ISPRS J. Photogramm. Remote Sens.* **2011**, *66*, 473–483. [CrossRef]
59. Rossum, G.V. *Python Tutorial*; Technical Report CS-R9526; Centrum voor Wiskunde en Informatica (CWI): Amsterdam, The Netherlands, 1995.

60. Inglada, J.; Christophe, E. The Orfeo Toolbox Remote Sensing Image Processing Software. In Proceedings of the 2009 IEEE International Geoscience and Remote Sensing Symposium, Cape Town, South Africa, 12–17 July 2009; p. 736.
61. Corman, S.S.; Roman, C.T.; King, J.W.; Appleby, P.G. Salt Marsh Mosquito-Control Ditches: Sedimentation, Landscape Change, and Restoration Implications. *J. Coast. Res.* **2012**, *28*, 874–880. [CrossRef]
62. *ESRI 2016—ArcGIS Desktop: Release 10.5*; Environmental Systems Research Institute: Redlands, CA, USA, 2016.
63. R Core Team. *R: A Language and Environment for Statistical Computing*; Version 3.5.0; R Foundation for Statistical Computing: Vienna, Austria, 2018.
64. Johnston, R.J.; Grigalunas, T.A.; Opaluch, J.J.; Mazzotta, M.; Diamantedes, J. Valuing Estuarine Resource Services using Economic and Ecological Models: The Peconic Estuary System Study. *Coast. Manag.* **2002**, *30*, 47–65. [CrossRef]
65. Vincent, R.E.; Burdick, D.M.; Dionne, M. Ditching and Ditch-Plugging in New England Salt Marshes: Effects on Hydrology, Elevation, and Soil Characteristics. *Estuaries Coasts* **2013**, *36*, 610–625. [CrossRef]
66. Watson, E.B. Wetland Elevation Changes: Expansion of Pannes. Sound Update, Newsletter of the Long Island Sound Study 2015. Available online: http://longislandsoundstudy.net/wp-content/uploads/2015/06/SoundUpdate_TidalWetlandsLoss_forweb.pdf (accessed on 15 January 2019).
67. Rasmussen, S.A.; Neil, A.J. *Marsh Elevation Points for Assateague Island National Seashore, Fire Island National Seashore, and Gateway National Recreation Area-Sandy Hook Unit*; Environmental Data Center: Kingston, RI, USA, 2017.
68. Thomlinson, J.R.; Bolstad, P.V.; Cohen, W.B. Coordinating Methodologies for Scaling Landcover Classifications from Site-Specific to Global: Steps Toward Validating Global Map Products. *Remote Sens. Environ.* **1999**, *70*, 16–28. [CrossRef]

© 2019 by the authors. Licensee MDPI, Basel, Switzerland. This article is an open access article distributed under the terms and conditions of the Creative Commons Attribution (CC BY) license (http://creativecommons.org/licenses/by/4.0/).

Review

Remote Sensing Applications in Monitoring of Protected Areas: A Bibliometric Analysis

Peili Duan [1,*], Yeqiao Wang [2,*] and Peng Yin [1,3]

1. School of Business, Ludong University, Yantai 264025, China; yinp438@nenu.edu.cn
2. Department of Natural Resources Science, University of Rhode Island, Kingston, RI 02881, USA
3. Institute of Marine Development, Ocean University of China, Qingdao 266100, China
* Correspondence: duanpl069@nenu.edu.cn (P.D.); yqwang@uri.edu (Y.W.)

Received: 22 December 2019; Accepted: 24 February 2020; Published: 28 February 2020

Abstract: The development of remote sensing platforms and sensors and improvement in science and technology provide crucial support for the monitoring and management of protected areas. This paper presents an analysis of research publications, from a bibliometric perspective, on the remote sensing of protected areas. This analysis is focused on the period from 1991 to 2018. For data, a total of 4546 academic publications were retrieved from the Web of Science database. The VOSviewer software was adopted to evaluate the co-authorships among countries and institutions, as well as the co-occurrences of author keywords. The results indicate an increasing trend of annual publications in the remote sensing of protected areas. This analysis reveals the major topical subjects, leading countries, and most influential institutions around the world that have conducted relevant research in scientific publications; this study also reveals the journals that include the most publications, and the collaborative patterns related to the remote sensing of protected areas. Landsat, MODIS, and LiDAR are among the most commonly used satellites and sensors. Research topics related to protected area monitoring are mainly concentrated on change detection, biodiversity conservation, and climate change impact. This analysis can help researchers and scholars better understand the intellectual structure of the field and identify the future research directions.

Keywords: remote sensing; protected areas; bibliometric analysis; VOSviewer

1. Introduction

In accordance with the International Union for Conversation of Nature (IUCN) [1,2], a protected area (PA) is defined as "a clearly defined geographical space, recognized, dedicated, and managed, through legal or other effective means, to achieve the long-term conservation of nature with associated ecosystem services and cultural values". In general, PAs are considered to be areas of land or sea, including national parks, national forests, natural reserves, conservation areas, wilderness areas, marine protected areas (MPAs), and wildlife refuges and sanctuaries, that are designated for the conservation of native biological diversity and natural and cultural heritage and significance [3]. Over the past century, the amount and coverage of both terrestrial and marine PAs have markedly increased [4]. As reported by the World Database on Protected Areas [5], as of July 2018, there were 238,563 designated PAs, covering about 14.9% and 7.3% of the Earth's land and ocean surface areas, respectively. PAs are central to nature conservation efforts with key environmental, social, cultural and economic functions throughout the world [3,6]. In addition, PAs play an important role in biodiversity conservation and ecosystem integrity [7–10].

Remote sensing refers to the art, science, and technology used for Earth system data acquisition through nonphysical contact sensors or sensor systems mounted on space-borne, airborne, and other types of platforms, data processing and interpretation from automated and visual analysis, information generation using computerized and conventional mapping facilities, and applications of the generated data and information to benefit society and meet its needs. Remote sensing can provide comprehensive geospatial information to map and study PAs at different spatial scales, e.g., high spatial resolution and large area coverage, different temporal frequencies (e.g., daily, weekly, monthly, or annual observations), different spectral properties (e.g., visible, near infrared, or microwave), and spatial contexts (e.g., the immediately adjacent areas of PAs vs. a broader background of land and water bases). Remote sensing is considered to be a cost-effective method to support the monitoring efforts of PAs and has played a vital role in protecting natural resources, ecosystems, and biodiversity [11,12].

In terms of the large-scale observation ability of remote sensing, the technology is becoming a common practice for monitoring the characteristics and change of land surface properties of PAs [13]. For example, remote sensing has been applied to the assessment of night-time lighting within and surrounding global terrestrial PAs and wilderness areas [14], continuous monitoring of the landscape dynamics of national parks by Landsat-based approaches [15–19], the evaluation of forest dynamics within and around the Olympic National Park using time-series Landsat observations [20], and monitoring the wildlife habitat changes in Kejimkujik National Park and the National Historic Site in southern Nova Scotia of the Canadian Atlantic Coastal Uplands Natural Region [21]. One particular advantage that remote sensing can provide for the inventory and monitoring of protected areas is information to better understand the past and current status, the changes that occurred under different impacting factors and management practices, the trends of changes in comparison with those in the adjacent areas, and the implications of changes on ecosystem functions [22–24]. Remote sensing has unique advantages in monitoring frontier lands, which are always in remote and difficult-to-reach locations and huge in their area coverage. Different types of remote sensing data have been applied in the study of frontier lands—for example, using hyperspectral and radar data to monitor forests in the Amazon [25–30], in Africa [31], and in Siberia [32–35], and for hydrologic change detection in the lake-rich Arctic region [36,37], along the coastal zones [38–41], and in MPAs [42].

There have also been several reviews on PA monitoring using remote sensing. For example, Nagendra et al. [43] provided a review of remote sensing for conservation monitoring by assessing PAs, habitat extent, habitat condition, species diversity, and threats. Kachelriess et al. [44] reviewed the application of remote sensing for MPAs management. Gillespie et al. [45] reviewed advances in the spaceborne remote sensing of terrestrial PAs. Willis [11] provided a review of the remote sensing change detection methods employed for the ecological monitoring of United States PAs. The existing reviews have mainly been focused on a certain type of PAs or the monitoring method. There have been no bibliometric analyses of remote sensing applications in the monitoring of PAs.

Bibliometric analysis, introduced by Pritchard (1969), is a mathematical and statistical approach to analyze pertinent literature and understand the global research trends in a specific area [46,47]. Bibliometric analysis methods are frequently used to provide quantitative analyses of academic literature [48], and have been applied to environmental engineering and science, soil science, ecology, food safety, new energy utilization, and other areas. A bibliometric analysis helps identify research gaps and directions in one certain area [49]. In recent years, studies have applied this method to evaluate the research trends of remote sensing and its application in different scientific fields [50–52]. For instance, Zhang et al. [53] combined the new index (geographical impact factor) and traditional bibliometric methods to study the global research trends in remote sensing studies. Viana et al. [54] performed a bibliometric analysis to appraise the publication, research trends, and characteristics regarding the application of remote sensing data in human health. Wang et al. used the bibliometric method to study the research status and trends in the remote sensing of crop growth monitoring in China [55]. However, no attempt has been made to evaluate the inventory and monitoring of PAs in the literature using bibliometric analysis methods. In recent years, the number of publications on the remote sensing monitoring of PAs has been increasing. Hence, it is necessary to summarize the current status and development trend in this field. With the help of bibliometric methods, researchers can better understand the current number of publications, what journals these documents are published in, what the influential countries and institutions in this field are, how the research direction of this discipline is developing, etc.

Using a bibliometric approach, this article analyzes the relevant literature specialized in remote sensing applications in PAs. The aims of this work are to (1) summarize the variation in the characteristics of document types, total publication output, subject categories, and source journals; (2) analyze the publication output and international collaboration by countries, institutions, and authors; and (3) reveal the common research topics of PA monitoring research based on a keyword analysis. This research can help us understand the progress in this field and identify the relevant research and application directions.

2. Methodology and Data Collection

The bibliometric indicators analyzed in this research include a number of publications, subject categories, source journals, countries, and institutions, which were all obtained directly from the Web of Science. The Web of Science database can offer various statistics on retrieved papers, including the author, series name, conference name, country/region, document type, editor, fund funding institution, authorization number, group author, language, institution, publication year, research direction, source publication name, and the Web of Science category. Another function of web of science is to "Create a Citation Report", which can directly generate the total quoted frequency of the retrieved documents, the total quoted frequency of the removed self-cited documents, the quoted documents, the quoted documents of the removed self-cited documents, the average times a document has been cited, and the H-index of each item.

Co-authorship among countries and institutions was also analyzed in this research. Co-authorship mainly analyzes the co-signature of authors in the published paper. If the authors sign their names together in the paper, they are considered to have a cooperative relationship. At present, co-authorship analysis not only focuses on an analysis of researchers, but also includes an analysis of the cooperation between countries and institutions. In the case of the co-authorship analysis, the link strength between countries and institutions indicates the number of publications that two affiliated countries and institutions have co-authored, whereas the total link strength indicates the total strength of the co-authorship links of a given country or institution with other countries and institutions. Similarly, in the case of the co-occurrence analysis, the link strength between the author keywords indicates the number of publications in which two keywords occur together. In order to investigate the development of remote sensing in the field of protection area monitoring, we determined the keywords related to satellite, sensor and remote sensing monitoring method from the results of the co-occurrence keywords.

In this study, the VOSviewer software was utilized to visualize the co-authorship collaboration networks of countries and institutions and produce a keywords co-occurrence analysis. Invented by Van Eck and Waltman (Leiden University) in the Netherlands, VOSviewer is freely available -text mining software for generating bibliometric maps and analyzing trends in the scientific literature [56]. The outstanding feature of VOSviewer is its strong graphic display ability, which is especially suitable for analyzing large-scale sample data. This visualization effect is better than that of other similar analysis software, and the analysis function is more comprehensive. VOSviewer is a robust tool that uses clustering algorithms and functionalities based on the strength of the connections among items to facilitate the analyses of the network. [57]. The VOSviewer software uses a circle and label to represent an element, in which the circle size represents the importance, and circles with the same color belong to the same cluster.

Bibliometric maps are created by VOSviewer. These maps include items such as countries, institutions, and author keywords in the present study. The connection or relation between two items is named a link. The strength of a link indicates the number of publications that two countries or institutions have co-authored in the case of co-authorship links, or the number of publications in which two author keywords occur together in the case of co-occurrence links [46,58]. In the VOSviewer, there are two methods used to calculate link strength: full counting and fractional counting.

Full counting means that a co-authored publication is counted with a full weight of one for each co-author, which implies that the overall weight of a publication is equal to the number of authors of the publication. Fractional counting means that a co-authored publication is assigned fractionally to each of the coauthors, with the overall weight of the publication being equal to one. As analyzed by Perianes-Rodriguez et al. (2016), a fractional counting approach is preferable to full counting [59]. Therefore, we chose fractional counting to calculate the link strength.

VOSviewer uses a clustering algorithm to cluster the literature network, which is similar to the network clustering method of Modularity, specifically the maximization formula:

$$V(c_1,\ldots c_n) = \frac{1}{2m} \sum_{i<j} \delta(c_i, c_j) w_{ij} \left(c_{ij} - \gamma \frac{c_i c_j}{2m} \right) \quad (1)$$

$$w_{ij} = \frac{2m}{c_i} c_j \quad (2)$$

where c_i is the cluster of element I, and γ is the resolution of clustering. By adjusting its size, different resolution clustering can be obtained. The larger γ is, the more clustering will be obtained, and the finer the classification will be.

In VOSviewer, the number of clusters is determined by the option "choose threshold". In the case of a co-authorship analysis, the threshold is the minimum number of documents of a country or an institution. In the case of co-occurrence analysis, the threshold is the minimum number of occurrences of a keyword. We can choose the threshold according to our own needs.

The VOSviewer software has been widely used in bibliometric analyses in many fields. For example, Santos et al. used VOSviewer to map knowledge networks on female entrepreneurship [60]. Sainaghi et al. mapped the co-citation network of journals and authors on the foundations of hospitality performance measurement research using VOSviewer [61]. Sweileh et al. used VOSviewer to visualize map-based bibliometric indicators for the global research output on antimicrobial resistance among uropathogens [62].

The relevant documents were retrieved from the Science Citation Index Expanded (SCI-Expanded) and Social Science Citation Index (SSCI) of the Web of Science database, which is a multidisciplinary database of Thomson Reuters [63]. The following keywords were used to search all archived documents: TS (Topic) = "protected area*" or "natural reserve*" or "conservation area*" or "national park*" or "national forest*" or "marine protected area*" or "wilderness area*" or "frontier land*" or "natural monument*" or "biodiversity conservation" and "remote sensing". The publications that contained any of those keywords or variants (with*) in their titles, abstracts, and keyword lists were included [48]. The information on title, authors, institution, abstract, keywords, and cited references was downloaded. We set the starting time of this study as 1991, considering that the number of publications under the subject of remote sensing applications in PAs and relevant studies increased significantly after 1991 in professional journals and publications. This is illustrated in Figure 1. The data collection was conducted on 16 November 2019. Until 2018, a total of 4546 records were retrieved as the data for this analysis. Among these records, 3994 papers were focused on the remote sensing monitoring of terrestrial PAs, while the other 552 papers were on MPAs.

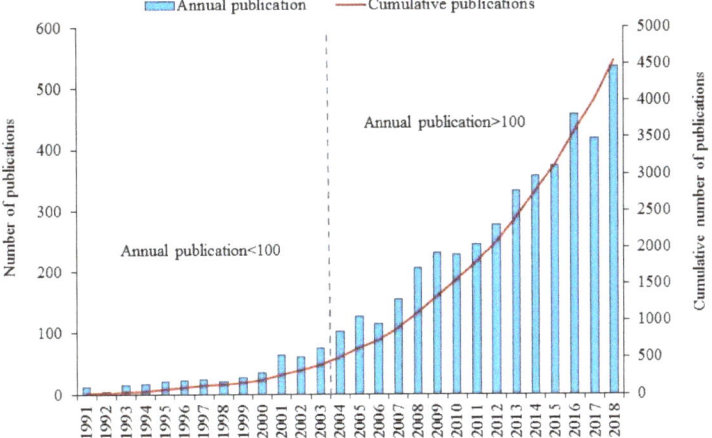

Figure 1. Annual publication and cumulative number, 1991–2018.

3. Results

3.1. General Characteristics and Trends of Publication Outputs

The trend for publications from 1991 to 2018 is illustrated in Figure 1. In general, the number of publications has shown an increasing trend over the years, with small fluctuations between individual years. According to the dates, the evolution of the published article output can be divided into three stages. The first stage extends from 1991 to 2003, with a relatively slow growth period. The second stage features a steady growth period from 2004 to 2011. The third stage is a fast growth period from 2012 to 2018.

The sample documents covered a total of 108 subject categories. The research domain covered a wide variety of themes and disciplines. The top 10 subject categories with more than 200 documents are displayed in Figure 2. The results indicate that environmental sciences ranked first with 1524 publications, followed by remote sensing with 1062 publications, ecology with 946 publications, and imaging science and photographic technology with 652 publications. Multidisciplinary geosciences, physical geography, forestry, biodiversity conservation, water resources, and meteorological and atmospheric sciences were also relevant subject categories.

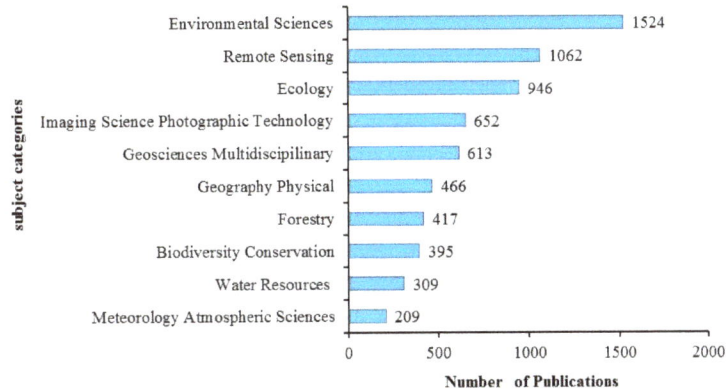

Figure 2. Top 10 subject categories in the field of the remote sensing monitoring of protected areas (PAs).

For the source journals, 739 different journals published papers related to remote sensing for PA monitoring. Table 1 shows the top 20 journals in terms of total relevant publications. Remote Sensing of Environment ranked first, with 256 articles covering 5.63% of the total publications. Remote Sensing ranked second, with 174 articles accounting for 3.83%, while the International Journal of Remote Sensing ranked third, with 153 articles accounting for 3.37%. The ISPRS Journal of Photogrammetry and Remote Sensing, Forest Ecology and Management, and the International Journal of Applied Earth Observation and Geoinformation (ranked 4th, 5th, and 6th places) accounted for 2.38%, 2.11%, and 2.02%, respectively.

Table 1. Top 20 main source journals in the research field.

Rank	Name	Country	Number	Percentage%
1	Remote Sensing of Environment	USA	256	5.63
2	Remote Sensing	Switzerland	174	3.83
3	International Journal of Remote Sensing	UK	153	3.37
4	ISPRS Journal of Photogrammetry and Remote Sensing	Netherlands	108	2.38
5	Forest Ecology and Management	Netherlands	96	2.11
6	International Journal of Applied Earth Observation and Geoinformation	Netherlands	92	2.02
7	Ecological Indicators	Netherlands	78	1.72
8	Environmental Monitoring and Assessment	Netherlands	70	1.54
9	Biological Conservation	UK	69	1.52
10	Applied Geography	UK	62	1.36
11	PLOS One	USA	58	1.28
12	Ecological Applications	USA	53	1.17
13	Journal of the Indian Society of Remote Sensing	USA	47	1.03
14	International Journal of Wildland Fire	Australia	44	0.97
15	Current Science	India	43	0.95
16	Environmental Management	USA	43	0.95
17	Environmental Research Letters	UK	43	0.95
18	Journal of Environmental Management	UK	42	0.92
19	Landscape Ecology	Netherlands	42	0.92
20	Biodiversity and Conservation	Netherlands	39	0.86

3.2. Countries, Institutions, and International Collaboration

According to the retrieved results, the papers covered a total of 153 different countries (or territories, hereafter referred to as "countries" for simplification). The geographical distribution of the top 20 productive countries for the overall study period is shown in Figure 3. The USA ranked first with a dominant output of 1655 papers or a share of 36.41%. China had 619 papers (13.62%) and UK had 479 (10.54%), ranking second and third, respectively. Other top ranked countries are Germany (7.92%), India (7.90%), Australia (7.11%), Canada (6.64%), and Italy (5.65%).

The co-authorship analysis studied a network of the main countries, which is plotted in Figure 4. These countries published more than 60 papers. There were four main clusters formed in the network (Table 2). The USA showed 62,644 citations and a link strength of 634, the UK showed 14,335 citations with a link strength of 241, and China showed 12,906 citations with a link strength of 265, which surpassed all the other clusters. The strongest link strength was evidenced by the USA and China, with a 151.93 link strength, followed by the USA and Canada with a 64.89 link strength, the USA and the UK with a 58.69 link strength, the USA and Germany with a 49.93 link strength, the USA and Australia with a 46.48 link strength, and the USA and Brazil with a 43.59 link strength.

According to the results, 4451 institutions contributed to the analyzed publications. The top 15 research institutions with the largest number of documents are listed in Table 3. By far the most productive institution was the Chinese Academy of Sciences in China, with 296 publications. The University of Maryland was in second place with 118 publications. The Chinese Academy of Sciences also ranked first in number of citations, followed by NASA, University of Maryland, and the U.S. Forest Service.

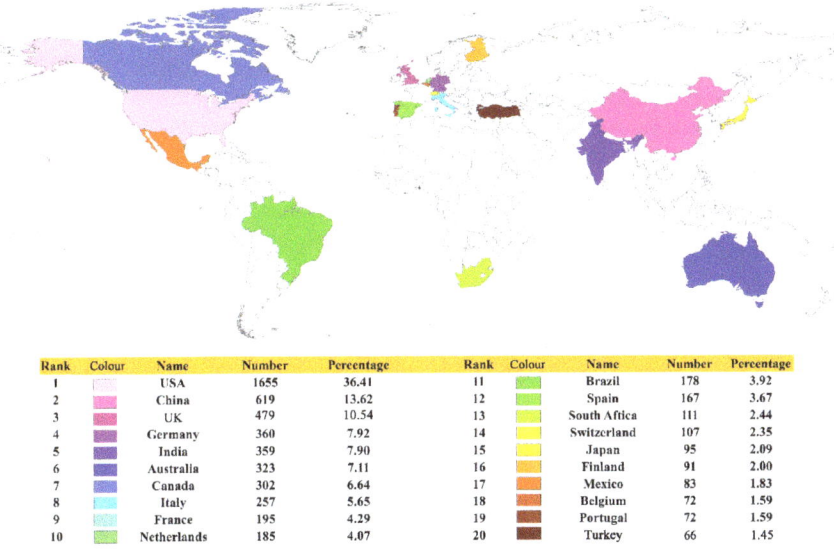

Rank	Colour	Name	Number	Percentage	Rank	Colour	Name	Number	Percentage
1		USA	1655	36.41	11		Brazil	178	3.92
2		China	619	13.62	12		Spain	167	3.67
3		UK	479	10.54	13		South Africa	111	2.44
4		Germany	360	7.92	14		Switzerland	107	2.35
5		India	359	7.90	15		Japan	95	2.09
6		Australia	323	7.11	16		Finland	91	2.00
7		Canada	302	6.64	17		Mexico	83	1.83
8		Italy	257	5.65	18		Belgium	72	1.59
9		France	195	4.29	19		Portugal	72	1.59
10		Netherlands	185	4.07	20		Turkey	66	1.45

Figure 3. The geographic distribution of the top 20 productive countries.

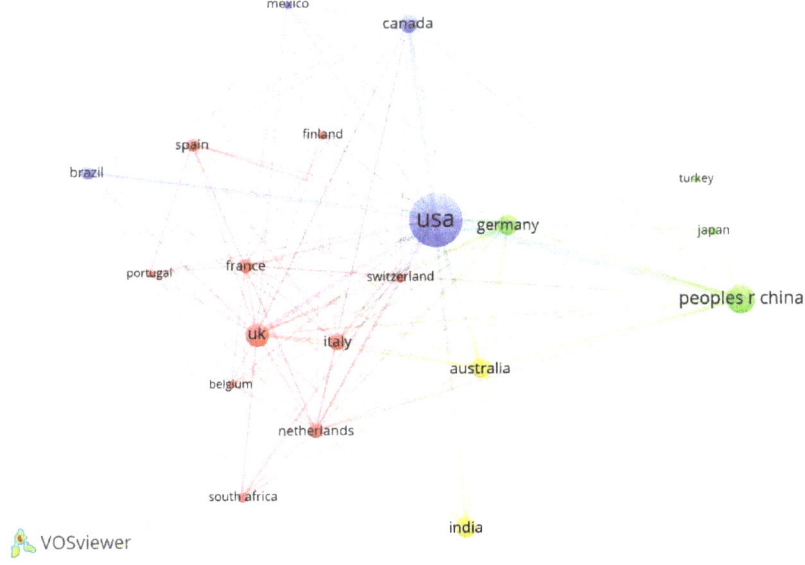

Figure 4. Co-authorship cooperation between productive countries. Each node represents a country. The size of the nodes reveals the citations of the countries, while the thickness of the lines between them shows the strength of collaboration.

Table 2. 5 main clusters for country collaboration.

Cluster	Country	Citations	Link Strength
1	UK	14,335	241
	Italy	8924	133
	Netherlands	7298	128
	France	6838	124
	Spain	4732	96
	Switzerland	4361	70
	South Africa	2715	59
	Finland	2596	40
	Belgium	1980	50
	Portugal	1288	54
2	USA	62,644	634
	Canada	11,665	140
	Brazil	4926	94
	Mexico	2026	58
3	China	12,906	265
	Germany	11,147	200
	Japan	2562	45
	Turkey	513	8
4	Australia	10031	164
	India	6772	73

Table 3. Top 15 institutions based on total publications.

Rank	Organization	Country	Number	Percentage %	Citations
1	Chinese Academy of Sciences	China	296	6.51	7279
2	University of Maryland	USA	118	2.60	5683
3	U.S. Forest Service	USA	110	2.42	5376
4	NASA	USA	108	2.38	6668
5	U.S. Geological Survey	USA	84	1.85	2842
6	University of Chinese Academy of Sciences	China	63	1.39	1222
7	Beijing Normal University	China	61	1.34	1703
8	University of Queensland	Australia	59	1.30	2347
9	University of Wisconsin	USA	56	1.23	3535
10	Oregon State University	USA	54	1.19	2367
11	University of Florida	USA	53	1.17	1529
12	Caltech	USA	51	1.12	2086
13	University of British Columbia	Canada	51	1.12	2765
14	University of Oxford	UK	48	1.06	1381
15	Natural Resources Canada	Canada	47	1.03	1636

An institutional cooperation network based on the VOSviewer software for the construction of scientific maps is presented in Figure 5. This figure presents the four clusters of collaboration among the prolific institutions with 35 or more publications. The largest cluster (red) contains nine institutions. All the institutions in the red cluster belong to the USA. The green and blue clusters both contain five institutions. Two of the institutions in the green cluster belong to the Netherlands, and the remaining three are from Australia, the UK and the USA. The blue cluster is composed of three Chinese institutions and two American institutions. The fourth cluster (yellow) includes three institutions from Canada. It can be seen that the cooperation between institutions is mainly focused within the same country or neighboring countries.

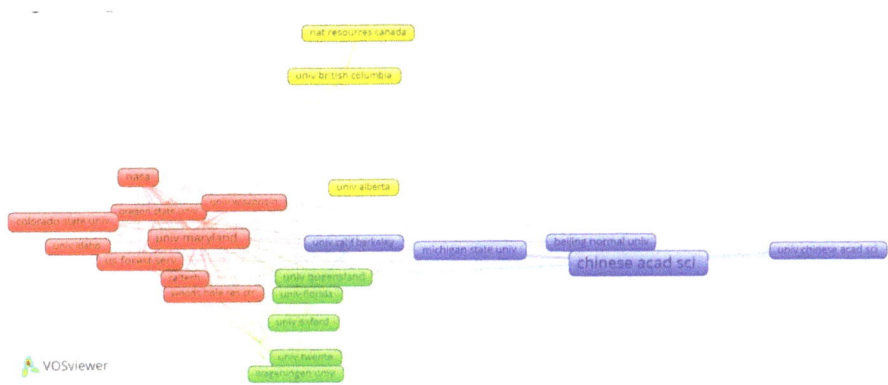

Figure 5. Co-authorship cooperation between productive institutions. The colors represent clusters of institutions, the size of frames represents the number of articles published by these institutions, and the lines represent the strength of cooperation among institutions.

3.3. Common Interests in Research Topics

Keywords, a core element of papers, offer a highly summarized form of a paper's contents. In order to understand the focus areas and development trends of one field, it is necessary to systematically analyze the selection of keywords in relevant studies [64]. Table 4 shows the 20 most frequently used author keywords from 1991 to 2018, including "remote sensing", "GIS", "Landsat", "deforestation", "LiDAR", "conservation", and "biodiversity", for research on PA monitoring that is concentrated on deforestation and biodiversity conservation.

A statistical analysis of the changes in the author keywords between different stages is beneficial for a comparative analysis of the changes in common research subjects and the development process of PA monitoring studies [19,65,66]. Table 4 separates the development of PA monitoring research into three stages, namely 1991–2003, 2004–2011, and 2012–2018. "Remote sensing" and "Landsat" were the most frequently used author keywords and appeared in the top 20 in all three periods. The "MODIS" and "LiDAR" keywords increased in frequency of appearance from 1991 to 2011 and increased further in 2012–2018, which indicates that the platform played a significant important role in PA monitoring. Comparing the three different stages, the keywords rankings changed considerably. The keyword "climate change" began to appear in the top 10 during 2012–2018, which suggests that more attention was being given to climate change on PA research. The research focus of each stage is as follows. The early stage of research focuses on landscape ecological change and human disturbance. The middle stage focuses on the change detection of land cover and land use caused by deforestation. The late stage focuses on the impact of climate change on PAs.

In order to trace the trend of the remote sensing data used in PAs research, the most frequently selected keywords related to satellites and sensors were counted. The top ten are Landsat, MODIS, LiDAR, SPOT, AVHRR, ASTER, IKONOS, PALSAR, Sentinel (Sentinel-1 and Sentinel-2), and WorldView, with low, moderate, or high-resolution sensors. The annual publications of the top ten satellites and sensors are shown in Figure 6. In terms of quantity, Landsat was the most frequently used satellites and sensors type, with 1078 papers, followed by MODIS with 439 papers and LiDAR, with 370 papers. In addition, with the continuous development of remote sensing technology, some new platforms and satellites have emerged and have been applied to monitor PAs in recent years. For example, there were 35 papers on the UAV monitoring of PA, and 26 papers on small satellites from 2001 to 2008.

Table 4. Top 20 author keywords in different stages, 1991–2018, 1991–2003, 2004–2011, and 2012–2018. F(%)—frequency of author keywords and their percentage of total publications in the corresponding stage. The total publications numbered 4546 in 1991–2018, 393 in 1991–2003, 1405 in 2004–2011, and 2748 in 2012–2018.

Rank	1991–2018 Keywords	F (%)	1991–2003 Keywords	F (%)	2004–2011 Keywords	F (%)	2012–2018 Keywords	F (%)
1	remote sensing	1933 (42.52)	remote sensing	135 (34.35)	remote sensing	659 (46.90)	remote sensing	1139 (41.45)
2	Landsat	1078 (23.71)	Landsat	112 (28.50)	Landsat	353 (25.12)	Landsat	613 (22.31)
3	MODIS	439 (9.66)	GIS	48 (12.21)	GIS	159 (11.32)	MODIS	310 (11.28)
4	GIS	425 (9.35)	landscape ecology	12 (3.05)	MODIS	118 (8.40)	LiDAR	293 (10.66)
5	LiDAR	370 (8.14)	MODIS	11 (2.80)	LiDAR	72 (5.12)	GIS	218 (7.93)
6	deforestation	161 (3.54)	conservation	11 (2.80)	biodiversity	49 (3.49)	deforestation	115 (4.18)
7	conservation	146 (3.21)	biodiversity	9 (2.29)	NDVI	42 (2.99)	conservation	97 (3.53)
8	biodiversity	136 (2.99)	modeling	8 (2.04)	deforestation	39 (2.78)	biodiversity	78 (2.84)
9	NDVI	124 (2.73)	land-cover	8 (2.04)	conservation	38 (2.70)	climate change	78 (2.84)
10	protected area	106 (2.33)	deforestation	7 (1.78)	protected area	36 (2.56)	NDVI	68 (2.47)
11	land-use	105 (2.31)	fragmentation	7 (1.78)	land-cover	34 (2.42)	protected area	66 (2.40)
12	land-cover	104 (2.29)	land-use	7 (1.78)	land-use	32 (2.28)	land-use	62 (2.26)
13	climate change	99 (2.18)	disturbance	6 (1.53)	wetland	32 (2.28)	land-cover	60 (2.18)
14	change detection	95 (2.09)	mapping	6 (1.53)	change detection	31 (2.21)	change detection	59 (2.15)
15	wetland	87 (1.91)	forest	6 (1.53)	tropical forest	29 (2.06)	land-use change	54 (1.97)
16	land-use change	83 (1.83)	LiDAR	5 (1.27)	land-cover change	29 (2.06)	wetland	52 (1.89)
17	monitoring	82 (1.80)	habitat fragmentation	5 (1.27)	monitoring	28 (1.99)	random forest	52 (1.89)
18	land-cover change	77 (1.69)	fire	5 (1.27)	land-use change	23 (1.64)	ecosystem service	51 (1.89)
19	forest	68 (1.50)	classification	5 (1.27)	forest	22 (1.57)	monitoring	51 (1.86)
20	soil erosion	62 (1.36)	satellite remote sensing	5 (1.27)	hyperspectral	22 (1.57)	REDD	44 (1.60)

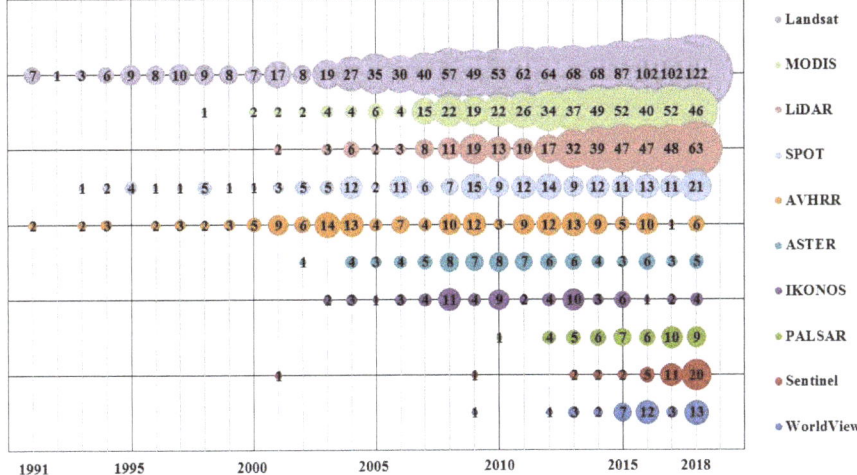

Figure 6. Annual publications of the main satellites and sensors in the research field.

Based on the co-occurrence analysis, the remote sensing monitoring methods are also counted in Table 5. The remote sensing monitoring methods mainly include classification, time-series analysis, model methods, object-oriented method, visual analysis, direct comparisons, and hybrid methods [67,68]. The classification method holds the first position with 526 papers and 11.57% of the total publications, followed by time-series analysis (288, 6.34%) and model method (159, 3.50%).

Table 5. The main remote sensing monitoring methods used for protected areas (PAs).

Rank	Methods	Number	Percentage %
1	classification	526	11.57
2	time-series analysis	288	6.34
3	model method	159	3.50
4	object-oriented method	131	2.88
5	visual analysis	95	2.09
6	direct comparison	72	1.58
7	hybrid methods	57	1.25

Figure 7 shows a co-occurrence network analysis of the keywords, which can be used to identify the research front in terms of topical trends for PA monitoring. In this analysis, the minimum number of occurrences of a keyword is 30 times for titles and abstracts in all publications. The research theme of PA monitoring has been categorized into six colored clusters, which were analyzed as follows. The red cluster with the highest number of keywords (12) is led by "land cover"; In addition, "land use", "monitoring", "mapping", "hyperspectral", and "classification" are also the main keywords of this cluster. Most keywords in this cluster are associated with studies on land use and land cover classification using hyperspectral remote sensing data. The blue cluster, with 11 keywords, has "Landsat", "MODIS", "NDVI", "climate change", "change detection", and "wetland" as its main related keywords, which appear in the relevant research on the habitat mapping and change detection of PAs, as well as the impact of climate change. The green cluster (11 keywords) focuses on the keywords: "deforestation", "LiDAR", "REDD", "biomass", "forest inventory", "tropical forest", "forest management", and "carbon". The keywords of this cluster are closely related to estimating forest biomass and carbon storage in PAs using LiDAR data. The yellow cluster has 10 keywords; the most frequently used is "remote sensing" followed by "conservation", while "biodiversity", "protected areas", and "fragmentation" are ranked 3rd–5th, respectively. Most keywords in this cluster relate to the use of remote sensing to support biodiversity conservation in PAs. The number of keywords in the purple cluster is four, including "land-use change", "land-cover change", "ecosystem service", and "landscape metrics". This cluster is related to the analysis of land-use/land-cover change and ecosystem service evaluation by remote sensing and landscape metrics. The orange cluster includes only three keywords. The keyword "GIS" appears most frequently, with a total of 387 occurrences. The other two keywords are "soil erosion" and "RUSLE (The Revised Soil Loss Equation)". This cluster has connections with keywords related to the study of soil erosion and its spatial distribution in PAs using the GIS analysis method.

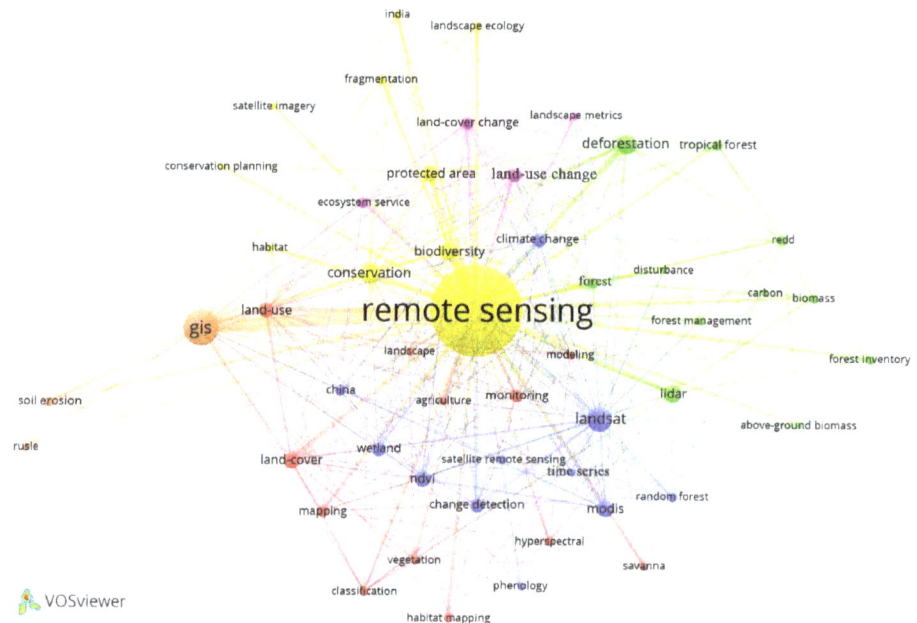

Figure 7. Keywords co-occurrence network. Each node represents a keyword, the size of the node indicates the number of occurrences of the keyword, and the line thickness of the two nodes represents the degree of connection.

4. Discussion

In this paper, by retrieving the relevant literature on remote sensing monitoring protected areas, we revealed hidden knowledge underlying this significant body of research. The number of publications shows a trend of continuous growth, demonstrating that more and more scholars have paid attention to this research field. From the perspective of subject categories, environmental sciences ranked first, followed by remote sensing and ecology, which shows that the remote sensing monitoring of PAs is a field closely related to environmental science, remote sensing, and ecology. The top three journals are all well-known journals in the field of remote sensing, include Remote Sensing of Environment, the International Journal of Remote Sensing, and the ISPRS Journal of Photogrammetry and Remote Sensing.

For country of origin, the USA is in the leading position. Moreover, the top 20 countries are mostly European countries. When considering institutions, the Chinese Academy of Sciences published the largest number of papers. The United States has the largest number of research institutions in the top 15, accounting for more than half of them. Through the co-authorship analysis of countries and institutions, this study determined that the USA was at the center of international cooperation, and the cooperation among national research institutions was relatively close, while its international cooperation was relatively less prevalent, which is not conducive to the long-term development of remote sensing monitoring for PAs. Countries and institutions should strengthen their knowledge exchanges and cooperation to more effectively discuss research trends and research status in the research field by holding relevant academic forums and conferences.

The analysis results showed that studies have mainly concentrated on terrestrial PAs, while literature on MPA monitoring is relatively less common. Future research should make full use of new monitoring technology and methods to establish a long-term, scientific, and systematic monitoring system and thereby provide a data-based foundation for evaluating the effectiveness of MPAs. Based on the changes in keywords, it can be seen that remote sensing monitoring of PAs mainly focused on vegetation classification, landscape pattern analysis, biodiversity protection, and the monitoring of changes in PAs. Future research trends will focus on the impact of climate change on PAs.

Considering temporal variation, the use of Landsat, MODIS, and LiDAR show a clear fluctuating and increasing trend. LiDAR and SAR have been increasingly used to monitor and evaluate the landscape in recent years. Different satellites and sensors are now applied in different fields and at different scales of PAs. For example, Landsat products include the Thematic Mapper (TM), Enhanced Thematic Mapper Plus (ETM+), and Operational Land Imager (OLI), which can be used to monitor vegetation dynamics and assess land-cover/land-use change. However, MODIS sensors are more appropriate for vegetation phenology and forest fire monitoring and can provide high temporal resolution time series data at the landscape, regional, and global spatial scales. LiDAR makes it possible to estimate tree height, biomass, and leaf area index in large areas of the world [69,70]. SAR facilitates the estimation of forest biomass and tree height at small and medium scale [71]. Furthermore, SPOT or QuickBird may be used for species or specific vegetation change monitoring [72–74]. AVHRR sensors are mainly used to analyze the impact of climate change on vegetation coverage in PAs. The high spatial resolution of ASTER can also be used to study land- cover/land-use change in PAs [75]. Other high-resolution satellites, such as IKONOS and WorldView, focus on mapping vegetation types or habitat associated with endangered fauna [76,77].

In recent years, with the rapid development of satellites, sensors, and techniques, the applications of remote sensing have been broadly employed in the monitoring and management of PAs. The relevant research results for improving the level of monitoring in PAs, formulating differentiated regional protection policies, and guiding sustainable development play an important role. According to this bibliometric analysis, research on the remote sensing monitoring of PAs has mainly focused on the inventory and classification of vegetation, change detection, habitat degradation, the impact of climate change, and biodiversity conservation. Among the various methods, classification, time series analysis, and model methods were the most frequently used types for PA monitoring. In the foreseeable future, there will be more new methods to monitor PAs. For example, big data approaches are being adopted to process large amounts of remote sensing data [78–80].

There are still some limitations to this study. Firstly, the single database that we used does not index all scientific journals and theme books, which could exclude some relevant articles. For example, some gray literature on this topic from government agencies, nature conservancies and other non-profits might have been excluded. Expanding the search across multiple databases, such as Scopus and Google Scholar, will help reduce omissions in the analysis. Secondly, setting 1991 as the starting time may omit some earlier studies. However, most articles relevant to remote sensing applications in PAs were published in professional journals from 1991 onwards. Therefore, we believe that using 1991 as the starting point is still representative and appropriate. Thirdly, the VOSviewer software has some functional restrictions. Another consideration is that other bibliometric analysis tools, such as Citespace, could be applied in combination with VOSviewer in the future to more extensively cover the published research on the remote sensing of PAs. In the meantime, we acknowledge that it is almost impossible to include all remote sensing applications for PAs by limiting the search to include "remote sensing" alone. Other terms and descriptions, such as "land-cover monitoring", "landscape configuration and composition", "habitat analysis", "biodiversity conservation", and "bathymetry assessment", among other examples, could be very relevant to studies in protected areas with remote sensing applications, but could be missed in the analysis. This is particularly true for monitoring changing terrestrial and marine environments under impacts from the natural and anthropogenic disturbances of protected areas. This challenge might be resolved when searches for bibliometric analysis are able to include into the full contents of published articles through the use of improved technologies, such as big data and artificial intelligence, instead of using limited and selected combination of keywords.

5. Conclusions

This paper evaluated the global research and publication trends in the remote sensing of PAs monitoring from 1991 to 2018. This analysis comprised eight main aspects: document types, publication output, subject categories, source journals, countries, organizations, and keywords. The results showed that since 2004, the number of publications increased rapidly and steadily. Environmental Sciences was the largest subject category. The highest number of papers was published in the two journals on Remote Sensing of Environment and Remote Sensing. The USA published the most in application of remote sensing technology in PA monitoring. Institutions affiliated to the USA have a massive number of publications and strong international collaboration in such type of explorations. Landsat, MODIS, and LiDAR are the most commonly used satellites and sensors. Most of the research was focused on classification, time-series analysis, and model methods. Keywords selections indicate that "Landsat", "deforestation", "LiDAR", "conservation", and "biodiversity" are among the most common subjects in the remote sensing of PAs. Studies on PA monitoring using remote sensing are mainly focused on change detection, biodiversity conservation, and the impact of climate change. In the future, we should continue to pay attention to the development trends and hot spots for the remote sensing monitoring of PAs. Researchers from all countries should strengthen the international exchanges of ideas and actively promote international research cooperation.

Author Contributions: P.D. designed the research, collected the data, analyzed the results, and wrote the first draft. Y.W. conceived the research and reviewed the manuscript. P.Y. processed and visualized the data. All authors have read and agreed to the published version of the manuscript.

Funding: This work was funded by the Humanity and Social Science Foundation of the Ministry of Education (19YJCZH229) and the China Postdoctoral Science Foundation (2018M632719).

Acknowledgments: We appreciate the insightful and constructive comments and suggestions from the anonymous reviewers that helped improve the quality and presentation of the manuscript.

Conflicts of Interest: The authors declare no conflict of interest.

References

1. UNEP-WCMC. *State of the World's Protected Areas: An Annual Review of Global Conservation Progress*; UNEP-WCMC: Cambridge, UK, 2008.
2. Dudley, N. *Guidelines for Applying Protected Area Management Categories*; IUCN: Gland, Switzerland, 2008.
3. Wang, Y. *Remote Sensing of Protected Lands*; CRC Press: Boca Raton, FL, USA, 2011; ISBN 978-1-4398-4187-7.
4. Watson, J.E.M.; Dudley, N.; Segan, D.B.; Hockings, M. The performance and potential of protected areas. *Nature* **2014**, *515*, 67–73. [CrossRef]
5. UNEP-WCMC; IUCN; NGS. *Protected Planet Report 2018*; UNEP-WCMC: Cambridge, UK; IUCN: Gland, Switzerland; NGS: Washington, DC, USA, 2018; ISBN 978-92-807-3721-9. Available online: https://www.iucn.org/theme/protectedareas/publications/protected-planet-report (accessed on 21 September 2019).
6. Erol, S.Y.; Kuvan, Y.; Yildirim, H.T. The general characteristics and main problems of national parks in Turkey. *Afr. J. Agr. Res.* **2011**, *6*, 5377–5385.
7. Bruner, A.G.; Gullison, R.E.; Rice, R.E.; Da Fonseca, G.A.B. Effectiveness of Parks in Protecting Tropical Biodiversity. *Science* **2001**, *291*, 125–128. [CrossRef] [PubMed]
8. Adhikari, S.; Southworth, J. Simulating Forest Cover Changes of Bannerghatta National Park Based on a CA-Markov Model: A Remote Sensing Approach. *Remote Sens.* **2012**, *4*, 3215–3243. [CrossRef]

9. Watson, J.E.M.; Venter, O.; Lee, J.; Jones, K.R.; Robinson, J.G.; Possingham, H.P.; Allan, J.R. Protect the last of the wild. *Nature* **2018**, *563*, 27–30. [CrossRef] [PubMed]
10. Leverington, F.; Costa, K.L.; Pavese, H.; Lisle, A.; Hockings, M. A Global Analysis of Protected Area Management Effectiveness. *Environ. Manag.* **2010**, *46*, 685–698. [CrossRef] [PubMed]
11. Willis, K.S. Remote sensing change detection for ecological monitoring in United States protected areas. *Biol. Conserv.* **2015**, *182*, 233–242. [CrossRef]
12. Wiens, J.; Sutter, R.; Anderson, M.; Blanchard, J.; Barnett, A.; Aguilar-Amuchastegui, N.; Avery, C.; Laine, S. Selecting and conserving lands for biodiversity: The role of remote sensing. *Remote Sens. Environ.* **2009**, *113*, 1370–1381. [CrossRef]
13. Lu, X.; Zhou, Y.; Liu, Y.; Yannick, L.P. The role of protected areas in land use/land cover change and the carbon cycle in the conterminous United States. *Glob. Chang. Biol.* **2017**, *24*, 617–630. [CrossRef]
14. Fan, L.; Zhao, J.; Wang, Y.; Ren, Z.; Zhang, H.; Guo, X. Assessment of Night-Time Lighting for Global Terrestrial Protected and Wilderness Areas. *Remote Sens.* **2019**, *11*, 2699. [CrossRef]
15. Wang, Y.; Mitchell, B.R.; Nugranad-Marzilli, J.; Bonynge, G.; Zhou, Y.; Shriver, G. Remote sensing of land-cover change and landscape context of the National Parks: A case study of the Northeast Temperate Network. *Remote Sens. Environ.* **2009**, *113*, 1453–1461. [CrossRef]
16. Gross, J.E.; Goetz, S.J.; Cihlar, J. Application of remote sensing to parks and protected area monitoring: Introduction to the special issue. *Remote Sens. Environ.* **2009**, *113*, 1343–1345. [CrossRef]
17. Jones, D.A.; Hansen, A.J.; Bly, K.; Doherty, K.; Verschuyl, J.; Paugh, J.I.; Carle, R.; Story, S.J. Monitoring land use and cover around parks: A conceptual approach. *Remote Sens. Environ.* **2009**, *113*, 1346–1356. [CrossRef]
18. Kennedy, R.E.; Yang, Z.; Cohen, W.B. Detecting trends in forest disturbance and recovery using yearly Landsat time series: 1. LandTrendr—Temporal segmentation algorithms. *Remote Sens. Environ.* **2010**, *114*, 2897–2910. [CrossRef]
19. Szantoi, Z.; Brink, A.; Buchanan, G.; Bastin, L.; Lupi, A.; Simonetti, D.; Mayaux, P.; Peedell, S.; Davy, J. A simple remote sensing based information system for monitoring sites of conservation importance. *Remote Sens. Ecol. Conserv.* **2016**, *2*, 16–24. [CrossRef]
20. Huang, C.; Schleerweis, K.; Thomas, N.; Goward, S.N. Forest Dynamics within and around Olympic National Park Assessed Using Time Series Landsat Observations (Chapter 4). In *Remote Sensing of Protected Lands*; CRC Press: Boca Raton, FL, USA, 2012; pp. 75–94.
21. Zorn, P.; Ure, D.; Sharma, R.; O'Grady, S. Using earth observation to monitor species-specific habitat change in the Greater Kejimkujik National Park Region of Canada. In *Remote Sensing of Protected Lands*; CRC Press: Boca Raton, FL, USA, 2012; pp. 95–110.
22. Hansen, A.J.; DeFries, R. Land use change around nature reserves: Implications for sustaining biodiversity. *Ecol. Appl.* **2007**, *17*, 972–973. [CrossRef] [PubMed]
23. Hansen, A.J.; DeFries, R. Ecological mechanisms linking protected areas to surrounding lands. *Ecol. Appl.* **2007**, *17*, 974–988. [CrossRef]
24. Clark, J.; Wang, Y.; August, P.V. Assessing current and projected suitable habitats for tree-of-heaven along the Appalachian Trail. *Philos. Trans. Roy. Soc. B* **2014**, *369*, 20130192. [CrossRef]
25. Sheng, Y.; Alsdorf, D. Automated ortho-rectification of Amazon basin-wide SAR mosaics using SRTM DEM data. *IEEE Trans. Geosci. Remote Sens.* **2005**, *43*, 1929–1940. [CrossRef]
26. Arima, E.Y.; Walker, R.T.; Sales, M.H.R.; Souza, C.M.; Perz, S.G. The fragmentation of space in the Amazon basin: Emergent road networks. *Photogram. Eng. Remote Sens.* **2008**, *74*, 699–709. [CrossRef]
27. Walsh, S.J.; Messina, J.P.; Brown, D.G. Mapping & modeling land use/land cover dynamics in frontier settings. *Photogram. Eng. Remote Sens.* **2008**, *74*, 677–679.
28. Mena, C.F. Trajectories of land-use and land-cover in the northern Ecuadorian Amazon: Temporal composition, spatial configuration, and probability of change. *Photogram. Eng. Remote Sens.* **2008**, *74*, 737–751. [CrossRef]
29. Wang, C.; Qi, J.; Cochrane, M. Assessment of tropical forest degradation with canopy fractional cover from Landsat ETM+ and IKONOS imagery. *Earth Interact.* **2005**, *9*, 1–18. [CrossRef]
30. Wang, C.; Qi, J. Biophysical estimation in tropical forests using JERS-1 SAR and VNIR Imagery: II-aboveground woody biomass. *Int. J. Remote Sens.* **2008**, *29*, 6827–6849. [CrossRef]
31. Ayebare, S.; Moyer, D.; Plumptre, A.J.; Wang, Y. Remote Sensing for Biodiversity Conservation of the Albertine Rift in Eastern Africa. In *Remote Sensing of Protected Lands*; CRC Press: Boca Raton, FL, USA, 2012; pp. 183–201.

32. Sun, G.; Ranson, K.J.; Kharuk, V.I. Radiometric slope correction for forest biomass estimation from SAR data in Western Sayani mountains, Siberia. *Remote Sens. Environ.* **2002**, *79*, 279–287. [CrossRef]
33. Bergen, K.M.; Zhao, T.; Kharuk, V.; Blam, Y.; Brown, D.G.; Peterson, L.K.; Miller, N. Changing regimes: Forested land cover dynamics in central Siberia 1974–2001. *Photogram. Eng. Remote Sens.* **2008**, *74*, 787–798. [CrossRef]
34. Kharuk, V.I.; Ranson, K.J.; Im, S.T. Siberian silkmoth outbreak pattern analysis based on SPOT VEGETATION data. *Int. J. Remote Sens.* **2009**, *30*, 2377–2388. [CrossRef]
35. Sherman, N.J.; Loboda, T.V.; Sun, G.; Shugart, H.H. Remote sensing and modeling for assessment of complex Amur (Siberian) Tiger and Amur (Far Eastern) Leopard Habitats in the Russian Far East. In *Remote Sensing of Protected Lands*; CRC Press: Boca Raton, FL, USA, 2012; pp. 379–407.
36. Stow, D.A.; Hope, A.; McGuire, D.; Verbyla, D.; Gamon, J.; Huemmrich, F.; Houston, S.; Racine, C.; Sturm, M.; Tape, K.; et al. Remote sensing of vegetation and land-cover change in Arctic tundra ecosystems. *Remote Sens. Environ.* **2004**, *89*, 281–308. [CrossRef]
37. Sheng, Y.; Shah, C.A.; Smith, L.C. Automated image registration for hydrologic change detection in the lake-rich arctic. *IEEE Geosci. Remote Sens. Lett.* **2008**, *5*, 414–418. [CrossRef]
38. Lu, Z.; Kwoun, O. RADARSAT-1 and ERS interferometric analysis over southeastern coastal Louisiana: Implication for mapping water-level changes beneath swamp forests. *IEEE Trans. Geosci. Remote Sens.* **2008**, *46*, 2167–2184. [CrossRef]
39. Lu, Z.; Dzurisin, D.; Jung, H.S. Monitoring Natural Hazards in Protected Lands Using Interferometric Synthetic Aperture Radar (InSAR). In *Remote Sensing of Protected Lands*; CRC Press: Boca Raton, FL, USA, 2012; pp. 439–471.
40. Campbell, A.; Wang, Y. Examining the Influence of Tidal Stage on Salt Marsh Mapping using High Spatial Resolution Satellite Remote Sensing and Topobathymetric LiDAR. *IEEE Trans. Geosci. Remote Sens.* **2018**, *56*, 5169–5176. [CrossRef]
41. Campbell, A.; Wang, Y. High spatial resolution remote sensing for salt marsh mapping and change analysis at Fire Island National Seashore. *Remote Sens.* **2019**, *11*, 1107. [CrossRef]
42. Friedlander, A.M.; Wedding, L.M.; Caselle, J.E.; Costa, B.M. Integration of Remote Sensing and in situ Ecology for the Design and Evaluation of Marine Protected Areas: Examples from Tropical and Temperate Ecosystems. In *Remote Sensing of Protected Lands*; CRC Press: Boca Raton, FL, USA, 2012; pp. 245–279.
43. Nagendra, H.; Lucas, R.; Honrado, J.P.; Jongman, R.H.G.; Tarantino, C.; Adamo, M.; Mairota, P. Remote sensing for conservation monitoring: Assessing protected areas, habitat extent, habitat condition, species diversity, and threats. *Ecol. Indic.* **2013**, *33*, 45–59. [CrossRef]
44. Kachelriess, D.; Wegmann, M.; Gollock, M.; Pettorelli, N. The application of remote sensing for marine protected area management. *Ecol. Indic.* **2014**, *36*, 169–177. [CrossRef]
45. Gillespie, T.W.; Willis, K.S.; Ostermann-Kelm, S. Spaceborne remote sensing of world's protected areas. *Prog. Phys. Geogr.* **2015**, *39*, 388–404. [CrossRef]
46. Khudzari, J.M.; Kurian, J.; Tartakovsky, B.; Raghavan, G.S.V. Bibliometric analysis of global research trends on microbial fuel cells using Scopus database. *Biochem. Eng. J.* **2018**, *136*, 51–60. [CrossRef]
47. Zou, X.; Yue, W.L.; Vu, H.L. Visualization and analysis of mapping knowledge domain of road safety studies. *Accid. Anal. Prev.* **2018**, *118*, 131–145. [CrossRef]
48. Chen, D.; Liu, Z.H.; Luo, Z.H.; Webber, M.; Chen, J. Bibliometric and visualized analysis of emergy research. *Ecol. Eng.* **2016**, *90*, 285–293. [CrossRef]
49. Geng, Y.; Chen, W.; Liu, Z.; Chui, A.S.F.; Han, W.; Liu, Z.; Zhong, S.; Qian, Y.; You, W.; Cui, X. A bibliometric review: Energy consumption and greenhouse gas emission in the residential sector. *J. Clean. Prod.* **2017**, *159*, 301–316. [CrossRef]
50. Hu, K.; Qi, K.; Guan, Q.; Wu, C.; Yu, J.; Qing, Y.; Zheng, J.; Wu, H.; Li, X. A Scientometric Visualization Analysis for Night-Time Light Remote Sensing Research from 1991 to 2016. *Remote Sens.* **2017**, *9*, 802. [CrossRef]
51. Zhang, Y.; Giardino, C.; Li, L. Water Optics and Water Colour Remote Sensing. *Remote Sens.* **2017**, *9*, 818. [CrossRef]
52. Cui, X.; Guo, X.; Wang, Y.; Wang, X.; Zhu, W.; Shie, J.; Lin, C.; Gao, X. Application of remote sensing to water environmental processes under a changing climate. *J. Hydrol.* **2019**, *574*, 892–902. [CrossRef]
53. Zhuang, Y.; Liu, X.; Nguyen, T.; He, Q.; Hong, S. Global remote sensing research trends during 1991–2010: A bibliometric analysis. *Scientometrics* **2013**, *96*, 203–219. [CrossRef]

54. Viana, J.; Santos, J.V.; Neiva, R.M.; Souza, J.; Duarte, L.; Teodoro, A.C.; Freitas, A. Remote Sensing in Human Health: A 10-Year Bibliometric Analysis. *Remote Sens.* **2017**, *9*, 1225. [CrossRef]
55. Wang, L.; Zhang, G.; Wang, Z.; Liu, J.; Shang, J.; Liang, L. Bibliometric Analysis of Remote Sensing Research Trend in Crop Growth Monitoring: A Case Study in China. *Remote Sens.* **2019**, *11*, 809. [CrossRef]
56. Van, E.N.J.; Waltman, L. Software survey: VOSviewer, a computer program for bibliometric mapping. *Scientometrics* **2010**, *84*, 523–538. [CrossRef]
57. Romanelli, J.P.; Fujimoto, J.T.; Ferreira, M.D.; Milanez, D.H. Assessing ecological restoration as a research topic using bibliometric indicators. *Ecol. Eng.* **2018**, *120*, 311–320. [CrossRef]
58. Van, E.N.J.; Waltman, L. *Manual for VOSviewer Version 1.6.10*; Universiteit Leiden: Leiden, The Nertherland, 2019.
59. Perianes-Rodriguez, A.; Waltman, L.; Van Eck, N.J. Constructing bibliometric networks: A comparison between full and fractional counting. *J. Informetr.* **2016**, *10*, 1178–1195. [CrossRef]
60. Santos, G.; Marques, C.S.; Ferreira, J.J. A look back over the past 40 years of female entrepreneurship: Mapping knowledge networks. *Scientometrics* **2018**, *115*, 953–987. [CrossRef]
61. Sainaghi, R.; Köseoglu, M.A.; d'Angella, F.; Tetteh, I.L. Foundations of hospitality performance measurement research: A co-citation approach. *Int. J. Hosp. Manag.* **2019**, *79*, 21–40. [CrossRef]
62. Sweileh, W.M.; Al-Jabi, S.W.; Zyoud, S.H.; Sawalha, A.F.; Abu-Taha, A.F. Global research output in antimicrobial resistance among uropathogens: A bibliometric analysis (2002–2016). *J. Glob. Antimicrob. Resist.* **2018**, *13*, 104–114. [CrossRef] [PubMed]
63. Birch, T.; Reyes, E. Forty years of coastal zone management (1975–2014): Evolving theory, policy and practice as reflected in scientific research publications. *Ocean Coast. Manag.* **2018**, *153*, 1–11. [CrossRef]
64. Liu, F.; Lin, A.; Wang, H.; Peng, Y.; Hong, S. Global research trends of geographical information system from 1961 to 2010: A bibliometric analysis. *Scientometrics* **2016**, *106*, 751–768. [CrossRef]
65. He, M.; Zhang, Y.; Gong, L.H.; Zhou, Y.; Song, X.F.; Zhu, W.P.; Zhang, M.M.; Zhang, Z. Bibliometrical analysis of hydrogen storage. *Int. J. Hydrog. Energy* **2019**, *44*, 28206–28226. [CrossRef]
66. Zhang, Y.; Zhang, Y.; Shi, K.; Yao, X. Research development, current hotspots, and future directions of water research based on MODIS images: A critical review with a bibliometric analysis. *Environ. Sci. Pollut. Res.* **2017**, *24*, 1–14. [CrossRef]
67. Zhang, H.; Huang, M.; Qing, X.; Li, G.; Tian, C. Bibliometric analysis of global remote sensing research during 2010–2015. *ISPRS Int. J. Geo Inf.* **2017**, *6*, 332. [CrossRef]
68. Gong, J.; Shi, H.; Ma, G.; Zhou, Q. A review of multi-temporal remote sensing data change detection algorithms. *Int. Arch. Photogramm. Remote Sens. Spat. Inf. Sci.* **2008**, *37*, 757–762.
69. Chen, B.; Pang, Y.; Li, Z.; North, P.; Rosette, J.; Sun, G.; Suárez, J.; Bye, I.; Lu, H. Potential of Forest Parameter Estimation Using Metrics from Photon Counting LiDAR Data in Howland Research Forest. *Remote Sens.* **2019**, *11*, 856. [CrossRef]
70. Saukkola, A.; Melkas, T.; Riekki, K.; Sirparanta, S.; Peuhkurinen, J.; Holopainen, M.; Hyyppä, J.; Vastaranta, M. Predicting Forest Inventory Attributes Using Airborne Laser Scanning, Aerial Imagery, and Harvester Data. *Remote Sens.* **2019**, *11*, 797. [CrossRef]
71. Austin, J.M.; Mackey, B.G.; Niel, K.P.V. Estimating forest biomass using satellite radar: An exploratory study in a temperate Australian Eucalyptus forest. *For. Ecol. Manag.* **2003**, *176*, 575–583. [CrossRef]
72. Wolter, P.T.; Townsend, P.A.; Sturtevant, B.R. Estimation of forest structural parameters using 5 and 10 meter SPOT-5 satellite data. *Remote Sens. Environ.* **2009**, *113*, 2019–2036. [CrossRef]
73. Langner, A.; Hirata, Y.; Saito, H.; Sokh, H.; Leng, C.; Pak, C.; Rašic, R. Spectral normalization of SPOT 4 data to adjust for changing leaf phenology within seasonal forests in Cambodia. *Remote Sens. Environ.* **2014**, *143*, 122–130. [CrossRef]
74. Viedma, O.; Torres, I.; Pérez, B.; Moreno, J.M. Modeling plant species richness using reflectance and texture data derived from QuickBird in a recently burned area of Central Spain. *Remote Sens. Environ.* **2012**, *119*, 208–221. [CrossRef]
75. Scharsich, V.; Mtata, K.; Hauhs, M.; Lange, H.; Bogner, C. Analysing land cover and land use change in the Matobo National Park and surroundings in Zimbabwe. *Remote Sens. Environ.* **2017**, *194*, 278–286. [CrossRef]
76. Navarrete, J.; Ramírez, M.I.; Pérez-Salicrup, D.R. Logging within protected areas: Spatial evaluation of the monarch butterfly biosphere reserve, Mexico. *For. Ecol. Manag.* **2011**, *262*, 646–654. [CrossRef]

77. Reshitnyk, L.; Costa, M.; Robinson, C.; Dearden, P. Evaluation of WorldView-2 and acoustic remote sensing for mapping benthic habitats in temperate coastal Pacific waters. *Remote Sens. Environ.* **2014**, *153*, 7–23. [CrossRef]
78. Tsai, Y.H.; Stow, D.; Chen, H.L.; Lewison, R.; An, L.; Shi, L. Mapping Vegetation and Land Use Types in Fanjingshan National Nature Reserve Using Google Earth Engine. *Remote Sens.* **2018**, *10*, 927. [CrossRef]
79. Lee, J.; Cardille, J.A.; Coe, M.T. BULC-U: Sharpening Resolution and Improving Accuracy of Land-Use/Land-Cover Classifications in Google Earth Engine. *Remote Sens.* **2018**, *10*, 1455. [CrossRef]
80. Mahdianpari, M.; Salehi, B.; Mohammadimanesh, F.; Homayouni, S.; Gill, E. The First Wetland Inventory Map of Newfoundland at a Spatial Resolution of 10 m Using Sentinel-1 and Sentinel-2 Data on the Google Earth Engine Cloud Computing Platform. *Remote Sens.* **2019**, *11*, 43. [CrossRef]

 © 2020 by the authors. Licensee MDPI, Basel, Switzerland. This article is an open access article distributed under the terms and conditions of the Creative Commons Attribution (CC BY) license (http://creativecommons.org/licenses/by/4.0/).

MDPI
St. Alban-Anlage 66
4052 Basel
Switzerland
Tel. +41 61 683 77 34
Fax +41 61 302 89 18
www.mdpi.com

Remote Sensing Editorial Office
E-mail: remotesensing@mdpi.com
www.mdpi.com/journal/remotesensing